ADVANCES IN NUCLEAR PHYSICS

VOLUME 8

Contributors to This Volume

J. L. Friar
Department of Physics
Brown University
Providence, Rhode Island

A. Gal
Racah Institute of Physics
The Hebrew University of Jerusalem
Jerusalem, Israel

J. W. Negele
Laboratory for Nuclear Science and Department of Physics
Massachusetts Institute of Technology
Cambridge, Massachusetts

Donald W. L. Sprung
Physics Department
McMaster University
Hamilton, Ontario, Canada

M. K. Srivastava
Physics Department
University of Roorkee
Roorkee, India

ADVANCES IN NUCLEAR PHYSICS

Edited by

Michel Baranger

Center for Theoretical Physics
Massachusetts Institute of Technology
Cambridge, Massachusetts

Erich Vogt

Department of Physics
University of British Columbia
Vancouver, B.C., Canada

VOLUME 8

Springer Science+Business Media, LLC

The Library of Congress cataloged the first volume of this title as follows:

Advances in nuclear physics. v. 1–   1968–
New York, Plenum Press.

   v. 24 cm. annual.

   Editors: 1968–   M. Baranger and E. Vogt.

1. Nuclear physics—Period.   ɪ. Baranger, Michel, ed. ɪɪ. Vogt,
Erich W., 1929–   ed.

QC173.A2545   539.7′05   67–29001

Library of Congress   ₍5₎

Library of Congress Catalog Card Number 67-29001
ISBN 978-1-4757-4400-2   ISBN 978-1-4757-4398-2 (eBook)
DOI 10.1007/978-1-4757-4398-2

© 1975 Springer Science+Business Media New York
Originally published by Plenum Press, New York in 1975.
Softcover reprint of the hardcover 1st edition 1975

ARTICLES PUBLISHED IN EARLIER VOLUMES

Volume 1
The Reorientation Effect • *J. de Boer and J. Eichler*
The Nuclear SU_3 Model • *M. Harvey*
The Hartree-Fock Theory of Deformed Light Nuclei • *G. Ripka*
The Statistical Theory of Nuclear Reactions • *E. Vogt*
Three-Particle Scattering – A Review of Recent Work on the Nonrelativistic Theory • *I. Duck*

Volume 2
The Giant Dipole Resonance • *B. M. Spicer*
Polarization Phenomena in Nuclear Reactions • *C. Glashausser and J. Thirion*
The Pairing-Plus-Quadrupole Model • *D. R. Bes and R. A. Sorensen*
The Nuclear Potential • *P. Signell*
Muonic Atoms • *S. Devons and I. Duerdoth*

Volume 3
The Nuclear Three-Body Problem • *A. N. Mitra*
The Interaction of Pions with Nuclei • *D. S. Koltun*
Complex Spectroscopy • *J. B. French, E. C. Halbert, J. B. McGrory, and S. S. M. Wong*
Single Nucleon Transfer in Deformed Nuclei • *B. Elbek and P. O. Tjøm*
Isoscalar Transition Rates in Nuclei from the (α, α') Reaction • *A. M. Bernstein*

Volume 4
The Investigation of Hole States in Nuclei by Means of Knockout and Other Reactions •
Daphne F. Jackson
High-Energy Scattering from Nuclei • *Wiesław Czyż*
Nucleosynthesis of Charged-Particle Reactions • *C. A. Barnes*
Nucleosynthesis and Neutron-Capture Cross Sections • *B. J. Allen, J. H. Gibbons.*
and R. L. Macklin
Nuclear Structure Studies in the $Z = 50$ Region • *Elizabeth Urey Baranger*
An s-d Shell-Model Study for $A = 18$-22 • *E. C. Halbert, J. B. McGrory, B. H. Wildenthal,*
and S. P. Pandya

Volume 5
Variational Techniques in the Nuclear Three-Body Problem • *L. M. Delves*
Nuclear Matter Calculations • *Donald W. L. Sprung*
Clustering in Light Nuclei • *Akito Arima, Hisashi Horiuchi, Kuniharu Kubodera,*
and Noboru Takigawa

Volume 6

Volume 7

ARTICLES PLANNED FOR FUTURE VOLUMES

PREFACE

Review articles on three topics of considerable current interest make up the present volume. The first, on Λ-hypernuclei, was solicited by the editors in order to provide nuclear physicists with a general description of the most recent developments in a field which this audience has largely neglected or, perhaps, viewed as a novelty in which a bizarre nuclear system gave some information about the lambda-nuclear intersection. That view was never valid. The very recent developments reviewed here—particularly those pertaining to hypernuclear excitations and the strangeness exchange reactions—emphasize that this field provides important information about the models and central ideas of nuclear physics.

The off-shell behavior of the nucleon–nucleon interaction is a topic which was at first received with some embarrassment, abuse, and neglect, but it has recently gained proper attention in many nuclear problems. Interest was first focused on it in nuclear many-body theory, but it threatened nuclear physicists' comfortable feeling about nonrelativistic potential theory, and many no doubt hoped that it would remain merely an esoteric diversion within the many-body cult. In the editors' opinion, this subject is now eminently respectable and a review of it indeed timely.

The third topic, nuclear charge distributions, is one which almost every nuclear physicist believed had been well in hand for some years. Here it was comforting to know that the Fermi distribution fitted much experimental data and that both electron scattering and muonic X-rays were providing information about the two parameters, surface thickness and radius. Now we find, with some chagrin, that the data is richer and subject to more systematic analysis if one wants to extract all one can from experiment.

For each of these articles we are fortunate in having authors who have contributed greatly to the events which have made their topics timely and important.

M. BARANGER

July 31, 1975   E. VOGT

PREFACE TO VOLUME 1

The aim of *Advances in Nuclear Physics* is to provide review papers which chart the field of nuclear physics with some regularity and completeness. We define the field of nuclear physics as that which deals with the structure and behavior of atomic nuclei. Although many good books and reviews on nuclear physics are available, none attempts to provide a coverage which is at the same time continuing and reasonably complete. Many people have felt the need for a new series to fill this gap and this is the ambition of *Advances in Nuclear Physics*. The articles will be aimed at a wide audience, from research students to active research workers. The selection of topics and their treatment will be varied but the basic viewpoint will be pedagogical.

In the past two decades the field of nuclear physics has achieved its own identity, occupying a central position between elementary particle physics on one side and atomic and solid state physics on the other. Nuclear physics is remarkable both by its unity, which it derives from its concise boundaries, and by its amazing diversity, which stems from the multiplicity of experimental approaches and from the complexity of the nucleon–nucleon force. Physicists specializing in one aspect of this strongly unified, yet very complex, field find it imperative to stay well-informed of the other aspects. This provides a strong motivation for a comprehensive series of reviews. Additional motivation arises from outside the community of nuclear physicists, through the inevitable occurrence of the nucleus as an accessory or as a tool in other fields of physics, and through its importance for terrestrial and stellar energy sources.

We hope to provide a varied selection of reviews in nuclear physics with a varied approach. The topics chosen will range over the field, the emphasis being on physics rather than on theoretical or experimental techniques. Some effort will be made to include regularly topics of great current interest which need to be made accessible by adequate reviews. Other reviews will attempt to bring older topics into clearer focus. The aim will be to attract the interest of both the active research worker and the research student. Authors will be asked to direct their article toward the maximum number of readers by separating clearly the technical material from the more basic aspects of the subject and by adopting a pedagogical point of view rather than giving a simple recital of recent results.

Initially, the *Advances* are scheduled to appear about once a year with approximately six articles per volume. To ensure rapid publication of the papers, we shall use the "stream" technique, successfully employed for series in other fields. A considerable number of planned future articles constitute the source of the stream. The flow of articles from the source takes place primarily to suit the convenience of the authors, rather than to include any particular subset of articles in a given volume. Any attempt at a systematic classification of the reviews would result in considerable publication delays. Instead, each volume is published as soon as an appropriate number of articles have been completed; but some effort is made to achieve simultaneity, so that the spread in completion dates of the articles in a given volume is much less than the interval between volumes.

A list of articles planned for future volumes is given on page v. The prospective articles together with those in this first volume still fall far short of our long-range aims for coverage of the field of nuclear physics. In particular, we definitely intend to present more articles on experimental topics. We shall eagerly receive and discuss outside suggestions of topics for additional papers, and especially suggestions of suitable authors to write them.

The editors owe a great deal to the authors of the present volume for their cooperation in its rapid completion, and to many colleagues who have already given advice about the series. In embarking on this venture, we have had the support of Plenum Press, a relatively new publisher in the field of physics, and of its vice-president, Alan Liss, who has an almost unmatched background in physics publications.

M. BARANGER

E. VOGT

October 15, 1967

CONTENTS

Chapter 3

THEORETICAL AND EXPERIMENTAL DETERMINATION OF NUCLEAR CHARGE DISTRIBUTIONS

J. L. Friar and J. W. Negele

Chapter 1

STRONG INTERACTIONS
IN Λ-HYPERNUCLEI

A. Gal

Racah Institute of Physics
The Hebrew University of Jerusalem, Israel

1. INTRODUCTION

Hypernuclei are nuclei in which at least one of the nuclear constituents is a hyperon, namely a baryon heavier than a nucleon and having a nonzero value of strangeness. In this review, discussion is limited to the study of Λ-hypernuclei $^A_\Lambda Z$, where Z denotes the number of protons, Λ means that there is *one* Λ-hyperon ($M = 1115$ MeV, $J^\pi = \frac{1}{2}^+$, $I = 0$, $S = -1$), and A is the total number of baryons (protons + neutrons + lambda). Double Λ-hypernuclei $^A_{\Lambda\Lambda} Z$, containing two lambdas, have been shown to exist in a stable form for at least two species. However, experimental data for these exciting species is severely restricted at present to ground-state binding energies and we shall not discuss them thoroughly here.

Σ-hypernuclei and Ξ-hypernuclei are expected to be observed in some favorable situations where due to charge conservation the generally allowed *strong interaction* exothermic decays

$$\Sigma N \to \Lambda N \qquad (1a)$$

$$\Xi N \to \Lambda\Lambda \qquad (1b)$$

are forbidden. No established configuration of this kind has yet been reported.

Major known hypernuclear decay modes, in which strangeness is released, are the ones induced by the basic strangeness-nonconserving weak decay modes:

$$\Lambda \to N\pi \tag{2a}$$

$$\Lambda N \to NN \tag{2b}$$

Of course, in the presence of other nucleons there might exist many-body decay modes obtainable form Eqs. (2a) and (2b) by adding the same number of nucleons on both sides and which cannot be explained by invoking the processes (2a) and (2b) in the framework of a many-body calculation. At present there is no information bearing on this subject. What is known, however, is that the basic pionic decay, Eq. (2a), is responsible for the larger fraction of decays of very light hypernuclei with lifetimes similar to that of a *free* Λ-particle, namely $\sim 10^{-10}$ sec. In medium and heavy hypernuclei the dominant decay mode (nonmesic) is probably given by Eq. (2b). The lifetimes involved are believed to be of a similar order of magnitude to that quoted above for mesic decays of light hypernuclei. For the latter to occur, the probability for producing a low-monumentum nucleon [as required by (2a)] is strongly suppressed by the presence of other nucleons (Pauli principle). As a result, mesic decays are suppressed by orders of magnitude (Lag+ 64).

The emerging picture is that on the time scale characteristic of strong interactions ($\sim 10^{-22}$ sec), levels of Λ-hypernuclei are stable in the same sense that ordinary nuclear levels are considered stable; weak and electromagnetic decay modes may be treated in perturbation theory and usually these modes do not interfere with or affect the deduction of strong interaction properties of hypernuclei in a direct way. This does not mean, however, that these decay modes do not test models for the structure of hypernuclear levels. In fact, some progress has been made along such lines. The study of hypernuclei, besides providing new and complementary information on the ΛN interaction which would otherwise be almost impossible to obtain, tests many of the models, ideas, and characteristics of nuclear physics by considering the Λ as a strongly interacting nuclear probe.

In discussing hypernuclear data and properties the scope of present-day experimentation in this field must not be overlooked. Stopping K^- (and to a lesser extent the accompanying Σ^- capture) in nuclear emulsions has for many years been the primary source for producing and detecting hypernuclei. The basic interaction is given by the strangeness exchange process:

$$K^- N \to \pi \Lambda \tag{3a}$$

$$K^- N \to \pi \Sigma \tag{3b}$$

where the produced Σ [Eq. (3b)] undergoes a strong decay [Eq. (1a)]. Because of the very short recoil track of medium and heavy nuclear fragments formed in hypernuclear weak decays, this method has proved useful only for the determination of light hypernuclear species ($A \lesssim 15$). Moreover, in most of the known cases information is provided only about hypernuclear ground states (binding energy). One identifies hypernuclei in emulsion by their decay, rather than production, which for light species occurs predominantly in the ground state; electromagnetic deexcitation rates are much faster than hypernuclear weak decay modes [Eqs. (2a) and (2b)]. In this way, binding energies of hypernuclear ground states, together with some special decay modes bearing some evidence on spins of these ground states, have been established for light hypernuclei. The most recent compilation (Jur+ 73) of measured binding energies is shown in Table I, where the quantity B_Λ stands for the Λ-separation energy ($c = 1$):

$$M(^A_\Lambda Z) = M(^{A-1}Z) + M(\Lambda) - B_\Lambda \tag{4}$$

as measured in these decay modes.

Interaction of K^--mesons with heavy emulsion, namely silver and bromine, leads to the formation of so-called spallation hypernuclei, whose decay is accompanied by nuclear evaporation and fragmentation. These spallation hypernuclei are believed to have mass values roughly between $A = 60$ and $A = 100$ and the deduced (Lag+ 64, Lem+ 65) B_Λ value for this wide range is given by

$$B_\Lambda(\text{heavy}) \approx 25 \pm 2 \text{ MeV} \tag{5}$$

In Section 3 we shall discuss the theoretical expectation for this number as a limiting value D_Λ for a very heavy hypernucleus.

Some illuminating work has been done with helium bubble chambers (Blo+ 63, Key+ 68). Here it proved possible to establish hypernuclear species ($^3_\Lambda$H, $^4_\Lambda$H, and $^4_\Lambda$He) directly by production reactions. The observed production rate is about 1% per stopped K^-.

Last but most promising are some recent counter experiments. One type of such esperiment (Bam+ 71, 73) is intended for observation of hypernuclear γ-rays with a deduction of the appropriate excitation energies. K^--mesons interact with a nuclear target either from rest or in flight, leading to an excited hypernuclear level. A counter detection of γ-rays emitted from these excited states together with observation of angular correlations with the initial production-pion or the final decay-pion may then yield the required information as to the relevant excited hypernuclear quantum

TABLE I

Compilation of Λ Binding Energies from Experiments Performed by the K^- European Collaboration (Jur+ 73)

In the last column the known spin values for the corresponding hypernuclear ground states are given. The values for $_\Lambda^3$H, $_\Lambda^4$H, $_\Lambda^4$He, $_\Lambda^8$Li, and $_\Lambda^{12}$B, where the spin assignment is nontrivial, have been experimentally deduced from two-body hypernuclear weak decay modes in a way which is described by Dalitz (Dal 69a). All known spin values satisfy $J = |J_N - \frac{1}{2}|$.

$_\Lambda^A Z$	Number of events	$B_\Lambda \pm \Delta B_\Lambda$, MeV	J
$_\Lambda^3$H	204	0.13 ± 0.05	$\frac{1}{2}$
$_\Lambda^4$H	155	2.04 ± 0.04	0
$_\Lambda^4$He	279	2.39 ± 0.03	0
$_\Lambda^5$He	1784	3.12 ± 0.02	$\frac{1}{2}$
$_\Lambda^6$He	31	4.25 ± 0.10	—
$_\Lambda^7$He	14	$(^a)$	$\frac{1}{2}$
$_\Lambda^7$Li	226	5.58 ± 0.03	—
$_\Lambda^7$Be	35	5.16 ± 0.08	$\frac{1}{2}$
$_\Lambda^8$He	6	7.16 ± 0.70	—
$_\Lambda^8$Li	787	6.80 ± 0.03	1
$_\Lambda^8$Be	68	6.84 ± 0.05	—
$_\Lambda^9$Li	8	8.53 ± 0.15	—
$_\Lambda^9$Be	222	6.71 ± 0.04	$\frac{1}{2}$
$_\Lambda^9$B	4	7.88 ± 0.15	—
$_\Lambda^{10}$Be	1	9.30 ± 0.26	—
$_\Lambda^{10}$B	10	8.89 ± 0.12	—
$_\Lambda^{11}$B	73	10.24 ± 0.05	—
$_\Lambda^{12}$B	87	11.37 ± 0.06	1
$_\Lambda^{13}$C	34^b	11.22 ± 0.08^{b}	$\frac{1}{2}$
$_\Lambda^{15}$N	14	13.59 ± 0.15	—

^{a} See Section 5.11.
^{b} See Cant 74, where a value 11.69 ± 0.12 MeV is derived.

numbers. This type of experiment is particularly well suited for determination of low-lying (a few MeV) hypernuclear levels in light elements, where the number of competing nuclear lines is not too large.

Another type of experiment consists in detecting the production π^- in the reaction

$$K^- \, {}^A Z \rightarrow \pi^- \, {}_\Lambda^A Z^* \tag{6}$$

where $^4_\Lambda Z^*$ stands for an excited state of the marked hypernucleus. Measuring the momentum of the outgoing pion yields the corresponding hypernuclear excitation energy directly. The resolution required from a suitable pion spectrometer is in the range of a few MeV/c at most. The reaction (6) (stopped K^- or in-flight, such as to minimize the nuclear momentum transfer) has been recently applied (Boh+ 70, Jur+ 72, Bon+ 73) to observe some special excited hypernuclear levels in $^{12}_\Lambda$C and $^{16}_\Lambda$O lying around and above 10 MeV excitation. The nature of these special levels as well as expectations for similar levels in heavier hypernuclei will be discussed in Section 5.

In this review article we do not attempt a discussion of experimental methods used in hypernuclear physics. Rather, we shall concentrate on the present status of phenomenology and theory of strong interactions in hypernuclei. Thus models for the structure of s-shell and p-shell hypernuclei will be discussed and a link to the basic ΛN force, as exhibited in the few pioneering experiments (Ale+ 68, Sec+ 68) on low-energy Λp scattering, will be sought. A short discussion of the Λ well depth parameter related to the measured binding energy [Eq. (5)] of very heavy hypernuclei is included. The concluding sections are devoted to excited hypernuclear levels, in particular those observed recently in the strangeness exchange reaction (6). The personal bias of the author is probably reflected in the references chosen for presentation and discussion, a fact which no apology could remedy.

Several other important topics have been reviewed in great detail in the literature. Their inclusion here was not expected to bring about new insight into or relationship with the immediate development of hypernuclear research. Such topics include exploration of particular pionic decay modes in emulsion and their usefulness for some determinations of hypernuclear spins, investigation of nonmesonic decay modes and their relation to ΛN hypernuclear correlations, rare weak decay modes such as induced by the basic

$$\Lambda \to p e^- \nu_e \tag{7a}$$

$$\Lambda \to p \mu^- \nu_\mu \tag{7b}$$

processes, and the production rates and cascades observed in propane, Freon, and hydrogen–neon chambers. The latest comprehensive review of hypernuclear physics was presented by Dalitz (Dal 69a), who has also given a more concise but updated presentation of the main problems of this field (Dal 73).

2. ΛN INTERACTION AND Λ-NUCLEUS PROPERTIES

It is customary in nuclear physics to state the purpose of study in the following way: Given the nucleon–nucleon interaction, whether phenomenologically from scattering experiments or theoretically from strong interaction meson theory (or preferably from a consistent agreement of both these methods), one ought to be able to derive nuclear ground-state energies, separation energies, excitation energies, transition probabilities, nuclear moments, and numerous other nuclear level characteristics. In practice, however, this program has proved a bit ambitious; it is true that there exist numerous methods for introducing *effective* nucleon–nucleon interaction parameters which lead to remarkable fits between limited experimental information in a particular region of the periodic table and a theory (e.g., shell model) built on these interaction parameters. However, the link between the basic NN interaction and the effective one to be used in nuclear calculations is not always available and its establishment requires particularly hard theoretical work in some of the regions of the periodic table. Lacking this link, most nuclear calculations satisfy their author's sense of consistency by dealing with effective interactions only. A situation which is qualitatively similar, though not as versatile as in the nuclear case, prevails in hypernuclear calculations. The scope of experimental information is severely restricted at present predominantly to binding energies (separation energies) of hypernuclear ground states of light species. Below we shall outline calculations of these B_{Λ} values, which essentially deal with *effective* ΛN interactions.

Two questions arise as a consequence of these calculations: (i) To what extent are the different available calculations consistent with each other, yielding *one* simple set of effective-interaction matrix elements? (ii) Is there a derivable link between the basic ΛN interaction and these effective-interaction matrix elements? The situation here is considerably worse than in nuclei. ΛN scattering information is scarce at present, consisting mainly of *total Λp* cross sections in the s-wave region. These scarcely allow a separation between triplet and singlet s-wave forces. Additional scattering and polarization experiments should be performed at low and medium Λ–nucleon energies. It is almost impossible at the moment to test meson-theoretic models for the ΛN interaction for their capability of deriving the few observed ΛN scattering parameters. The former contain enough parameters, such as hard core radii and some poorly determined coupling constants, to allow a fit to the latter. Whether or not these fits are meaningful is a question which in our opinion has to await accumulation of more detailed ΛN scattering information before a unique answer can be given. In this chapter we shall not concern ourselves deeply with the second question posed above. Current notions of the nature of the ΛN interaction will be briefly reviewed and then we shall address ourselves to the first question, namely to obtaining a consistent picture of binding in light hypernuclei in terms of effective quantities.

2.1. Λp Cross Sections and ΛN Potential

Λp low-energy scattering experiments (Ale+ 68, Sec+ 68) have been performed in a hydrogen chamber for laboratory kinetic energy in the range 120–320 MeV/c. Besides measuring total cross sections in this interval (Fig. 1), the polar–equatorial ratio as well as the backward–forward ratio

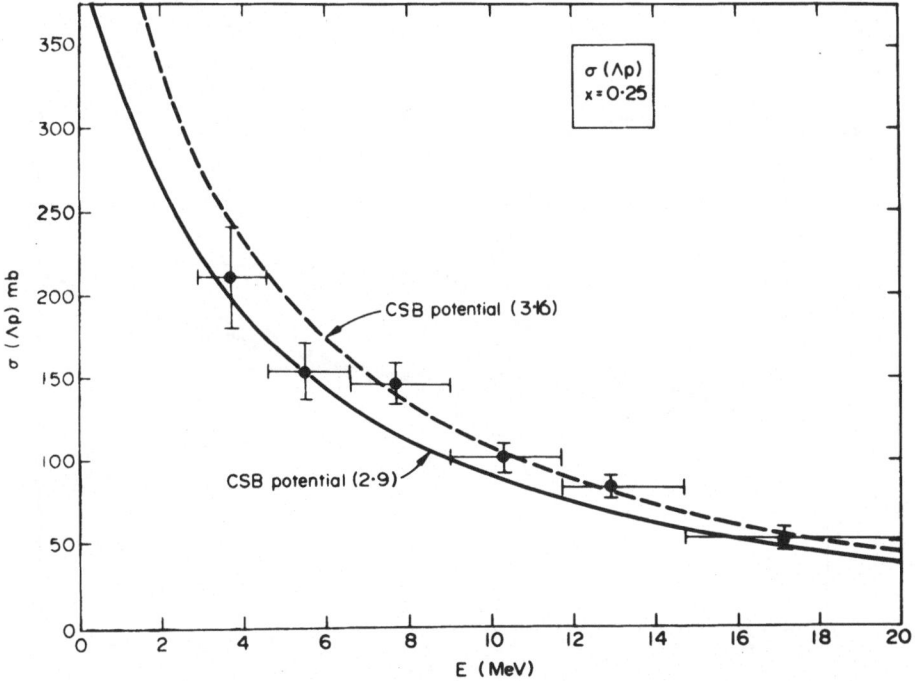

Fig. 1. **Total cross section for** Λp **scattering, taken from Alexander** *et al.* **(Ale+ 68) and Sechi-Zorn** *et al.* **(Sec+ 68) as a function of c.m. kinetic energy** E. The solid line is obtained using the potentials of DHT (DHT 72) with a hard core of radius 0.45 fm and intrinsic range of 2.0 fm, employing a spin–spin CSB component [DHT, Eq. (3.16)], and the dashed line corresponds to taking a spin-independent CSB component [DHT, Eq. (2.9)]. The strength of the Λp interaction in odd-parity states relative to that for even-parity states is given by the ratio $(1 - 2x)$:1. [From (DHT 72).]

(Fig. 2) were determined. From the deviation of the latter from unity it is evident that at the higher part of the momentum interval some p-waves enter the scattering process, although most of the scattering is still described in terms of s-waves. The effective range expansion for s-wave scattering gives the total cross section

$$\sigma = \frac{3\pi}{[-(1/a_t) + \frac{1}{2}r_{0t}k^2]^2 + k^2} + \frac{\pi}{[-(1/a_s) + \frac{1}{2}r_{0s}k^2]^2 + k^2} \qquad (8)$$

in terms of four low-energy parameters, a_s and a_t being singlet and triplet scattering lengths, respectively, and r_{0s} and r_{0t} the corresponding effective ranges. It is impossible, without further constraints, to determine the value of these four parameters, but only to correlate among them. Such a strong (a_s, a_t) correlation is shown in Fig. 3, where χ^2 contours for fitting σ by

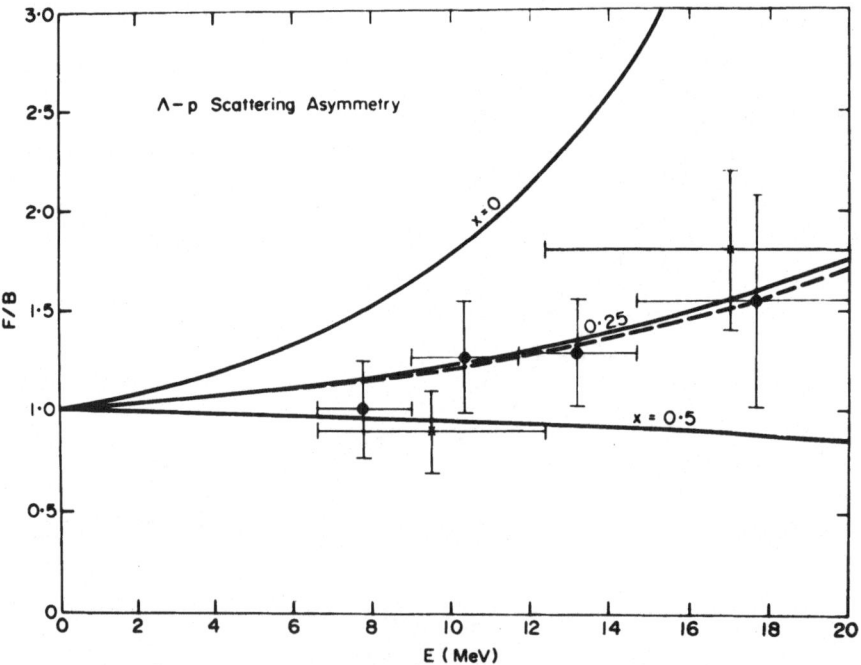

Fig. 2. Forward–backward ratio F/B for the Λp scattering data of Alexander *et al.* (Ale+ 68) and Sechi-Zorn *et al.* (Sec+ 68) as a function of c.m. kinetic energy E. The solid lines are obtained using a spin–spin CSB component, while the dashed line is obtained using a spin-independent CSB component in the potentials of DHT (DHT 72). These potentials are of the form $(1 - x + xP_{\Lambda N})v(r)$, where $v(r)$ denotes the s-wave potential specified by a hard core radius of 0.45 fm and an intrinsic range of 2.0 fm. [From (DHT 72).]

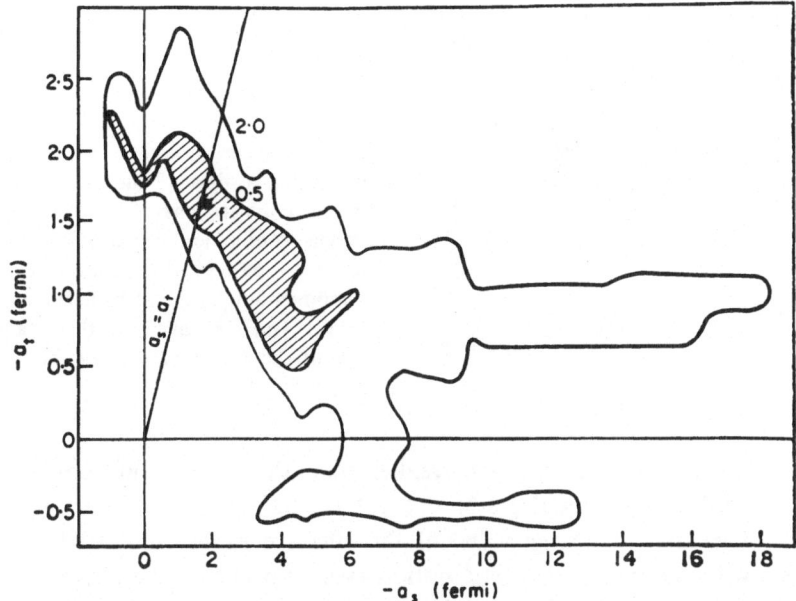

Fig. 3. Mapping of the likelihood function in the (a_s, a_t) plane for the low-energy Λp elastic scattering data of Alexander *et al.* (Ale+ 68). Point f is the best value obtained in a four-parameter fit. The shaded region includes all sets (a_s, a_t) that lie within one standard deviation of the best fit. The curve marked 2.0 encloses all points lying within four standard deviations of the best fit. [From (Ale+ 68).]

Eq. (8) are plotted for fixed ("best" in some sense) values for r_{0s} and r_{0t}. If the spin dependence of the ΛN s-wave interaction is small relative to its central part, it is sensible in the absence of depolarization experiments to attempt a determination of this spin dependence by means of a fit to an appropriately chosen hypernuclear spin-flip excitation and obtain from Eq. (8) the average central parameters a and r_0,

$$[(4\pi/\sigma) - k^2]^{1/2} = -(1/a) + \tfrac{1}{2}r_0 k^2 \qquad (9)$$

by a fit linear in k^2. Such a fit is shown in Fig. 4, yielding the values

$$a = -1.8 \text{ fm}, \qquad r_0 = 3.2 \text{ fm} \qquad (10)$$

Note, however, that the fits depicted in Fig. 3 allow quite a considerable spin dependence, so that the values in (10) for the scattering length may prove illusory.

The sign of the scattering lengths obtained is negative, in accordance with the experimental failure to reproduce a bound ΛN system. The ΛN

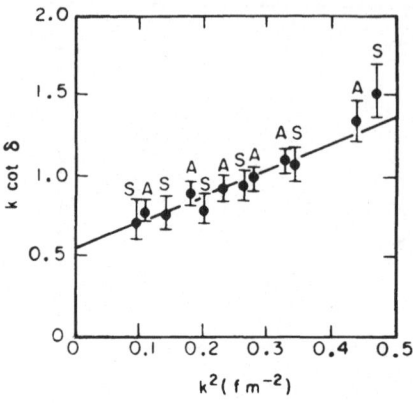

Fig. 4. The quantity $k \cot \delta = [(4\pi/\sigma) - k^2]^{1/2}$ plotted against k^2 for the Λp total scattering cross sections $\sigma(\Lambda p)$ reported by Alexander *et al.* (Ale + 68) and Sechi-Zorn *et al.* (Sec + 68). Here k denotes the Λp c.m. momentum. The best fit linear in k^2, $-(1/a) + \frac{1}{2}r_0 k^2$, is given by the straight line, with the values of a and r_0 given in (10). [From (LD 72).]

force, however, cannot be much weaker than the nucleon–nucleon force since it produces a bound Λnp system ($^3_\Lambda$H), the lightest bound hypernuclear system. The relatively large value of the effective range [see (10)], when coupled with the meson-theoretic notion (see below) that the range of the ΛN interaction should be somewhat shorter than that of the NN interaction, gives rise to the belief in an appreciable *hard core* in the ΛN interaction, the radius of which is estimated by various authors in the range 0.4–0.6 fm. It is customary then to construct a central ΛN interaction potential $V(r)$ (we suppress for the moment a possible spin–spin term, which does not affect the present argument) by fitting its strength and range (beyond the hard core radius) to the s-wave scattering parameters, say those in (10). Such a potential gives rise to scattering in higher partial waves. It is found, however, that the measured forward-to-backward ratio at the beginning of the p-wave scattering region is too low to be explained by an L-independent potential. A considerable suppression in the p-wave attraction is required, which is demonstrated by the different fits shown in Fig. 2. Of course, noncentral components of the ΛN force, such as the spin–orbit component, may show up in this p-wave region and give rise to the observed effect, but it is impossible on the basis of the present experimental data to analyze such noncentral ΛN interactions. Detailed proposals (AGG 67, Lon 71) have been made in the literature for Λ–^{4}He low-energy scattering experiment where a spin–orbit interaction might show up more favorably than in other experiments.

Detailed calculations (DHT 72; HT 67a, 67b, 68) of B_Λ values of s-shell hypernuclei have so far assumed spatial S wave functions which are completely symmetric in the coordinates of the nucleons and for which the p-wave component of ΛN potential has almost no effect. The only effective component of the potential for these wave functions is s-wave. Hence the use of a central potential for the ΛN system which fits the low-energy Λp cross sections is justified in these model calculations, regardless of the above remarks on a possible p-wave suppression in the ΛN interaction. In Section 4 we shall outline results of such calculations in the s shell, where a Λp potential of intrinsic range 2.0 fm and hard core of radius 0.45 fm is used. The strengths of this potential in the triplet and singlet s state are then determined by the B_Λ analysis and the resulting potential fits the Λp low-energy cross sections (DHT 72).

2.2. Meson-Theoretic Models for the ΛN Interaction

One-pion exchange between Λ and N is forbidden by charge independence since the latter is violated by the $\Lambda\Lambda\pi$ vertex. We thus expect the range of the ΛN force to be somewhat shorter than that of the NN interaction, the latter having the one-pion-exchange potential (OPEP) as its outermost part. Two-pion (and higher) exchange is allowed and in particular some of the multipion resonances, K-meson, and $K\pi$ resonances can and probably do contribute to the ΛN interaction. Most meson-theoretic calculations derive one-boson-exchange (OBE) potentials for the ΛN interaction. Isoscalar nonstrange bosons, such as $\eta(0^-)$, $\omega(1^-)$, and a somewhat hypothetical $\sigma(0^+)$, representing a broad enhancement in the s-wave part of the $\pi\pi$ cross section, are used in these calculations. In addition, unlike the case of NN interaction, it is possible to exchange strangeness in the ΛN system by mediation of $K(0^-)$ and $K^*(1^-)$ mesons. Some of the inherent uncertainties of OBE calculations are as follows: (i) There is a reliance on symmetry schemes [$SU(3)$, quark model] to relate some of the experimentally unknown, or poorly known, coupling constants (e.g., $\Lambda\Lambda\omega$) to known ones. (ii) The one-boson description cannot be expected to completely produce all the effects associated with a (partly nonresonating) multipion exchange. In particular, the use of an isoscalar scalar σ-meson is suspected to be a gross oversimplification of reality. (iii) These OBE potentials are singular at the origin, in the vicinity of which many other effects may enter the description of the ΛN interaction. The most obvious prescription is to use a cutoff procedure by choosing an interaction radius $r_c \sim 0.4$ fm below which a hard core is invoked. Smearing procedures may be applied instead, to obtain a soft core repulsion. However, there is an uncertainty as to how correctly one treats the short-range ΛN interaction region by cutting out a considerable fraction of these already short-range OBE potentials.

Nevertheless, OBE models of the ΛN interaction, even those ignoring the nearby ΣN channel (Del 68, DP 65), give some qualitative guidelines to the nature and strength of the various components of this interaction. As for the NN interaction, a strong central repulsion originates from ω exchange, which might have justified the use of a hard core. This is accompanied by an attraction provided by σ exchange. A tensor force is not expected to show up strongly in the ΛN system since ϱ and π exchanges, most significant for the nuclear tensor force, are forbidden due to charge independence, and the effective $\Lambda\Lambda\omega$ vertex for tensor interaction is probably very close to zero (GSD 71). Exchanges of η, K, and K^* give rise to a weak tensor component in view of the medium and small values for the coupling constants involved. A strong component of spin–orbit interaction, similar to the case of NN, is expected for the ΛN system. This originates from exchanges of ω-, K^*-, K-, and σ-mesons, where K^* exchange contributes only to a *symmetric* $[(\mathbf{s}_\Lambda + \mathbf{s}_N) \cdot \mathbf{l}_{\Lambda N}]$ force and K exchange gives rise to a significant *antisymmetric* $[(\mathbf{s}_\Lambda - \mathbf{s}_N) \cdot \mathbf{l}_{\Lambda N}]$ force. The latter cannot appear for charge-independent NN interactions due to the permutation symmetry between the two nucleons. However, the mass of the Λ-particle differs from the mass of the nucleon, an effect giving rise to K and σ contributions to an antisymmetric ΛN spin–orbit force. The latter is also derived (BDI 70) from the asymmetric way in which a boson (with two types of coupling) is coupled to both these baryons, as is the case for the ω coupling. In passing we may note that the sign predicted by these OBE models for the symmetric spin–orbit interaction is the same as for the appropriate NN component and that their strengths are comparable (GSD 71, LD 71). Here we only mention in passing a possible quadratic spin–orbit force, although OBE models also give rise to such a component in the ΛN interaction.

In hypernuclei the question arises as to whether or not the Λ–nucleus interaction can be sufficiently described by an effective Λ–nucleon potential (e.g., derived from OBE models). Nearby (to Λ) baryonic states may have to be introduced explicitly. The small excitation mass, $\Delta \approx 80$ MeV of the Σ triplet, having the same strangeness as the Λ-particle, indicates the necessity of introducing Λ–Σ coupling in hypernuclei. Indeed, the earliest calculations on the Λ–nucleon interaction (SI 62, 63) were based on a two-pion-exchange (TPE) model where the intermediate strangeness is carried by Σ (in order to satisfy charge independence) as given by Fig. 5(a), p. 15. One considers a two by two hyperon–nucleon interaction matrix

$$V_{YN \to YN} = \begin{pmatrix} V_{\Lambda N, \Lambda N} & V_{\Lambda N, \Sigma N} \\ V_{\Sigma N, \Lambda N} & V_{\Sigma N, \Sigma N} \end{pmatrix} \tag{11}$$

which operates on a two-component wave function

$$\psi = \begin{pmatrix} \psi_A \\ \psi_\Sigma \end{pmatrix} \tag{12}$$

It is clear that the Σ channel can be eliminated in favor of an *effective AN* potential

$$V_{\text{eff}} = V_{AA} + V_{A\Sigma}[1/(E - \Delta - T_\Sigma - V_{\Sigma\Sigma})]V_{\Sigma A} \tag{13}$$

where for brevity the notation $V_{YY'}$ was introduced for $V_{YN,Y'N}$. The effective AN potential is energy dependent; however, for a limited energy range in the vicinity of E_0, where $E_0 \ll \Delta$, this dependence is a weak one. Similarly, for the A–nucleus interaction one can envisage an *effective A–nucleus* potential in the form of Eq. (13) where the hidden subscript N now means "nucleus" and an antisymmetrization projection is to be inserted in the numerator of the Green's function for the nuclear intermediate states.

Calculations in the literature have attempted to relate the effective AN potential to basic exchange processes in the YN system. For a thorough review consult de Swart *et al.* (Swa+ 71). Up to second order in the potentials and with a neglect of multiple, other than two-pion, exchanges the effective AN potential is given by

$$V_{\text{eff}} = (AN \mid T^{(1)}_{\text{OBE}} \mid AN) + (AN \mid T^{(2)}_{\text{TPE}} \mid AN)$$
$$- (AN \mid T^{(1)}_{\text{OPE}} \mid \Sigma N)G_0(E - \Delta)(\Sigma N \mid T^{(1)}_{\text{OPE}} \mid AN) \tag{14}$$

where $T^{(1)}$ is a scattering amplitude associated with the OPE pole or OBE pole, and $T^{(2)}$ is that associated with the two-pion cut. Special care should be exerted in order to avoid double counting, since some of the one-boson exchanges, such as σ exchange, may be effectively included in the two-pion-exchange amplitude; the evaluation of the latter is somewhat ambiguous.[†] The main feature of the TPE calculations is that the long-range strong tensor component characteristic of OPEP averages out to give a mostly central component after this OPE contribution is iterated twice.

A treatment of both TPEP and OBEP for the YN system has been given by Fast *et al.* (FHS 69) where fitting (adjustable hard core radii) to YN scattering data around the ΣN thereshold is enforced. In particular

---

[†] Recently a more proper treatment of the two-pion-exchange potential has been offered (Ris 72) which makes use of dispersion techniques in decomposing the $N\bar{N} \rightarrow \pi\pi$ and $A\bar{A} \rightarrow \pi\pi$ amplitudes.

the measured Λp cross section seems to acquire an enhancement just below the ΣN channel, as observed in $K^-d \to \Sigma N\pi^- \to \Lambda p\pi^-$ from rest (Tan 69) and in flight (CLM 68, Ale+ 69), where the Λp effective mass strongly peaks only a few MeV below the Σ^+n threshold in the $^3S_1$ channel. From this fitting Fast *et al.* (FHS 69) obtain the following values for the *s*-wave low-energy ΛN scattering parameters:

$$a_t = -1.5 \pm 0.05 \text{ fm}, \qquad a_s = -1.7 \pm 0.5 \text{ fm} \tag{15a}$$

$$r_t = 2.0 \pm 0.05 \text{ fm} \qquad r_s = 2.5 \, {}^{+1.0}_{-0.5} \text{ fm} \tag{15b}$$

where the small errors in the triplet parameters reflect only the uncertainty in the position of the (supposed) resonance but not any theoretical uncertainty. More recent calculations (Swa+ 71) favor values of

$$a_t = -1.7 \text{ fm}, \qquad a_s = -1.8 \text{ fm} \tag{16}$$

for a reasonable fit to the Σ^-p reaction data.

Most recent comprehensive work concentrates on OBEP without regard to (ambiguous) TPE contributions. The latter are believed to be largely reproduced by invoking known and speculated bosons which serve as multipion resonances. We mention here two such works. Brown *et al.* (BDI 70, 72) construct an OBEP model for the Λp interaction which includes the ΣN channel explicitly. They reproduce scattering lengths and effective ranges which fit low-energy Λp cross sections by utilizing exchanges of nonets of pseudoscalar, vector, and scalar bosons outside a channel-independent hard core radius of 0.46 fm. Many of the parameters of their model are related either directly to measured quantities or by symmetry schemes to such quantities. They obtain Λp cross sections around and above the ΣN threshold which are in good agreement with experiment. The $^3S_1-^3D_1$ partial cross sections are dominant just below the ΣN threshold in a nonresonant manner, which explains the observed enhancement in Λp cross sections in that region. The main contributor in their model to the coupling between the ΛN and the ΣN channel is the strong tensor component in $V_{\Lambda\Sigma}$ arising primarily from OPE. Nagels *et al.* (NRS 73) performed a combined analysis of low-energy NN and YN scattering data in terms of the OBE model for the lowest-lying meson nonets, the pseudoscalar, the vector, and the scalar meson nonets. A strong central attraction, partly to be cancelled by the (maybe somewhat large) ω coupling, is supplied by their broad unitary (mostly) singlet ε ($m = 670$ MeV, $\Gamma = 500$ MeV). From a fit to the data they obtain values for $F/(F + D)$ ratios and coupling constants for the pseudoscalar and vector mesons which are in good agreement with other independent data. For the scalar meson nonet, where such information is scarce, they thus determine the mixing angle, $F/(F + D)$ ratio, and coupling constants.

2.3. The Σ Channel in Hypernuclei and Three-Body ΛNN Forces

In the following sections, where the ΛN interaction in hypernuclei is considered, we will have to consider the possibility that an *explicit* use of a Σ channel is necessary and that a strong $^3S_1$–$^3D_1$ transition is induced by the relatively long-range tensor OPE YN potential. On the other hand, if the Σ channel is eliminated, ΛNN *three-body forces* should be added to the description of Λ–nucleus interaction. These are not necessarily genuine three-body forces since they partly arise from an elimination of the Σ–nucleus channel. A typical diagram is depicted in Fig. 5(b), where the two pions are exchanged between the Λ-particle and *different* nucleons in the nucleus. The intermediate hyperonic state, represented by a blob in the figure, can be a Σ as well as any of the $I = 1$, $S = -1$ Σ^* resonances. The first such resonance is a $\pi\Lambda$ $p_{3/2}$ resonance $\Sigma^*(1385)$, analogous to $N^*(1236)$ in the πN system. Discussion of ΛNN three-body forces in the literature has concentrated on TPE, where the intermediate blob is given by Σ and $\Sigma^*(1385)$ (BLN 67). More complicated forces of a shorter range can be obtained

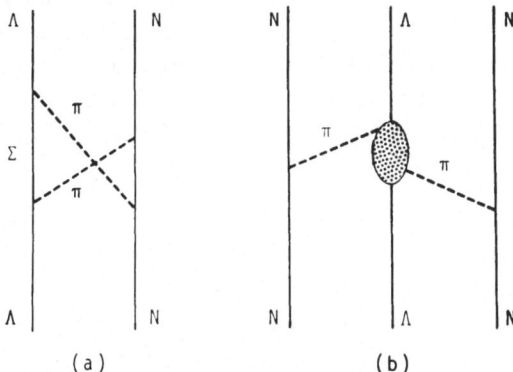

(a)　　　　　　(b)

Fig. 5. Two graphs which contribute to two-pion-exchange processes in hypernuclei. Graph (a) gives rise to a ΛN hypernuclear interaction potential which in most cases may be identified with the corresponding ΛN free-pair interaction potential. Graph (b) corresponds to two-pion-exchange hypernuclear ΛNN interaction potential. The blob in the middle line represents any $S = -1$, $I = 1$ baryonic excitation such as Σ, $\Sigma^*(1385)$, etc.

by πK, $\pi\eta$, $\sigma\sigma$, etc. exchanges. Physically, the inclusion of the TPE ΛNN force in a realistic Λ–nuclear calculation is important since this force senses regions of the ΛN configuration which are almost asymptotic for any two-body ΛN interaction. In order for the diagram of Fig. 5(b) to be operative, each of the nucleons must be localized within a fairly large distance of $\hbar/m_\pi c \approx 1.4$ fm from the Λ. Short-range repulsive NN correlations, however, render the use of the TPE ΛNN force derived from the diagram of Fig. 5(b) rather dubious at short NN distances, where this force becomes highly singular.

We would like to follow in a more systematic way the appearance of three-body ΛNN effective forces in hypernuclei. Equation (13) for the effective Λ–nucleus interaction after the Σ channel has been explicitly removed becomes

$$V_{\text{eff}}(\Lambda\text{–nucleus}) = \sum_{i=1}^{A} V_{\Lambda i,\Lambda i} + \left(\sum_{i=1}^{A} V_{\Lambda i,\Sigma i} \right)$$
$$\times \frac{\mathscr{A}}{E - \Delta - T_{\Sigma\text{–nucleus}} - V_{\Sigma\text{–nucleus}}} \left(\sum_{j=1}^{A} V_{\Sigma j,\Lambda j} \right) \quad (17)$$

where $\mathscr{A}$ is an antisymmetrization projection for the intermediate nuclear states. However, if V_{eff} is evaluated, as it should be, between states antisymmetric in the nuclear coordinates, the symbol $\mathscr{A}$ can be omitted since both $V_{\Lambda\Sigma}$ and $V_{\Sigma\Lambda}$ are here written in a manifestly symmetric (in the nucleons) way. The Σ–hypernuclear intermediate states $|\varphi_k\rangle$ each draws a contribution $(E - \Delta - \varepsilon_k)^{-1}$ from the Green's function, where ε_k is the energy of this intermediate $|\varphi_k\rangle$ relative to the Σ–nucleus threshold. *If* the $\Lambda\Sigma$ coupling favors excitations heavily centered in a narrow energy region $|\Delta\varepsilon_k| \ll \varepsilon_k$, we can take an average value of these excitations, $\varepsilon_k \to \bar{\varepsilon}$, and make use of closure. The second term on the right-hand side of Eq. (17) then reads

$$\frac{1}{E - \Delta - \bar{\varepsilon}} \left(\sum V_{\Lambda i,\Sigma i} \right)\left(\sum V_{\Sigma j,\Lambda j} \right)$$
$$= \frac{1}{E - \Delta - \bar{\varepsilon}} \left(\sum_{i=1}^{A} V_{\Lambda i,\Sigma i}V_{\Sigma i,\Lambda i} + \sum_{i \neq j}^{A} V_{\Lambda i,\Sigma i}V_{\Sigma j,\Lambda j} \right) \quad (18)$$

The first sum in the square brackets is a *two-body ΛN* interaction to be added to the first sum on the right-hand side of Eq. (17). The *effective hypernuclear two-body* interaction (for a ΛN pair imbedded in a nucleus) is therefore given by

$$V_{\Lambda\Lambda} + [1/(E - \Delta - \bar{\varepsilon})]V_{\Lambda\Sigma}V_{\Sigma\Lambda} \quad (19)$$

which differs from the effective two-body ΛN interaction given in Eq. (13) by essentially having a specific average intermediate Σ–hypernuclear excitation energy $\bar{\varepsilon}$ instead of a two-body ΣN characteristic excitation.

One speaks therefore of a modified or effective ΛN interaction when the latter occurs in a nuclear medium.

The second sum in the square brackets of Eq. (18) is an *effective three-body interaction*. The coupling between Λ and nucleons 1 and 2 is given here by

$$\frac{1}{E - \Delta - \bar{\varepsilon}} \left(V_{\Lambda 1, \Sigma 1} V_{\Sigma 2, \Lambda 2} + V_{\Lambda 2, \Sigma 2} V_{\Sigma 1, \Lambda 1} \right) \tag{20}$$

If one takes for $V_{\Lambda N, \Sigma N}$ and $V_{\Sigma N, \Lambda N}$ the static OPE forms

$$V_{\Lambda N, \Sigma N}(\mathbf{r}) = V_{\Sigma N, \Lambda N}(\mathbf{r})$$
$$= f_{NN\pi} f_{\Lambda \Sigma \pi} \mu 3^{-1/2} \mathbf{\tau}_N \cdot \mathbf{\Phi} [\mathbf{\sigma}_N \cdot \mathbf{\sigma}_Y + T(r) S_{NY}(\hat{\mathbf{r}})] Y(r) \tag{21}$$

where

$$Y(x) = \exp(-\mu x)/\mu x, \qquad T(x) = 1 + (3/\mu x) + [3/(\mu x)^2] \tag{22}$$

$$S_{12}(\hat{\mathbf{x}}) = 3(\mathbf{\sigma}_1 \cdot \hat{\mathbf{x}})(\mathbf{\sigma}_2 \cdot \hat{\mathbf{x}}) - \mathbf{\sigma}_1 \cdot \mathbf{\sigma}_2 \tag{23}$$

μ is the pion mass, and $\mathbf{\Phi}$ is a unit isotopic vector for the Σ-particle, then one obtains the following results:

(i) The term $V_{\Lambda N, \Sigma N} \cdot V_{\Sigma N, \Lambda N}$ of Eq. (19) for the ΛN interaction:

$$\tfrac{1}{3}\mu^2 f_{NN\pi}^2 f_{\Lambda \Sigma \pi}^2 \{3 - 2\mathbf{\sigma}_\Lambda \cdot \mathbf{\sigma}_N + 2T(r) S_{\Lambda N}(\hat{\mathbf{r}})$$
$$+ T^2(r)[6 + 2\mathbf{\sigma}_\Lambda \cdot \mathbf{\sigma}_N - 2S_{\Lambda N}(\hat{\mathbf{r}})]\} Y^2(r) \tag{24}$$

which is dominated by the term containing the tensor factor $T(r)$ twice, namely

$$\tfrac{1}{3}\mu^2 f_{NN\pi}^2 f_{\Lambda \Sigma \pi}^2 Y^2(r) T^2(r)[6 + 2\mathbf{\sigma}_\Lambda \cdot \mathbf{\sigma}_N - 2S_{\Lambda N}(\hat{\mathbf{r}})] \tag{25}$$

This expression obviously vanishes for the singlet ΛN interaction, whereas for the $^3S_1$ configuration an overall [note the $(E - \Delta - \bar{\varepsilon})^{-1}$ factor in Eq. (19)] diagonal attraction is obtained. In spite of the fact that the central part of $V_{\Lambda \Sigma}$ is considerably weaker than the tensor part, the average excitation of the former, $\bar{\varepsilon}_{\text{central}}$, involved in the closure application is certainly lower than $\bar{\varepsilon}_{\text{tensor}}$. Bodmer *et al.* (BRM 70) find that for Yukawa shapes of the same range the tensor force is more effective in bringing in high momentum components than is the central force: $\bar{\varepsilon}_{\text{tensor}} \sim 5\bar{\varepsilon}_{\text{central}}$.

This holds true since the tensor component couples the ΛN s wave function $u_s{}^\Lambda$ to a ΣN d wave function $u_d{}^\Sigma$ which, because of the centrifugal barrier, has poorer overlap with the product $u_s{}^\Lambda(r)V(r)$ than a ΣN s-wave does. One has to invoke quite high excitations in order to push $u_d{}^\Sigma(r)$ inside, thereby improving the overlap. As a result, the central–central component of Eq. (24) for the effective ΛN s-wave interaction, proportional to $(3 - 2\sigma_\Lambda \cdot \sigma_N)Y^2(r)$, should actually weigh, with respect to the tensor–tensor component proportional to $(6 + 2\sigma_\Lambda \cdot \sigma_N)Y^2(r)T^2(r)$, about five times more than Eq. (24) suggests. Yet, in actual calculations, this central–central component makes an almost negligible contribution and hence $V_{\Lambda\Sigma}$ in Eq. (21) may for all practical purposes be taken as a tensor interaction. The situation is different for coupling ΛN and ΣN channels via ϱ exchange since the central part of such a $V_{\Lambda\Sigma}$ interaction is strong.

(ii) The three-body Λ, 1, 2 term of Eq. (20) assumes for $E = \bar{\varepsilon} = 0$ (in the spirit of the static approximation) the following form:

$$-\frac{2\mu^2}{9\Delta} f_{NN\pi}^2 f_{\Lambda\Sigma\pi}^2 \tau_1 \cdot \tau_2 \{\sigma_1 \cdot \sigma_2 + T(r_{\Lambda 1})S_{12}(\hat{\mathbf{r}}_{\Lambda 1}) + T(r_{\Lambda 2})S_{12}(\hat{\mathbf{r}}_{\Lambda 2})$$

$$+ T(r_{\Lambda 1})T(r_{\Lambda 2})[9(\hat{\mathbf{r}}_{\Lambda 1} \cdot \hat{\mathbf{r}}_{\Lambda 2})(\sigma_1 \cdot \hat{\mathbf{r}}_{\Lambda 1})(\sigma_2 \cdot \hat{\mathbf{r}}_{\Lambda 2}) - S_{12}(\hat{\mathbf{r}}_{\Lambda 1}) - S_{12}(\hat{\mathbf{r}}_{\Lambda 2})$$

$$- \sigma_1 \cdot \sigma_2]\} Y(r_{\Lambda 1})Y(r_{\Lambda 2}) \tag{26}$$

This coincides exactly with the three-body ΛNN force derived with a Σ pole contribution only to $\pi\Lambda$ scattering. Bhaduri et al. (BLN 67) also evaluated the $\Sigma^*(1385)$ contribution, which they find to possess the same functional form as above (since both Σ and Σ^* show up in the p-wave $\pi\Lambda$ region) with about 60% of the Σ contribution strength. If we write the terms in front of the curly brackets in (26) in the form $-\frac{1}{3}C_p\tau_1 \cdot \tau_2$, then

$$C_p = +\tfrac{2}{3}(\mu^2/\Delta)f_{NN\pi}^2 f_{\Lambda\Sigma\pi}^2 + (\Sigma^* \text{ contribution}) \tag{26'}$$

and C_p is of the order of 1 MeV. The s-wave $\pi\Lambda$ contribution is believed to be particularly small, similar to the suppression in the s-wave πN interaction.

In the following sections we will discuss the various functional forms contained in expression (26) for the three-body force as appropriate to s- and p-shell hypernuclei. Here we note only that this force is independent of the Λ spin operator σ_Λ which is a consequence of the symmetry $1 \leftrightarrow 2$ of expression (20) (and can be derived in a general way from the diagram in Fig. 5b, too). The strongest components of the TPE ΛNN force involve terms of tensorial character which may require for their evaluation a more precise handling of d-wave hypernuclear bonds than has hitherto been done. These terms, accompanied by a radial strength proportional to $T(r_{A1})$ $\cdot T(r_{A2})Y(r_{A1})Y(r_{A2})$, become quite singular at short AN distances, where the effect of a cutoff or a smearing may lead to considerable uncertainties in their evaluation.

It was shown by Nogami (Nog 69) for the lowest order of TPE in nuclear matter that the obvious restrictions on intermediate nuclear states due to the Pauli principle can be lifted once the ΣN channel is removed and three-body ANN forces of the type discussed above appear. In this sense the three-body force is a device by the introduction of which intermediate states for the two-body interaction in nuclear matter have no exclusion restriction imposed upon them.

2.4. Charge Symmetry Breaking of ΛN Interactions

With Ap cross sections measured to some extent, the question arises whether or not the An interaction is equal to the Ap interaction. A direct comparison between the two is not feasible at present. The An scattering parameters may in some cases be deduced from certain reactions by invoking final-state interaction. However, theoretical uncertainties in reaction calculations make a precise comparison of Ap interactions with An interactions almost impossible. A less indirect method is to look for B_A values of hypernuclear members of an isospin multiplet. If isospin were an exact nuclear symmetry, then in the absence of AN charge symmetry breaking (CSB) these B_A values should be equal to each other. However, in real hypernuclei, deviations from such equality do not provide conclusive evidence for AN CSB since such deviations could arise from isospin impurities of the nuclear core (including Coulomb effects, a Thomas–Ehrman shift, a genuine nuclear isospin-violating force, or a possible nuclear CSB). The first established evidence for AN CSB came from the difference $\Delta B_A(_{A}^{4}\text{He}-_{A}^{4}\text{H}) = 0.34 \pm 0.08$ MeV (Table I) in the A binding energies of the $A = 4$, $I = \frac{1}{2}$ members. An evaluation of the induced (by the charge-symmetric part of the AN interaction) nuclear CSB effects, such as the extra Coulomb repulsion of the two protons in $_{A}^{4}\text{He}$, leads to an estimate (DP 66, DV 64, HT 68) of $(\Delta B_A)_{AN,\text{CSB}=0} < 0$ with a probable value of about -0.25 MeV. Thus, unless other, unknown at present, effects are discovered, a value of about $(\Delta B_A)_{AN,\text{CSB}} \sim 0.60$ MeV should be considered as arising from the difference between the Ap interaction and the An interaction in the $A = 4$ hypernuclei. In judging whether or not such a seemingly big value can be attributed to AN CSB interaction, the following estimate may prove helpful. The Λ–^{3}H potential energy is of the order of 20 MeV (DHT 72), namely about 6.5 MeV per ΛN pair. For the dominant symmetric (for the nuclear core) S configuration, CSB ΛN effects can arise only from *one* ΛN pair.[†] Hence the ratio of the charge-symmetry-breaking component to the charge-symmetric component of the ΛN force in $^4_\Lambda$H is about $0.5 \times 0.65/6.5 = 0.05$ [the 0.5 factor originates from the fact that the contributions to $(\Delta B_\Lambda)_{\mathrm{CSB}}$ from $^4_\Lambda$H and $^4_\Lambda$He are equal in magnitude, whereas here attention is confined to $^4_\Lambda$H]. This is a relatively high ratio to be expected for CSB in nuclear processes, but as we shall see, certain enhancement factors are expected in the case of the ΛN interaction which cause the ΛN CSB interaction to be considerably stronger than an ordinary electromagnetic effect would lead to (about less than 1%).

Other partners of hypernuclear isospin multiplets have been identified in the p shell. Thus $^7_\Lambda$Be and $^7_\Lambda$He belong to an $I = 1$ multiplet; the other member $^7_\Lambda$Li* (in the following an asterisk stands for an excited state) could not yet have been detected since it is not a ground state. However, the B_Λ distribution for assigned $^7_\Lambda$He events is particularly broad (Jur+ 73), possibly due to an isomeric state $^7_\Lambda$He*, which makes it almost impossible at present to determine the B_Λ value for the appropriate ground state with meaningful precision. The $^8_\Lambda$Li and $^8_\Lambda$Be form an $I = \frac{1}{2}$ isomultiplet. Here the near equality of Λ binding energies, 6.80 ± 0.03 and 6.84 ± 0.05 MeV, respectively, stands at present in (a real or apparent) contrast to the difference discussed above for $A = 4$. On the other hand, $^9_\Lambda$Li and $^9_\Lambda$B, belonging to an $I = 1$ isomultiplet, differ quite appreciably in their B_Λ values, 8.53 ± 0.15 and 7.88 ± 0.15 MeV. The last known isomultiplet is the $I = \frac{1}{2}$ ($^{10}_\Lambda$Be, $^{10}_\Lambda$B), with B_Λ values of 9.30 ± 0.26 and 8.89 ± 0.12 MeV. For all these p-shell hypernuclei the information bearing on CSB in the ΛN interaction has not yet been extracted.

On the theoretical side, CSB in the ΛN interaction may be attributed to a number of effects, the most important of which is probably isospin mixing of baryons and mesons (DP 66, DV 64). Dalitz and Von Hippel (DV 64) emphasized, in particular, the possibility of having a nonzero CSB coupling $\Lambda\Lambda\pi^0$ due to Λ, Σ^0 mixing. The idea is that both Λ and Σ^0, which differ in their quantum numbers only by their isospin, 0 and 1, respectively, are actually not pure eigenstates of isospin but involve admix-

---

[†] Thus the most general functional form for the ΛN CSB interaction in this case is $\sum_{i=1}^{3} \tau_{iz}(A + B\mathbf{\sigma}_\Lambda \cdot \mathbf{\sigma}_i)$, where A and B are radial matrix elements. For the A term the overall contribution comes from one of the neutrons, whereas for the B term it comes from the proton (since $\langle\mathbf{\sigma}_i\rangle$ is zero for a $^1S_0$ pair of neutrons).

tures of isospin 1 and 0, respectively. A model for the evaluation of these admixtures is provided by $SU(3)$ (DV 64) or by the quark model (GS 67) and the mixing is given by

$$| \Lambda \rangle = (\cos \alpha) | I = 0 \rangle + (\sin \alpha) | I = 1 \rangle \qquad (27a)$$

$$| \Sigma^0 \rangle = -(\sin \alpha) | I = 0 \rangle + (\cos \alpha) | I = 1 \rangle \qquad (27b)$$

$$\alpha \approx -\frac{(\Sigma^0 | \delta M | \Lambda)}{M(\Sigma^0) - M(\Lambda)}$$

$$= -\frac{1}{[M(\Sigma^0) - M(\Lambda)]\sqrt{3}} \begin{cases} \times [M(\Sigma^0) - M(\Sigma^+) - M(n) + M(p)] \\ \quad = -0.0134 \pm 0.0012 \quad [SU(3)] \\ \times \frac{1}{2}[M(\Xi^-) - M(\Xi^0) - M(\Xi^{*-}) \\ \quad + M(\Xi^{*0})] = -0.011 \pm 0.004 \quad \text{(quark)} \end{cases} \begin{matrix} (27c) \\ \\ (27d) \end{matrix}$$

The $\Lambda\Lambda\pi^0$ coupling is hence given to first order in α by

$$f_{\Lambda\Lambda\pi^0} = (\sin 2\alpha)f_{\Lambda\Sigma^0\pi^0} = (2/\sqrt{3})(1 - f)(\sin 2\alpha)f_{NN\pi^0} \qquad (28)$$

where $f \sim 0.4$ is the $SU(3)$ pseudoscalar $F/(F + D)$ ratio. Here $f_{NN\pi^0}^2 \approx 0.082$ is the pion–nucleon coupling constant. This coupling is particularly important since it provides the ΛN interaction with some OPE component of longer range than any of the charge-symmetric contributions to the ΛN interaction. It generates the following CSB potential:

$$V_{\Lambda N}^{\text{OPE}} = (2/\sqrt{3})(1 - f)(\sin 2\alpha)\tau_3^N V_{pp}^{\text{OPE}} \approx -0.019\tau_3^N V_{pp}^{\text{OPE}} \qquad (29)$$

where V_{pp}^{OPE} is the OPE proton–proton potential with variables as appropriate to the lambda–nucleon system:

$$V_{pp}^{\text{OPE}} = f_{NN\pi^0}^2 m_\pi \frac{1}{3}[\mathbf{\sigma}_\Lambda \cdot \mathbf{\sigma}_N + T(r)S_{\Lambda N}(\hat{\mathbf{r}})]Y(r) \qquad (30)$$

where τ_3^N is the third component of the nucleon's isotopic Pauli matrix, $+1$ for the proton and -1 for the neutron. The $SU(3)$ estimate for α, Eq. (27c), was employed.

Similarly, a ϱ^0 exchange becomes feasible for the ΛN interaction due to charge symmetry breaking. The functional form of this CSB force involves more terms than Eqs. (29) and (30) show. In particular, a spin-independent central term is present (DP 66).

A mixing between mesons, such as (π^0, η) and (ϱ^0, ω), is possible in principle. However, the appropriate contribution to CSB of the ΛN interaction vanishes in the limit of equal admixed meson masses. Thus (ϱ^0, ω)

mixing is of no practical interest for ΛN CSB, whereas (π^0, η) mixing contributes, but less than (Λ, Σ^0) mixing, to CSB.

Other, less significant, sources of CSB in the ΛN interaction include mass differences between Σ^+ and Σ^- and between a proton and a neutron in the intermediate states appropriate for the TPE ΛN interaction (DV 64).

If we consider the sign of the OPE CSB ΛN potential [Eq. (29)] for $^1S_0$ ΛN states as appropriate for the dominant component of the $A = 4$ hypernuclear wave function,[†] we find it negative for the relevant Λn pair in $_\Lambda^4$He and positive for the relevant Λp pair in $_\Lambda^4$H, in accordance with the measured ΔB_Λ value. This potential also provides the right order of magnitude for the $A = 4$ binding energy difference (DV 64). More on the contribution of CSB to ground-state energies as well as first excited state energies for $A = 4$ is found in Section 4.2.

In considering ΛN CSB effects in the p shell one should keep in mind the possibility that a dominant contribution to B_Λ differences may arise from the tensor term of Eq. (30). In the p shell this term contributes in first order, unlike the situation in the s shell for simple common wave functions. A preliminary evaluation (Gal 75) of this tensor contribution for the $A = 8$ ground state gives, however, a very small number, less than 0.05 MeV in magnitude, which is due to large cancellations. It remains to be seen whether or not this is the general trend for other isomultiplets of the p shell.

Expression (29) for the ΛN OPE CSB potential, as well as for potential forms arising from other mixings or breakings, is of a *vector* character in the nuclear isospace. Thus there is just one reduced matrix element which is supposed to reproduce CSB contributions to B_Λ values (δB_Λ) within a hypernuclear isomultiplet. These δB_Λ values are then effectively given to first order by $(\delta B_\Lambda)_{I_3} = a I_3$, where a is a constant. For $I > \frac{1}{2}$, relations should hold among the various (≥ 2) B_Λ values measurable in principle. However, at present, no example of a whole $I = 1$ (or larger) isomultiplet is known, since for light hypernuclei the middle member $(I_3 = 0)$ of such multiplet corresponds to an excited state of a hypernucleus whose ground state has $I = 0$. This is the situation for $A = 7, 9$, where the first $I = 1$ excited state of $_\Lambda^7$Li and $_\Lambda^9$Be is missing experimentally in isomultiplets partly composed of $(_\Lambda^7$He, $_\Lambda^7$Be$)$ and $(_\Lambda^9$Li, $_\Lambda^9$B$)$, respectively.

For some hypernuclei, off-diagonal contributions from the ΛN CSB interaction should be regarded as nonnegligible. Consider, for example,

---

[†] For $_\Lambda^4$He, for example, the two paired protons cannot obtain a contribution from Eq. (30) since the latter is proportional to σ_N. The experimental evidence is that the Λn pair is coupled to a total spin of zero; hence $\langle \sigma_\Lambda \cdot \sigma_N \rangle = -3$ for this case.

$^7_\Lambda$Li, where the ground state corresponds to $I = 0$ and where a first $I = 1$ level is expected roughly at about 3.5 MeV excitation as a result of the Λ coupling to the 0^+, $I = 1$ ^{6}Li level at 3.56-MeV excitation. This $I = 1$ level is the missing partner for the $A = 7$, $I = 1$ hypernuclear isomultiplet mentioned above. Our previous considerations would indicate that the first-order CSB contribution vanishes for both these states [the triangular condition is violated for the ground state and the "parity" condition $\binom{111}{000} = 0$ for the excited state]. However, since these two states are separated by a few MeV only, a mixing could effectively occur between them via the ΛN CSB interaction, Eq. (29). If δ is the value of the matrix element connecting these $I = 0$ and $I = 1$ hypernuclear states and Δ is their separation in excitation, then to first order in δ/Δ these levels "repel" each other by a quantity δ^2/Δ. It is obvious that for $A = 9$, where $\Delta \sim 17$ MeV, this constitutes a negligible effect. For $^3_\Lambda$H, however, where $\Delta \sim 2.3$ MeV, the effect may be of the order of 0.1 MeV. A more dynamically consistent calculation for $^3_\Lambda$H taking into account isospin mixing has been reported (HS 71). The possibility of admixing in the nuclear giant (dipole) resonance by the ΛN CSB interaction has not yet been explored for hypernuclear ground states. Although the excitation energies involved are not small (recall also that the Λ should be excited to a p state to compensate for the parity change inherent in the dipole state) the specific collectivity of the giant resonance might turn out to be easily excitable by a ΛN interaction of a relatively long range which acts in opposite ways on protons and neutrons.

3. THE WELL DEPTH D_Λ FOR Λ IN NUCLEAR MATTER

As mentioned in the introduction, the pionic decay modes of spallation hypernuclei produced as a result of K^--mesons interacting with Ag and Br nuclei in nuclear emulsion give upper limits on B_Λ values for these hypernuclei (Lag+ 64, Lem+ 65). Allowing for a small (of the order of a few MeV at most) kinetic energy of the Λ in a big hypernucleus, an upper limit for D_Λ, the well depth spanned by the Λ–nuclear interaction in nuclear matter, is established. In this way the following value is found:

$$D_\Lambda = 27 \pm 1.5 \text{ MeV} \tag{31}$$

It seems that in these experiments heavy hypernuclei in the mass number range of $A \sim 60$–100 are produced and that for these hypernuclei the differences in B_Λ values do not exceed a few MeV in spite of the large mass range involved. It would be very desirable in the near future to deduce particular ground-state B_Λ values by using the $K^- \to \pi^-$ reaction [Eq. (6)] in flight for heavy nuclear targets. As mentioned in the introduction, strangeness exchange reactions may favor populating particular coherent excitations in the residual hypernucleus. Nevertheless, a fair distribution over hypernuclear levels is expected which might depict the ground state at the very end of a visible tail of strengths.[†] In this way a pattern of convergence of B_Λ values pertaining to heavy elements may be established and a formula similar in principle to the nuclear Bethe–Weizsäcker mass formula might emerge.

The physical picture behind the concept of "well depth" and its association with binding energies is quite transparent (Wal 60). The Λ-particle is distinct from the nucleons; no Pauli principle is imposed for the available states of a single Λ. Hence in the Λ–nucleus potential well generated by the Λ interacting with nucleons, the Λ occupies the lowest s state. As A increases to infinity, the depth of this potential well presumably remains constant in view of the observed constancy of central nuclear densities [$A\varrho(0) \approx$ const] and the short range of the ΛN interaction relative to the nuclear size. Indeed, to first order for the Λ–nucleus potential

$$V(r) = A \int v(\mathbf{r} - \mathbf{r}')\varrho(\mathbf{r}')\, d^3r' \approx A\varrho(r) \int v(\mathbf{r}'')\, d^3r'' \tag{32}$$

where v is a ΛN (presumably effective; should be nonlocal) interaction potential. The radius of this effective Λ–nucleus potential is very close to that of $\varrho(r)$, namely $R = r_0 A^{1/3}$. It is easy to show that the kinetic energy of a particle in the ground state of such a well goes to zero with increasing A. In fact, by the uncertainty principle

$$E_{kin} = \left\langle \frac{p^2}{2M_\Lambda} \right\rangle \sim \left(\frac{\hbar\pi}{R} \right)^2 \bigg/ 2M_\Lambda = \frac{\hbar^2\pi^2}{2M_\Lambda r_0^2}\, A^{-2/3} \xrightarrow[A\to\infty]{} 0$$

so that for this simple picture

$$B_\Lambda \xrightarrow[A\to\infty]{} D_\Lambda - (\hbar^2\pi^2/2M_\Lambda r_0^2)A^{-2/3} \tag{33}$$

where, according to Eq. (32), the Λ well depth may be evaluated as

$$D_\Lambda = A\varrho(0) \int v_{\Lambda N}(\mathbf{r})\, d^3r \tag{34}$$

[†] This is actually the situation for production of $^{12}_\Lambda$C and $^{16}_\Lambda$O by this technique (Bon + 73) and it remains to be seen whether the generalization discussed here holds.

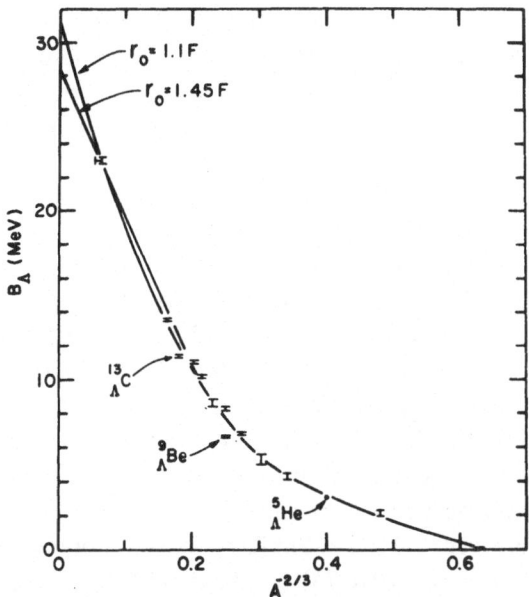

Fig. 6. The Λ separation energy B_Λ as a function of $A^{-2/3}$. The two curves are fitted to the values of B_Λ for the indicated values of r_0, so as to correspond for large A to a Λ kinetic energy given by an infinitely deep square well of radius $R = r_0 A^{1/3}$. [From (Bod 73).]

Equation (33), which holds strictly for an infinite square well, has been used to extrapolate from B_Λ values around $^{13}_{\Lambda}\text{C}$ to $A \to \infty$ by plotting the observed B_Λ values as a function of $A^{-2/3}$ and fitting them with the best straight line for $^{13}_{\Lambda}\text{C}$ and heavier nuclei. The resulting intercept with the ordinate then gives the value of D_Λ (RB 70). This is shown in Fig. 6 for two different choices of r_0. Another method (Dal 69a, DK69), which also takes into account to some extent surface effects (varying as $A^{-1/3}$) of the nuclear density, employs a fitted Fermi distribution for the nuclear density with a depth (as a result of folding in the ΛN potential) given by D_Λ. By requiring B_Λ values in the p shell and in heavy elements to be reproduced in such a Λ–nucleus potential well, one obtains a value for D_Λ. The outcome of these fitting procedures is the following value (BR 72):

$$D_\Lambda \approx 30 \pm 3 \text{ MeV} \tag{35}$$

with an upper limit of $D_\Lambda \lesssim 35$ MeV.

A simple calculation of the Λ well depth in nuclear matter gives, according to Eq. (34), a significant *overbinding*. For a central, smooth, spin-independent, ΛN potential $v(r)$ that fits the low-energy Λp scattering data, the volume integral of the interaction for a fixed range depends only weakly on the shape of the interaction and is given by $\sim 360 \pm 30$ MeV-fm³ for an intrinsic range $b = 1.5$ fm. Since $\Lambda \varrho(0) \approx 0.17$ fm⁻³, a value of $D_\Lambda \sim 60 \pm 5$ MeV is obtained, in gross disagreement with the phenomenological value given above. The use of a larger value for b, as indicated by scattering results, would increase the above value to 80 MeV. Allowing for some spin dependence of the ΛN potential, consistent with Λp scattering data, does not alleviate this discrepancy. Since the triplet ΛN interaction enters with a weight 3 relative to the singlet interaction, the discrepancy seems to lie with effects associated with the triplet ΛN interaction, as will be confirmed below. A calculation of the second-order contribution [the first order is given by Eq. (34)] indicates that the series may converge (BS 62, RB 70). However, the high-order terms do not seem large enough to bring the calculated value of the first-order term, 80 MeV, significantly closer to the "observed" 30 MeV. Similar results are obtained if an effective ΛN interaction [G-matrix (BR 73) or phase-shift approximation (BL 70)] is used in the sense of usual nuclear matter calculations. In the following we review results and various effects associated with a Λ-particle in nuclear matter as elaborated extensively by Bodmer and collaborators and described recently (BR 72, Bod 73). These authors use the reaction matrix (G) approach with the independent pair approximation for ΛN in nuclear matter. Linear effects in the nuclear density are then correctly taken into account, the well depth being given by

$$D_\Lambda = \sum_{\substack{k \leq k_F \\ k_0 = 0}} \langle \mathbf{k}_0, \mathbf{k} \mid G_{\Lambda N} \mid \mathbf{k}_0, \mathbf{k} \rangle \qquad (36)$$

Higher order effects in the density have recently been examined within the Bethe–Fadeev equations, which take into account ΛNN clusters as well as nuclear rearrangement effects. The additional contribution to D_Λ seems to be about 3 MeV (Dab 73a).

As stated before, the relatively large Λp effective ranges required by Λp low-energy scattering data indicate the existence of an appreciable ($c \sim 0.4$–0.6 fm) hard core in the ΛN s-wave interaction. It has been found (BR 73) that for fixed low-energy parameters a and r_0 the dependence of a calculated $D_\Lambda^{(s)}$ value (the superscript s stands for contribution from s-waves) on the hard core radius c is a strongly decreasing one, whereas the calculated $D_\Lambda^{(p)}$ increase only moderately with c. The overall effect on D_Λ of increasing the hard core radius is to decrease D_Λ, in agreement with

TABLE II

Self-Consistent s-State $\varLambda$ Well Depth $D_\varLambda^{(s)}$ as a Function of the Hard Core Radius c Employed by Bodmer and Rote (BR 73)

The $\varLambda N$ potential outside the hard core region is of the form $-v_0[\exp(-\mu r)]/\mu r$ adjusted to yield the same low-energy scattering parameters in the triplet and in the singlet $\varLambda N$ states: $a = -2$ fm, $r_0 = 3$ fm. The parameters employed in the nuclear matter calculation are $k_F = 1.4$ fm^{-1}, $\varDelta_N = 85.4$ MeV, and $M_N^* = 0.638 M_N$. For the range of $c = 0$–0.6 fm the p state $\varLambda$ well depth $D_\varLambda^{(p)}$ slowly rises from 17 to 22 MeV.

c, fm	0	0.2	0.3	0.43	0.6
$D_\varLambda^{(s)}$, MeV	60.0	59.5	56.7	50.1	37.7

the expectation that some of the high lying excitations are blocked for a $\varLambda N$ pair in the presence of a hard core. In Table II the dependence of $D_\varLambda^{(s)}$ on c is shown for $a = -2$ fm and $r_0 = 3$ fm. The attractive tail of the potential was fitted by a Yukawa shape.

It should be emphasized that the phenomenological method of deriving $D_\varLambda$ from $B_\varLambda$ values, by means of extrapolating a fit to Eq. (33) or employing Fermi distribution shapes for the $\varLambda$–nucleus interaction, suffers from some drawbacks when the $\varLambda N$ interaction involves an appreciable hard core. The latter increases the kinetic energy of the $\varLambda$-particle when interacting with nucleons, so that the expansion (33) may require further terms to simulate this effect (Wal 60). Effectively, however, a hard core component in the $\varLambda N$ interaction might be taken into account by keeping Eq. (33) with a slight decrease in the accepted value of r_0. One should then treat r_0 as a *parameter* to be fitted along with $D_\varLambda$ in Eq. (33). It is pretty clear that with a smaller value of r_0 a higher value for $D_\varLambda$ would be phenomenologically determined from the known $B_\varLambda$ values, but this is still a small effect. For example, for a hard core radius $c = 0.4$ fm and an initial choice of $r_0 = 1.1$ fm, the nuclear volume accessible to the interacting $\varLambda$ is decreased from $(4\pi/3)r_0^3 A$ to $(4\pi/3)(r_0^3 - c^3)A$, so that

$$r_0^2 \to r_0^2[1 - (c/r_0)^3]^{2/3} \sim r_0^2[1 - \tfrac{2}{3}(c/r_0)^3] \simeq r_0^2(1 - 0.075)$$

If for $r_0 = 1.1$ fm the kinetic energy of a $\varLambda$ in a hypernucleus of a heavy emulsion has been estimated by the appropriate fit depicted in Fig. 6 as about 8 MeV, for a hard core of $c = 0.4$ it would increase by less than 1 MeV, increasing by this quantity the deduced value for $D_\varLambda$, well within the uncertainty quoted in (35). In this context of discussing possible drawbacks of phenomenological extrapolations of B_A values, one should recall the important role played by B_A values around $^{13}_A$C in the fitting of a B_A curve as a function of $A^{-2/3}$; it is only for hypernuclear mass numbers above 13 that this curve levels off to approximately a straight line. However, for $^{13}_A$C there are certain surface effects which are unjustifiedly ignored and which on general grounds should give rise to a term proportional to $A^{-1/3}$ in any empirical formula for B_A.

Another objection, raised by Bhaduri et al. (Bha+ 68), has to do with effects nonlinear in the nuclear density. Expression (34) for D_A assumes contributions to the Λ well depth from a two-body ΛN interaction which couples to the nuclear density. However, as discussed earlier, the presence of the Σ channel leads to effective ΛNN forces which affect other properties of the nuclear medium than single-nucleon density. In particular these forces may contribute in a manner which depends crucially on NN correlations quadratic in the density. The strong noncentral terms of the TPE three-body interaction presumably derive their full strength from d-wave components of NN configurations in nuclear matter; a reliable evaluation of the latter is an ambitious project by itself.

Bodmer, Rote, and Mazza (BRM 70) investigated the effect of ΛN tensor forces for scattering and for Λ binding in nuclear matter. The tensor ΛN force couples a ΛN triplet s-wave to a triplet d-wave. The overall triplet potential is written as a sum

$$V_t(r) + V_T(r)S_{\Lambda N}(\hat{\mathbf{r}}) \tag{37}$$

of a central triplet interaction V_t and a tensor term with a shape V_T. The latter connects the s wave function u to the d wave function w:

$$u'' + k^2 u = V_t u + \sqrt{8}\, V_T w \tag{38a}$$

$$w'' + [k^2 - (6/r^2)]w = (V_t - 2V_T)w + \sqrt{8}\, V_T u \tag{38b}$$

To eliminate the d-channel in an approximate way, we argue that the main contribution of d-wave intermediate ΛN (free or in nuclear matter) states comes from a narrow energy region around $\bar{\varepsilon}_T$, similar to what was done earlier to eliminate a ΣN channel. Application of closure leads then to the *effective* ΛN triplet s-wave interaction

$$V_t(r) - [8V_T^2(r)/\bar{\varepsilon}_T] \tag{39}$$

where $\bar{\varepsilon}_T$ depends on the medium. For nuclear matter there is a restriction on available intermediate states. Single-particle energy gaps for Λ and N and the Fermi energy for N are exceeded by the relevant intermediate excitations. This leads quite generally to the relation

$$(\bar{\varepsilon}_T)_{\text{nuclear matter}} > (\bar{\varepsilon}_T)_{\text{scattering}} \tag{40}$$

so that some of the ΛN attraction due to operation of the ΛN tensor component is *suppressed* in nuclear matter relative to free scattering. However, due to the short range of the intrinsic ΛN tensor interaction, arising probably from K, η, and heavier mass exchanges, the appropriate average excitation $\bar{\varepsilon}_T$ is very high; for a Yukawa range of 0.4 fm, corresponding to K exchange, $\bar{\varepsilon}_T \gtrsim 3000$ MeV,[†] much higher than any of the obstructing energies mentioned above for excitation in nuclear matter. Hence for such a short range the suppression of the tensor ΛN interaction in nuclear matter is a minor effect. Bodmer *et al.* (BRM 70) find only a small reduction of about 4 MeV in the value expected for D_Λ. The effect of tensor coupling is then similar for scattering and for nuclear matter and an effective central potential may be used which is similar for both.

A suppression effect is more pronounced for the effective $\Lambda N \rightarrow \Sigma N$ coupling in nuclear matter (BR 71). As discussed earlier, the main coupling is between a triplet ΛN s-wave and a triplet ΣN d-wave. This time, however, the coupling contains an OPE component with a Yukawa range of about 1.4 fm, considerably larger than that for the ΛN tensor coupling. The relatively large range of the coupling means that lower intermediate d-wave excitation energies are involved, with the result that $\bar{\varepsilon}_{\Sigma N}$ is only about a few hundreds of MeV and the suppression in nuclear matter is quite appreciable. For "realistic" coupling constants $f^2_{\Lambda\Sigma\pi}$ and a moderate hard core radius, Bodmer and Rote (BR 71) find that the value for D_Λ should be reduced by 6–16 MeV due to suppression of the ΣN channel in nuclear matter.

It was argued in Section 2.1 that the ΛN p-wave interaction is considerably weaker than a L-independent interaction potential would predict. This is brought about by considering the measured forward–backward ratio in low-energy Λp scattering, as sketched in Fig. 2. On the theoretical side, p-wave ΛN phase shifts derived from a fit of the OBE model to YN reaction data prove to be exceptionally small (BR 73) and may even correspond to a slight net central repulsion. Since the p-wave ΛN interaction

---

[†] A central interaction of a similar range is characterized by a considerably smaller average excitation energy $\bar{\varepsilon}_c \sim 500$ MeV. This is due to the centrifugal barrier for the intermediate d-wave, which necessitates high excitation for penetration.

contributes about 20 MeV at full strength to D_Λ, a reduction of its strength by 50% corresponds to a decrease of about 10 MeV in the calculated value for D_Λ.

It is clear then that to obtain a considerable reduction in a calculated value for D_Λ, where the s-wave contribution alone is certainly bigger than 50 MeV for the ΛN interaction, by fitting low-energy scattering data (with a moderate hard core radius), a number of effects have to be invoked simultaneously, which may bring this calculated value down to the phenomenological value of about 30 MeV. These effects, investigated in some detail in the literature, include: (i) suppression of $\Lambda\Sigma$ coupling in nuclear matter—a reduction of about 10 MeV; (ii) suppression in the p-wave strength of the ΛN interaction—a reduction of about 10 MeV for a 50% suppression with respect to the s-wave interaction; (iii) suppression of the ΛN tensor component in nuclear matter—a reduction of about 5 MeV. A combination of *all* these reductions seems at present required in order to bring the calculated value of the Λ well depth into accord with the current phenomenological value, Eq. (35).

Finally we mention the effect of neutron excess in heavy hypernuclei on the Λ well depth. Actual stable nuclei above ^{40}Ca are characterized, among other things, by an increasingly larger neutron excess. In the preceding discussion we have tacitly assumed that the Λp interaction is the same as the Λn interaction and therefore we have constantly identified the measured Λp parameters with "ΛN scattering parameters." This assumption requires some modification in view of the charge symmetry breaking (see Section 2.4) expected for the ΛN interaction both from the measured difference $B_\Lambda(^4_\Lambda\text{He}) - B_\Lambda(^4_\Lambda\text{H})$ and the current theoretical notions about mixing of baryons and mesons. A rough way to estimate the effect of CSB on the calculation outlined for the Λ well depth is to assume Eq. (34) for the correction δD_Λ due to V_{CSB}. In first order in the difference $\Delta\bar{a} \equiv \bar{a}_n - \bar{a}_p$ (where $\bar{a} \equiv \frac{3}{4}a_t + \frac{1}{4}a_s$),

$$\delta D_\Lambda = -A\varrho(0)\frac{2\pi\hbar^2}{\mu_{\Lambda N}}\Delta\bar{a}\frac{N-Z}{2A} \tag{41}$$

For heavy nuclei, where $Z/A \sim 0.4$, $(N-Z)/2A \sim 0.1$ and the following estimate is obtained:

$$\delta D_\Lambda = -8.6\,\Delta\bar{a} \qquad (\delta D \text{ in MeV for } \Delta\bar{a} \text{ in fm}) \tag{42}$$

If the neutron–lambda spin-averaged scattering length is smaller in magnitude than the spin-averaged proton–lambda scattering length, i.e., the attraction in the Λn system is weaker than in the Λp system, the correction (42) is negative, decreasing the calculated value for D_Λ. For $\Delta \bar{a} = 0.3$ fm one obtains $\delta D_\Lambda \simeq -2.6$ MeV. It is our feeling, however, that Eq. (42) overestimates the above correction, but it is not clear to what extent this is indeed the case.

The neutron excess parameter $\eta \equiv (N - Z)/2A$ appears linearly for the charge-symmetry-breaking contribution. In addition, a quadratic term in η is expected to show up in an empirical formula for D_Λ, similar to the nuclear symmetry energy. This would be the first-order charge-symmetric contribution to the Λ well depth which is associated with neutron excess. The coefficient of such a term could in principle be determined by comparing B_Λ values of hypernuclei of the same mass number but having different values of isospin. At present, the only relevant information available is for p-shell hypernuclei, where $A = 7, 8, 9$ are the masses involved. However, one would like to possess more data for the elimination of other effects (spin, $\Lambda\Sigma$ suppression; see discussion on p-shell hypernuclei) intimately connected with isospin in the p shell before a definite conclusion can be reached.

At this stage it only remains to repeat what has been said throughout this chapter: More B_Λ values are required, mostly for medium and heavy mass numbers, where the present available information is limited to one rough number. With these B_Λ values, hopefully to be deduced by applying the (K^-, π^-) reaction in flight for heavy elements, a semiempirical hypernuclear mass formula could in principle be established. The various terms of such a hypothetical formula would probably include a kinetic energy term (the only one used to date), a surface term, a symmetry term, a charge-symmetry-breaking term, and possibly other terms related to the various suppression effects discussed above. The establishment of a hypernuclear mass formula would shed more light on the nuclear interaction of the Λ-particle, since more than one (D_Λ at present) parameter will emerge to bear on this interaction.

4. BINDING ENERGIES OF HYPERNUCLEAR SYSTEMS

Attempts to relate the measured B_Λ values of hypernuclei to ΛN interaction parameters have been extensively reported in the literature. In the s shell it is generally required that a set of ΛN parameters reproducing B_Λ values can be found so as to yield the low-energy measured Λp cross sections. Most of these calculations were constrained to deal with central (spin dependent, however) ΛN potentials $V_t(r)$ and $V_s(r)$, the idea being that these are the only operative components of the ΛN interaction for hypernuclear wave functions which have $L = 0$ for their spatial part. Of course, noncentral NN interactions, in particular the nuclear tensor force, are known to play a decisive role in building up higher L terms for the nuclear core. A ΛN weak tensor component could then benefit in first order form coupling a nuclear D-wave to a nuclear S-wave; a second-order contribution for which a Λ–nucleus d-wave component is built in is not really required for the ΛN tensor interaction to become operative in s-shell hypernuclei. However, realistic nuclear wave functions have so far not been used in s-shell hypernuclear calculations.

Of the three charge-symmetric B_Λ values known in the s shell (see Table I), namely for $A = 3$, 4, 5, only one proves essentially problematic in the following sense: For ΛN central interactions that fit low-energy scattering, an appreciable overbinding is found in calculations for $^5_\Lambda$He, namely instead of reproducing a B_Λ value close to 3 MeV, a value about 6 MeV or higher is calculated. This overbinding appears independent of the particular model employed (AGK 67, BNV 67, HT 67b) and has led to many speculations on the origin of such a discrepancy and how theoretically to envisage a mechanism for suppression of the basic ΛN interaction in $^5_\Lambda$He. In Section 4.2 we will discuss this problem in detail.

The situation is different in the p shell, where noncentral ΛN interaction components may contribute to hypernuclear binding energies in first order through "large" components of relevant wave functions. Here a microscopic approach is almost impossible and one invokes nuclear models for the nuclear core; the best studied one is the shell model. The interactions used are effective ones; it is hard to relate them directly to the basic ΛN interaction parameters because of certain renormalization effects. In Section 4.3 we shall discuss in detail shell model calculations in the p shell.

4.1. General Methods of Calculation

Here we would like to make a few obvious observations as to the nature of a B_Λ calculation.

4.1.1. A Rigid Core Approximation

In this approach, the first one to be used in hypernuclear calculations (DD 58), the assumption is made that the ΛN interaction does not change the properties of the nuclear core, described by a wave function Φ_0. Hence the core is considered as rigid; its internal degrees of freedom are described by Φ_0 and are governed therefore only by the pure nuclear Hamiltonian. The total hypernuclear wave function is $\Phi_0 \varphi(r)$, where r is the radius vector of the Λ with respect to the nuclear center of mass. The Λ s wave function $\varphi(r)$ is given by the solution of the two-body Λ–nucleus Schrödinger equation

$$[-(\hbar^2/2\mu_{\Lambda N})\Delta_{\mathbf{r}} + v(r)]\varphi(r) = -B_\Lambda \varphi(r) \qquad (43)$$

corresponding to the highest eigenvalue for B_Λ, where

$$v(r) = A \int \varrho(\mathbf{r}')V_{\Lambda N}(\mathbf{r} - \mathbf{r}') \, d^3r' \qquad (44)$$

and $\varrho(r)$ is the appropriate one-body nuclear density:

$$\varrho(r) = \int | \Phi_0(\mathbf{r}, \mathbf{r}_2, \ldots, \mathbf{r}_A) |^2 \, d^3r_2 \cdots d^3r_A \qquad (45)$$

Spin degrees of freedom are suppressed in these equations. It is possible to take them into account but the generalization detracts from the simplicity of the approach. Hence the application of the rigid core model has mainly been reported for spinless-core hypernuclei, such as $^5_\Lambda$He, $^7_\Lambda$Be, $^9_\Lambda$Be, and $^{13}_\Lambda$C (BM 65). When the rigid core model was applied to hypernuclei whose core nucleus possesses a nonzero value of spin, such as $^4_\Lambda$H, the further assumption was made that a separation between space and spin is possible for the nuclear part and that the nuclear spin couples directly to the Λ spin to give the total hypernuclear spin. The hypernuclear wave function corresponding to such an approach is of the form

$$\Phi_0(\mathbf{r}_1, \ldots, \mathbf{r}_A)\varphi(r)[\chi_0 \otimes \chi]_J \qquad (46)$$

where χ_0 and χ are nuclear and Λ spin wave functions, respectively, coupled together to a total hypernuclear spin J (DD 58).

The highest B_Λ value obtained by solving Eq. (43) gives a lower bound for the physical B_Λ value, the evaluation being variational in the following sense: Suppose a trial hypernuclear wave function is taken of the form $\Phi_0(\{\mathbf{r}_i\})\varphi(r)$, where Φ_0 *is* the nuclear core ground-state wave function

$$H_N\Phi_0 = -\varepsilon_0\Phi_0 \qquad (47)$$

The total binding energy E_0 is given by $\varepsilon_0 + B_\Lambda$ and is approached from below by the lowest expectation value of the total Hamiltonian $H = H_N + T_A + \sum V_i(\mathbf{r} - \mathbf{r}_i)$, for a wave function of the type $\Phi_0 \varphi(r)$:

$$E_0 \geq E = \max_{\varphi(r)} \langle \Phi_0 \varphi(r) \mid -H \mid \Phi_0 \varphi(r) \rangle$$

$$= \varepsilon_0 + \max_{\varphi(r)} \langle \varphi(r) \mid -T_A - v(r) \mid \varphi(r) \rangle$$

Hence

$$B_A = E_0 - \varepsilon_0 \geq \max_{\varphi(r)} \langle \varphi(r) \mid -T_A - v(r) \mid \varphi(r) \rangle \qquad (48)$$

and the best (approaching from below) value of a calculated B_A value is therefore obtained by finding a $\varphi(r)$ that minimizes the expectation value of $T_A + v$, in accordance with Eq. (43). We stress that had we not taken ε_0 to be the *observed* nuclear binding energy, the evaluated B_A would not provide in general a lower bound to the observed one, since the latter is defined as the difference between two such quantities, E_0 and ε_0, for each of which individually a maximum principle exists.

Actually the rigid core approximation is the first natural step toward the following expansion:

$$\Psi(\mathbf{r}, \{\mathbf{r}_i\}) = \sum_i \Phi_i(\{\mathbf{r}_i\}) \varphi_i(\mathbf{r}) \qquad (49)$$

where the summation extends over *all* nuclear-core eigenfunctions

$$H_N \Phi_i = -\varepsilon_i \Phi_i \qquad (50)$$

The unknown functions $\varphi_i(\mathbf{r})$ satisfy coupled equations for an eigenvalue $-E$:

$$T_A \varphi_j(\mathbf{r}) + v_{jj}(\mathbf{r}) \varphi_j(\mathbf{r}) + (E - \varepsilon_j) \varphi_j(\mathbf{r}) = -\sum_l v_{jl}(\mathbf{r}) \varphi_l(\mathbf{r}) \qquad (51)$$

where

$$v_{jl}(\mathbf{r}) = A \int \Phi_j^*(\mathbf{r}', \mathbf{r}_2, \ldots, \mathbf{r}_A) V_{AN}(\mathbf{r} - \mathbf{r}') \Phi_l(\mathbf{r}', \mathbf{r}_2, \ldots, \mathbf{r}_A)$$

$$\times d^3 r' \, d^3 r_2 \cdots d^3 r_A \qquad (52)$$

This set of equations can be truncated at some stage, leaving a finite number of coupled functions, $j = 0, 1, \ldots, k$. The largest E that solves this finite set is the "best" total binding energy evaluated within the space spanned by $\Phi_0, \ldots, \Phi_k$ and $E - \varepsilon_0$ therefore provides a variational lower bound for the physical B_A value in question; any further inclusion of new nuclear eigenfunctions in the set of equations dealt with can only improve this bound. We point out that for a spin-independent central ΛN interaction, the nuclear states $i \neq 0$ of Eq. (49) are connected with the nuclear ground state $i = 0$ only if these excited states involve an excitation to a higher nuclear shell. This necessarily involves exciting the Λ from a $1s$ state generated by v to a higher single-particle state. It is thus reasonable to expect good results form the rigid core approximation, not only for $^5_\Lambda$He, where the first nuclear excitations are above 20 MeV, but also for some of the p-shell hypernuclei, where all the low-lying nuclear core levels are well described in terms of a $(1p)^{4-5}$ configuration. If, on the other hand, appreciable spin-dependent ΛN interaction terms exist, in particular tensor and spin–orbit terms, the rigid core approximation is expected to become invalid for $A > 5$ unless special circumstances prevail.

4.1.2. Variational Calculations

Since the first variational calculation for $^3_\Lambda$H (DD 59), such calculations have been extensively reported by Herndon and Tang (HT 67a, b, 68) and recently by Dalitz, Herndon, and Tang (DHT 72) for hypernuclei in the s shell. In this approach a form is chosen for the total hypernuclear wave function Ψ, containing a few parameters which provide both short-range correlations connected with the repulsive baryon–baryon interaction at small distances and the correct asymptotic behavior for each baryon in the system. The maximum value of $(\Psi, -H\Psi)$ is sought as a function of the assumed parameters and is determined. Likewise, the maximum value of $(\Phi, -H_N\Phi)$ is found as a function of the parameters in the assumed form of the nuclear wave function Φ and an estimate (not a bound) for B_Λ is thus established:

$$B_\Lambda = \max_{\Psi}(\Psi \mid -H \mid \Psi) - \max_{\Phi}(\Phi \mid -H_N \mid \Phi) \qquad (53)$$

The reason for doing the nuclear calculation instead of simply subtracting the observed nuclear ground-state binding energy ε_0 from max $(\Psi, -H\Psi)$ is twofold. First, to check within the calculation, for NN forces that are generally deduced from scattering information, that the trial wave function is flexible enough to correctly approximate ε_0 and other relevant nuclear quantities (particularly size). Second, whatever serious drawbacks are inherent in the evaluation of the nuclear binding energy, these may to a large extent cancel in the subtraction procedure [Eq. (53)]. The problem, however, is quite complicated since for *realistic* nucleon–nucleon interactions it is difficult to obtain good evaluated values for both nuclear binding energy and nuclear size simultaneously. The correct nuclear size is tremendously important for obtaining [by folding, as above—Eq. (44)] a meaningful Λ–nucleus interaction; a compressed nucleus would lead to an over-attractive Λ–nucleus force, which would result in overshooting B_Λ appreciably. For this reason the simple rigid core approximation, in which the Λ–nucleus potential is constructed from an experimentally deduced nuclear density $\varrho(r)$, may give results which are considerably better than expected.

4.1.3. Hartree–Fock Calculations

In this approach it is assumed that a set of single-particle orbitals can be found for the nucleons and for the Λ in a self-consistent way, namely that these orbitals are eigenstates of a single-particle Hamiltonian obtained by generating single-particle N and Λ potentials from the same orbitals. The extension of the nuclear Hartree–Fock (HF) equations to the hypernuclear case is straightforward (BG 70, LGB 69). There exists of course a coupling between the equations for the nucleons and that for the Λ. The self-consistent Λ potential is constructed from the ΛN interaction and the nuclear potential also receives contribution from this interaction, so that the nuclear orbitals in the presence of the Λ are not the same as in its absence. Since the total hypernuclear wave function is of the form

$$\Psi = \det_{i=1,\ldots,A} \{\varphi_i\} \cdot \varphi_\Lambda \tag{54}$$

the HF procedure gives rise to a definite core-polarization correction over the simple product (rigid core approximation) given by Eq. (46); the nuclear HF wave function is given by:

$$\Phi = \det_{i=1,\ldots,A} \{\varphi_i^{(0)}\} \tag{55}$$

There are several drawbacks in the application of HF equations to hypernuclear B_Λ values. Center-of-mass correction for light systems, projection of a definite angular momentum, and isospin are effects involving a few MeV or less that are not considered or are ill treated in a standard HF calculation. It is not obvious that these effects cancel out for a B_Λ value, namely for a difference between two binding energies in the nuclear and hypernuclear cases, respectively. Unless such cancellations reduce the contributions from the effects mentioned above to an order of 0.5 MeV, it is difficult to ascribe a physical meaning to B_Λ values obtained in Hartree–Fock calculations.

4.1.4. Cluster Calculations

When binding is small between two (or more) hypernuclear constituent clusters relative to the separation energies involved within each of these clusters, it may be useful to consider the system as being actually composed of such basic clusters. The dynamical variables then exclude the internal degrees of freedom of the separate clusters and the latter are considered as basic entities, information for which can be obtained from other processes where they appear as either free or bound entities. For example, in the discussion of the rigid core approximation $^5_\Lambda$He is being treated as a two-body cluster whose constituents are Λ and α (^{4}He); the internal binding energies of the latter are considerably larger than the B_Λ value of about 3 MeV for the separation energy between the two constituents of this cluster. Three-body hypernuclear clusters discussed in the literature are $^6_\Lambda$He(α-n-Λ) (LR 67), $^7_\Lambda$Li (α-d-Λ) (MB 66), [†] $^9_\Lambda$Be (α-α-Λ), and $^{13}_\Lambda$C (α-α-α-Λ) (AMB 65; BA 64; Ray 72, 73). In these cluster calculations the potential between Λ and a nuclear cluster $^A Z$ is required to fit the B_Λ value of $^{A+1}_{\ \ \Lambda}Z$. For example, in the last two cases mentioned above the only Λ–nuclear potential required is the Λ–α potential and this is taken to fit B_Λ($^5_\Lambda$He). An obvious advantage of cluster calculations is that they take into account some of the readily available nuclear correlations with and without the Λ. The main disadvantage of these calculations is their failure to give a systematic pattern over a large range of hypernuclear masses. By their nature they are confined to some isolated events. In passing we note that calculations on available B_Λ values of *double* $\Lambda\Lambda$-hypernuclei ($_{\Lambda\Lambda}^{\ 6}$He and $_{\Lambda\Lambda}^{\ 10}$Be or $_{\Lambda\Lambda}^{\ 11}$Be) have been done (AB 67a, 67b; BA 65; DR 64) mostly in the framework of a cluster decomposition, i.e., α-Λ-Λ for $_{\Lambda\Lambda}^{\ 6}$He and α-α-Λ-Λ for $_{\Lambda\Lambda}^{\ 10}$Be.

4.1.5. Shell Model Calculations

Such calculations have been applied (GSD 71, 72; LHC 70) in the p shell for those hypernuclear species for which the core nucleus is describable in terms of an intermediate coupling (between LS and jj coupling). The main idea of a shell model calculation is that throughout the p shell the Λ wave function remains pretty much the same and hence the relevant

---

[†] Unjustified according to the criterion given above. The evidence for a cluster structure derives from the nuclear core ^{6}Li.

Λ–nuclear matrix elements can all be parametrized in terms of a relatively small number of reduced matrix elements. This basic assumption can be justified to a great extent by noting that two opposing effects operate on the spatial extension of the $1s$ Λ wave function: The increasing B_Λ value pushes the Λ wave function to smaller Λ–nucleus distances, since at the beginning of the p shell the Λ wave function extends appreciably outside the nuclear core due to the small B_Λ value of ${}^5_\Lambda$He (about 3 MeV). On the other hand, as A increases, the Λ–nucleus interaction volume increases, which pushes the Λ wave function to larger Λ–nucleus distances in order to keep the nuclear volume within the close proximity of the interacting Λ. These two effects cancel each other to a large extent, leaving the Λ wave function almost the same throughout the p shell, as shown in Fig. 9 (Section 4.3). From a shell model fit of the observed B_Λ values a set of effective ΛN matrix elements (and possibly ΛNN) is deduced. These contain matrix elements of the ΛN spin–orbit and tensor forces, which give rise to first-order contributions in the p shell. As we shall discuss later in greater detail, the shell model diagonalization takes into account the coupling of Λ to any of the low-lying nuclear states describable in terms of the nuclear $(1p)^{A-5}$ configuration, so that a distortion of the nuclear core by the Λ is allowed. The main drawback of a shell model calculation is that small variations with A in the size of the nuclear core may give rise to a caricature of effects connected with the Λ spin, such as spin–spin, part of the spin–orbit, and the tensor force matrix elements (BM 65). At present, available B_Λ values are associated with ground states, but in due time such values should become known for excited hypernuclear states and a restricted shell model calculation for *one* hypernuclear species will give rise to more reliable information of the ΛN effective matrix elements.

4.2. s-Shell Hypernuclei

Here we will report on B_Λ calculations in the s shell. Most of this section is devoted to a review of the series of papers by Herndon and Tang (HT 67a, b, 68), who together with Dalitz (DHT 72), recently summarized their variational calculations. Other calculations either reach similar results or are not sufficiently detailed to present a complete picture. The ΛN potential adopted by Dalitz, Herndon, and Tang (DHT 72) is central outside a hard core of radius 0.45 fm and has an intrinsic range of 2.0 fm. Charge-symmetry-breaking is allowed, to be constrained by the CSB effect in the $A = 4$ hypernuclei (see Section 2.4). Thus

$$U_{\Lambda N} = \tfrac{1}{4}(3 + \sigma_\Lambda \cdot \sigma_N)U_t(r_{\Lambda N}) + \tfrac{1}{4}(1 - \sigma_\Lambda \cdot \sigma_N)U_s(r_{\Lambda N}) + U_{\text{CSB}}(r_{\Lambda N}) \quad (56)$$

where

$$U_t(r) = \infty, \qquad U_s(r) = \infty \qquad r < d \qquad (57a)$$

$$U_i = \frac{1}{r} U_{0i} e^{-\lambda(r-d)}, \qquad i = s, t \quad r > d \qquad (57b)$$

with

$$d = 0.45 \text{ fm}, \qquad \lambda = 3.219 \text{ fm}^{-1} \qquad (57c)$$

For the CSB component, as well as for three-body ΛNN forces, several choices are considered and these are discussed later. The nucleon–nucleon interaction used by these authors is central, of a form similar to the charge-symmetric component of $U_{\Lambda N}$ [Eq. (56)], and contains, in addition, a Coulomb interaction for the protons. The same hard core radius is chosen for both ΛN and NN pairs. The other available NN parameters are chosen so as to give a good fit to the NN effective range parameters and satisfactory values for calculated binding energies and rms radii of the core nuclei ^{3}H and ^{4}He. Thus their nuclear calculation gives for the cores: $E(^2$H$) = -2.225$ MeV (exp., -2.225 MeV), $E(^3$H$) = -7.42 \pm 0.06$ MeV (exp., -8.48 MeV), and $E(^4$He$) = -28.31 \pm 0.19$ MeV (exp., -28.30 MeV). Note that no tensor component is used here for the NN interaction. The hypernuclear trial wave function used by Dalitz, Herndon, and Tang (DHT 72) is of the form $\Psi = \psi\chi$, where the $L = 0$ spatial component ψ is symmetric with respect to all nucleons

$$\psi = \prod_{i>j}^{A-1} g(r_{ij}) \prod_{i=1}^{A-1} f(r_{\Lambda i}) \qquad (58)$$

The functions g and f are constructed, up to some parameters whose value is to be determined by the variational calculation, to give the right asymptotic behavior for one baryon removed from the system. The spin–isospin function χ can be eliminated, once spatial expectation values for $L = 0$ are considered, by recalling that for a hypernucleus of spin J and nuclear spin J_N

$$\left(\psi\chi\left|\sum_i \tfrac{1}{4}\boldsymbol{\sigma}_\Lambda \cdot \boldsymbol{\sigma}_i \Delta(r_{\Lambda i})\right|\psi\chi\right) = (\psi \mid \Delta(r_{\Lambda 1}) \mid \psi) \cdot (\chi \mid \mathbf{s}_\Lambda \cdot \mathbf{J}_N \mid \chi)$$
$$= \tfrac{1}{2}\bar{\Delta}[J(J+1) - J_N(J_N+1) - \tfrac{3}{4}] \qquad (59)$$

where $\bar{\Delta}$ is the value of the above spatial matrix element. Hence we obtain for the charge-symmetric part of (56)

$$\left(\Psi_A\left|\sum_{i=1}^{A-1} U_{\Lambda i}\right|\Psi_A\right) = (A-1)(\psi_A \mid U_\Lambda(r_{\Lambda 1}) \mid \psi_A) \qquad (60)$$

where

$$U_A(r) = \infty \qquad\qquad r < d \qquad\qquad (61a)$$

$$U_A(r) = -U_{0A}e^{-\lambda(r-d)}, \qquad r > d \qquad\qquad (61b)$$

and

$$U_{03} = \tfrac{1}{4}U_{0t} + \tfrac{3}{4}U_{0s}, \qquad J = \tfrac{1}{2}, \quad {}^{3}_{\Lambda}\mathrm{H} \qquad\qquad (62a)$$

$$U_{04} = \tfrac{1}{2}U_{0t} + \tfrac{1}{2}U_{0s}, \qquad J = 0, \quad ({}^{4}_{\Lambda}\mathrm{H}, {}^{4}_{\Lambda}\mathrm{He}) \qquad\qquad (62b)$$

$$U_{05} = \tfrac{3}{4}U_{0t} + \tfrac{1}{4}U_{0s}, \qquad J = \tfrac{1}{2}, \quad {}^{5}_{\Lambda}\mathrm{He} \qquad\qquad (62c)$$

are the ΛN potential strengths appropriate to the deduced spin values of hypernuclear ground states in the s shell. For $A = 3, 4$, excited states with $J = \tfrac{3}{2}, 1$, respectively, may turn out to be particle-stable. Their analysis runs parallel to that of ground states, with the following modification of the appropriate ΛN potential strengths:

$$U_{03}^{*} = U_{0t}, \qquad\qquad J = \tfrac{3}{2}, \quad {}^{3}_{\Lambda}\mathrm{H}^{*} \qquad\qquad (62d)$$

$$U_{04}^{*} = \tfrac{5}{6}U_{0t} + \tfrac{1}{6}U_{0s}, \qquad J = 1, \quad ({}^{4}_{\Lambda}\mathrm{H}^{*}, {}^{4}_{\Lambda}\mathrm{He}^{*}) \qquad\qquad (62e)$$

DHT evaluate the various expectation values by the Monte Carlo method, kinetic energies as well as potential energies. For each hypernucleus this is done for a fixed value of the appropriate U_{0A}, and consequently the corresponding B_Λ value is determined. By using the measured values of $B_\Lambda({}^{3}_{\Lambda}\mathrm{H})$, $B_\Lambda({}^{4}_{\Lambda}\mathrm{H})$, and $B_\Lambda({}^{4}_{\Lambda}\mathrm{He})$, three parameters are determined: $U_{0t} = 412.35$ MeV, $U_{0s} = 432.83$ MeV, and the strength of an assumed spin–spin CSB component in the ΛN interaction designed to yield a difference of $(\Delta B_\Lambda)_{\mathrm{CSB}} = 0.46$ MeV[†] for the $A = 4$ hypernuclei. With these values the following ΛN effective-range parameters are obtained:

$$a_t^p = -1.77 \text{ fm}, \qquad r_t^p = 3.45 \text{ fm} \qquad\qquad (63a)$$

$$a_s^p = -1.83 \text{ fm}, \qquad r_s^p = 3.39 \text{ fm} \qquad\qquad (63b)$$

$$a_t^n = -1.61 \text{ fm}, \qquad r_t^n = 3.61 \text{ fm} \qquad\qquad (63c)$$

$$a_s^n = -2.45 \text{ fm}, \qquad r_s^n = 3.01 \text{ fm} \qquad\qquad (63d)$$

the Λp parameters of which are in a close proximity to the "best" values quoted earlier [see (10)].

[†] Extra Coulomb repulsion due to nuclear compression amounts to 0.20 MeV for ${}^{4}_{\Lambda}\mathrm{He}$ in their calculation. When added to the measured B_Λ value known then, one obtains 0.46 MeV for the ΛN charge-symmetry-breaking effect for $A = 4$.

When a spin-independent CSB component is assumed, the resulting ΛN effective-range parameters are

$$a_t{}^p = -1.94 \text{ fm}, \qquad r_t{}^p = 3.30 \text{ fm} \qquad (64a)$$

$$a_s{}^p = -2.45 \text{ fm}, \qquad r_s{}^p = 3.01 \text{ fm} \qquad (64b)$$

$$a_t{}^n = -1.47 \text{ fm}, \qquad r_t{}^n = 3.77 \text{ fm} \qquad (64c)$$

$$a_s{}^n = -1.83 \text{ fm}, \qquad r_s{}^n = 3.39 \text{ fm} \qquad (64d)$$

which for the proton part result in a satisfactory fit to the measured low-energy Λp cross sections, although the larger magnitude [compared to Eqs. (63a)–(63d)] of the Λp scattering lengths indicates a stronger Λp force.[†] We note that for the singlet quantities in Eqs. (63a)–(63d) and (64a)–(64d) the interchange $p \leftrightarrow n$ leads from one set to the other. This is clear from the observation that for a spin–spin CSB potential its contribution to B_Λ of $^4_\Lambda$H$(J = 0)$ comes from the singlet Λp state and within the variational calculation described above is constrained to give the same value as a spin-independent CSB potential would yield. The latter, however, receives a contribution for the state mentioned above from the Λn interaction, and since it is spin independent, the Λn singlet contribution for this potential equals the Λp singlet contribution for the former CSB potential. This relation between different possible sets of ΛN low-energy parameters on the one hand and different $J = 1$ excitation patterns for $A = 4$ on the other hand (since a spin-independent and spin–spin CSB, each fitted separately to the $J = 0$, $A = 4$ isodoublet, lead to quite different excitation energies for the $J = 1$ appropriate states) indicates the type of relationship existing between free ΛN s-wave scattering information and some excitation properties of particular hypernuclei. At present, it would seem premature to speculate on which of the ΛN low-energy sets of parameters, Eqs. (63a)–(63d) or (64a)–(64d), is closer to giving the correct excitation pattern for $A = 4$, since it is not yet established whether the 1.09-MeV γ-ray observed by Bamberger *et al.* (Bam+ 71, 73) corresponds to $^4_\Lambda$H* or $^4_\Lambda$He* and whether or not the second line, at 1.42 MeV, observed by them is of hypernuclear origin. On the calculational side the predicted excitation energies for the $A = 4$ species are very sensitive to the input $B_\Lambda(^3_\Lambda$H$)$. An

---

[†] Note that $\Delta\bar{a} \equiv \bar{a}_n - \bar{a}_p = 0.77$ fm for the spin-average scattering lengths of Eqs. (64a)–(64d), which, according to Eq. (42), implies a CSB *reduction* in the Λ well depth by 6.6 MeV.

increase of the latter by about 0.05 MeV leads to an increase by approximately 0.3 MeV in the excitation energies of the $J = 1$ states for $A = 4$. If $B_\Lambda(^3_\Lambda H)$ were decreased below 0.07 MeV, the $J = 1$ $^4_\Lambda H$ state would become the ground state, which is empirically excluded.

With U_{0s} and U_{0t} determined from $B_\Lambda(^3_\Lambda H)$ and $B_\Lambda(CS, A = 4)$ it is a simple matter to "predict" a B_Λ value for $^5_\Lambda He$. The value obtained by DHT, $B_\Lambda(^5_\Lambda He) = 5.46$ MeV, greatly exceeds the measured value of 3.12 $\pm$ 0.02 MeV. The same feature is characteristic of most other calculations in the field, which use completely different methods, such as a rigid core approximation (Dal 69a) and Hartree–Fock calculations (BG 70, GGW 69), provided ΛN interactions that fit the low-energy Λp scattering parameters are employed. This calculational overbinding of $^5_\Lambda He$ is well demonstrated in Fig. 7, where the calculated (by DHT) $B_\Lambda(^5_\Lambda He)$ is displayed as function of $B_\Lambda(^3_\Lambda H)$ and $B_\Lambda(CS, A = 4)$. It is clear that in spite of the strong dependence of the calculated $B_\Lambda(^5_\Lambda He)$ on the two measured B_Λ

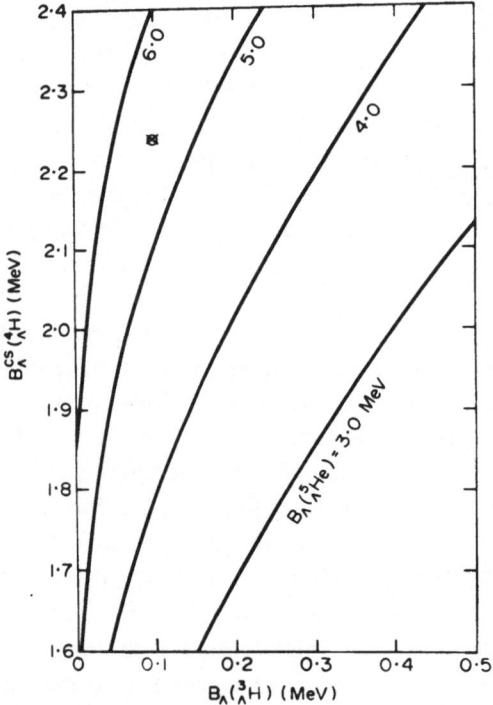

Fig. 7. The binding energy $B_\Lambda(^5_\Lambda He)$, calculated with two-body ΛN potentials with a hard core radius of 0.45 fm and intrinsic range of 2.0 fm, displayed as a function of the fitted values of $B_\Lambda(^3_\Lambda H)$ and $B_\Lambda(CS, ^4_\Lambda H)$. The cross corresponds to the values of these latter B_Λ values used by the authors. [From (DHT 72).]

values, in particular $B_A({}^3_A\text{H})$, the gap between the measured value of $B_A({}^5_A\text{He})$ and the calculated one is too wide to be bridged by more exact and refined measurements of these B_A values.

The problem of overbinding in ${}^5_A\text{He}$ has attracted many speculations as to its origin. Three possible mechanisms have been mentioned. (i) suppression of the ΛN tensor interaction in ${}^5_A\text{He}$ due to the zero spin of the nuclear core (Dal 67), (ii) suppression of the Σ channel (Bod 66), and (iii) repulsive three-body ΛNN forces (BLN 67, Gal 66). We will now discuss these suggestions.

Among the considered s-shell hypernuclei, ${}^5_A\text{He}$ obtains the largest contribution per ΛN pair [Eqs. (62a)–(62c)] from the triplet interaction. The presence of a significant tensor component in the ΛN interaction implies that in the triplet ΛN channel two waves, s and d, are mixed together. For an isolated ΛN pair the d-wave may be eliminated in favor of an approximate additional central interaction. When a ΛN s-wave interaction is fitted to correctly yield the triplet ΛN low-energy scattering parameters, it necessarily includes the original central as well as the additional (effective) central interaction. However, the argument continues, the additional central interaction, necessarily attractive, should be considerably suppressed in ${}^5_A\text{He}$ since it derives from a ΛN tensor force which is a *vector* in the nuclear labels ($\sim \mathbf{s}_A \cdot \mathbf{A}_i$, where $\mathbf{A}_i = [\mathbf{s}_i \otimes Y_2(\hat{\mathbf{r}}_i)]^{(1)}$). The expectation value of such a vector for a ${}^5_A\text{He}$ state based on a ${}^4\text{He}$ ground state vanishes since $J_N({}^4\text{He}) = 0$. Note that this argument does not require that ${}^4\text{He}$ be described by one LS component, the most symmetric one being ${}^1S_0$. An admixture of a ${}^5D_0$ component, for example, will not change the above conclusion as long as $J_N = 0$ holds for the core nucleus. Of course, the Λ may induce a distortion of the nuclear core allowing admixture of $J_N{}^\pi = 1^+$ nuclear excitations, particularly those containing a significant ${}^3D_1$ component. However, such nuclear levels have not been observed below 30 MeV excitation, which makes their contribution small. The possibility exists for the Λ to be excited to a $d_{3/2,5/2}$ state or to a $p_{1/2,3/2}$ state, but each of these possibilities for excitation of Λ via the ΛN tensor interaction must be accompanied by an appropriate excitation of the nuclear core to compensate for the change in Λ spin and/or parity. The energy gap for such states would involve at least 30 MeV (20 MeV for the nucleus and 10 MeV for the Λ, which is a conservative estimate). These considerations indicate that the second-order contribution of the ΛN tensor force might be suppressed in ${}^5_A\text{He}$. However, an approximate Hartree–Fock calculation by Law *et al.* (LGB 69) indicates that suppression effects as indicated above (for the ΛN tensor force generated by an exchange of K or η) can reduce the calculated value of $B_\Lambda(^5_\Lambda He)$ by 0.5 MeV at most, insufficient to explain the ~2.5 MeV overbinding. In $^3_\Lambda H$ and $^4_\Lambda H$ a tensor suppression does not initially seem an important factor. This is true since realistic ground-state core wave functions for these species contain, in addition to the dominant $^3S_1$ and $^2S_{1/2}$ components, appreciable admixtures (~0.20 in amplitude at least) of $^3D_1$ and $^4D_{1/2}$ components, respectively. Since $J_N \neq 0$ holds for ^2H and ^3H, a tensor contribution is allowed in *first order* through coupling of these admixtures to the appropriate dominant configurations. For example, in $^4_\Lambda H$ the matrix element

$$\langle ^2S_{1/2}, \, s_\Lambda = \tfrac{1}{2}; \, J = 0 \text{ or } 1 \,|\, \text{tensor} \,|\, ^4D_{1/2}, \, s_\Lambda = \tfrac{1}{2}; \, J = 0 \text{ or } 1\rangle \qquad (65)$$

does not vanish. This, together with the less important role that the ΛN triplet interaction plays in the $A = 3, 4$ hypernuclei (with respect to $^5_\Lambda He$), leads to $^5_\Lambda He$ as probably the special s-shell hypernucleus for which tensor suppression may constitute a significant effect. Even so, it would be very instructive to evaluate ΛN tensor contributions arising from realistic D-wave components in the nuclear core, such as indicated by Eq. (65).

In the *nuclear case*, it is known that the NN tensor force, of a large range due to OPE, is strongly suppressed in ^4He and nuclear matter. However, for the parallel hypernuclear problem presented here the appropriate ΛN tensor interaction is due to heavy meson exchange and its range is thus too small, at most 0.4 fm, to act coherently on an assembly of nucleons and a lambda situated at an average distance of roughly 2 fm from each other (Dal 67). The smallness of the tensor suppression is also borne out in the nuclear matter calculations of Bodmer *et al.* (BRM 70), where the intermediate states reached in the second-order ΛN tensor contribution lie much higher in energy than the Fermi energy and the gap in the single-nucleon spectra. This is a direct consequence of the short range expected for the ΛN tensor force and the subsequent incoherence pointed out above.

The other two mechanisms mentioned above are related to each other. As argued in Section 2.3, an elimination of the Σ channel (i) modifies the two-body free ΛN interaction and (ii) induces three-body ΛNN forces. The latter feature has been discussed earlier with the result that for s-shell hypernuclei, taking into account only s-wave bonds between pairs of the form Λi and Λj, this ΛNN interaction [Eq. (26)] assumes the form (BLN 67)

$$V(\Lambda; i, j) = C_p[1 + (3\cos^2\theta - 1)T(r_{\Lambda i})T(r_{\Lambda j})]Y(r_{\Lambda i})Y(r_{\Lambda j}) \qquad (66)$$

where

$$\cos\theta = \hat{\mathbf{r}}_{i\Lambda} \cdot \hat{\mathbf{r}}_{j\Lambda}$$

and in deriving Eq. (66) use was made of the relation $-\frac{1}{3}(\tau_i \cdot \tau_j)(\sigma_i \cdot \sigma_j)$ $= 1$ for spatially symmetric nuclear core wave functions as appropriate to the calculations described above. We note that the first term in the square brackets in Eq. (66) is a central term independent of angles and is weakly repulsive. Its largest effect in the s shell is expected for $^5_\Lambda$He due to the six nuclear bonds ($i < j$) available and the relative compactness of this system. With $C_p \sim 1$ MeV this central part of the ΛNN interaction contributes a repulsion of only about 0.25 MeV in $^5_\Lambda$He, a small amount to alleviate the overbinding problem. However, the second term of the ΛNN force [Eq. (66)] is angle dependent and draws strong contributions from small ΛN distances where the singular $T(r)$ factors make them significant, considerably more than those due to the central term. DHT evaluated this term as function of a cutoff parameter δ for the ΛN distances involved, starting from $\delta = d$, the hard core radius, and increasing it gradually. Below this cutoff value other competing mechanisms may dominate and the use of TPE might become useless. They find that for δ in the range 0.45–0.75 fm the contribution of the above three-body force to $^5_\Lambda$He (for the optimum wave function evaluated with two-body forces alone) is attractive, whereas for larger values of δ it becomes repulsive. The maximum repulsion, attained for $\delta \sim 1.2$ fm, is about 0.7 MeV for $^5_\Lambda$He, again insufficient for solving uniquely the overbinding problem. We point out that short-range NN and ΛN correlations may significantly affect the ΛNN interaction at small distances. In the case discussed above the function $T(r_{\Lambda i})T(r_{\Lambda j})$ peaks at the boundary $r_{\Lambda i} = r_{\Lambda j} = \delta$, where the cutoff is enforced. Since nucleons i and j cannot come closer to each other than the hard core radius d, one obtains $\cos\theta \leq 1 - (d^2/2\delta^2)$, which for $\delta \to d$ yields $\cos^2\theta < \frac{1}{4}$ for the most important θ segment, $[0, \pi/2]$, and hence $3\cos^2\theta - 1 \leq -\frac{1}{4}$, so that attraction follows from Eq. (66). It turns out that for the "best" choice of δ, a variational calculation, including ΛNN forces to first order, of $B_\Lambda(^3_\Lambda H)$ and $B_\Lambda(CS, A = 4)$ yields a B_Λ value for $^5_\Lambda$He which is only 0.3 MeV lower than the 5.46 MeV obtained without using TPE ΛNN forces.

It is worthwhile to note that a considerable part of the ΛNN TPE interaction [Eq. (26)], which is accompanied by the singular strength $T(r_{\Lambda i})T(r_{\Lambda j})Y(r_{\Lambda i})Y(r_{\Lambda j})$, is noncentral for the nucleons (i, j) involved and has not been considered so far. Thus an average on a Λ s wave function leads to an induced two-body nuclear force which contains a significant tensor component arising from noncentral ΛNN terms. Such a strong component may have appreciable matrix elements between the dominant $^1S_0$ configuration for ground-state ^{4}He and other "small" configurations, such as $^5D_0$, admixed by an intrinsic NN tensor force. Until such contributions to $B_\Lambda(^5_\Lambda\text{He})$ are estimated there will remain some doubt as to whether or not three-body ΛNN forces play a major role in alleviating the overbinding problem in $^5_\Lambda\text{He}$.

The second feature of an elimination of the Σ–nucleus channel is a modification of the effective ΛN interaction. Contrary to expectations, this modification does not directly arise from a partial blocking of intermediate Σ–nucleus states due to the Pauli principle, since the antisymmetry of intermediate nuclear states was considered in Section 2.3 leading, as in Eq. (20), to the appearance of three-body ΛNN forces. Rather, the modification discussed above is due to specific nuclear structure effects, such as single-nucleon gaps (as also shown by the work of Bodmer *et al.* reviewed in Section 3 for nuclear matter). This is indeed the case for a ΛN interaction in $^5_\Lambda\text{He}$: The transition for the system as a whole $(\Lambda + {}^4\text{He})_{I=0} \rightarrow (\Sigma + {}^4\text{He})_{I=1} \rightarrow (\Lambda + {}^4\text{He})_{I=0}$ is forbidden by isospin conservation, so that Σ–nuclear intermediate states must be of the form $\Sigma + {}^4\text{He}^*(I = 1)$ (Bod 66). However, excitation of ${}^4\text{He}$ is costly, with the lowest $I = 1$ excited states lying above 25 MeV excitation energy. On the other hand, for $^4_\Lambda\text{H}$ the nuclear ground state may take part in an intermediate Σ hypernuclear state and for $^3_\Lambda\text{H}$ the nearby virtual $(I = 1)$ state of ${}^2\text{H}$ can couple to the $(I = 0)$ nuclear ground state. This singles out $^5_\Lambda\text{He}$ for suppression of the Σ channel. Moreover, as pointed out by Gibson *et al.* (GW 72), the *observed* ${}^4\text{He}^*(I = 1)$ states all have negative parity [corresponding to a particle–hole $(1p)(1s)^{-1}$ nuclear excitation]; in order to conserve parity in the Σ–${}^4\text{He}^*$ intermediate state, the Σ should be created in a p-wave (at least) with respect to the nucleus. Besides involving a higher excitation energy, the amplitude for such a process is additionally suppressed by the centrifugal barrier for the Σ within the typical distance $[2m_\Sigma(m_\Sigma - m_\Lambda)]^{-1/2} \sim 0.5$ fm over which the Σ–nuclear wave function is of a nondecaying form.[†] To maximize the overlap between the Λ $1s$ wave function and the Σ wave function, the latter should also be of a $1s$ type. But such a choice requires, for parity and angular momentum conservation, a nuclear excitation of an $I = 1$, $J^\pi = 0^+$ or

---

[†] This argument actually holds for exchanges, such as scalar meson exchange, which can generally proceed through a $s_{1/2} \rightarrow s_{1/2}$ transition for the hyperon. On the other hand, OPE necessarily induces a $s_{1/2} \rightarrow p_{1/2}$ transition, because of the pseudoscalar nature of the pion, for both hyperon and nucleon. Since the dominant OPE tensor potential couples the initial ${}^3S_1$ hyperon–nucleon state (composed of $s_{1/2}$ and $s_{1/2}$ states) to the final ${}^3D_1$ hyperon–nucleon state (composed of $p_{1/2}$ and $p_{1/2}$ states), the centrifugal barrier has a similar effect for a *free* $\Lambda N \rightarrow \Sigma N$ transition. The argument in the text, reproduced from Gibson *et al.*, does not therefore provide for an obvious extra nuclear suppression for OPE, a case to which we will return below.

1^+, namely an excitation by $2\hbar\omega_N \sim 40$ MeV to the $1d$–$2s$ shell [or a $(1p)^2$ $(1s)^{-2}$ two particle–two hole excitation whose energy could in principle be lower than the above estimate]. This is quite costly in energy and results in a poorer overlap for the nucleons than before.

The above arguments for suppression of the Σ hypernuclear channel depend in a crucial way on the nature and range of the interaction process. If the latter is of a particularly short range, it requires nuclear high momentum components, which are more readily provided by highly excited nuclear states than by the nuclear ground state. In such a case the suppression due to the inability of the nuclear ground state to participate in an intermediate Σ hypernuclear state may prove to be a minor effect. The situation is probably different for the basic long-range OPE $\Lambda N \rightarrow \Sigma N$ transition characterized by a range of 1.4 fm, comparable to the interbaryon rms distance of about 2 fm in $^5_\Lambda$He. A distinction between $^5_\Lambda$He and the other s-shell hypernuclei comes about in the limit of a very large range of the exchange process with respect to the nuclear size. A *coherence* over ^{4}He is then reached for the exchange process with the dominant pion–nucleon p-wave $\boldsymbol{\sigma}_i \cdot \boldsymbol{\nabla}_i \tau_i$ couplings adding up to yield an effective π^4He^{4}He coupling of the form $\boldsymbol{\nabla} \cdot \sum_{i=1}^4 \boldsymbol{\sigma}_i \tau_i$. Since the dominant $^{11}S_0$ configuration of ^{4}He is totally symmetric in the spatial coordinates of the nucleons, its spin–isospin component belongs to the identity representation of $SU(4)$, one of whose generators is $\sum_{i=1}^4 \boldsymbol{\sigma}_i \tau_i$. As a generator, the latter cannot connect the $^{11}S_0$ state to any other state which necessarily belongs to a different irreducible representation of $SU(4)$. Operating on the $^{11}S_0$ state, the above operator vanishes, as appropriate to the identity representation. Hence, for the limit considered for the range of the exchange process and for the assumed structure of the ground state ^{4}He, the $\Lambda \rightarrow \Sigma$ transition through the p-wave coupling of pions to baryons is *completely suppressed in $^5_\Lambda$He*. This, we believe, is the physical content of statements on Σ channel suppression in $^5_\Lambda$He; the relatively high excitation energies involved affect suppression only in a secondary, though nonnegligible manner.[†] In contrast, for $^3_\Lambda$H and $^4_\Lambda$H the corresponding core nuclei do not belong in zeroth

---

[†] In the above "coherent" approximation, but allowing for additional configurations in ground-state ^{4}He, the most obvious of which is the $^{13}P_0 + (1p)^2(1s)^{-2}$ two particle–two hole excitation, the $\Lambda\Sigma$ transition proceeds mostly from this "small" nuclear component to a $^{33}P_0 - (1p)(1s)^{-1}$ particle–hole nuclear excitation (probably ^{4}He* at 29.5 MeV). Both these states are connected by the $SU(4)$ generator $\sum \boldsymbol{\sigma}_i \tau_i$ within the irreducible 15-dimensional representation of $SU(4)$. As explained in the text, a $1p$ excitation for Σ is required as well. Hence the contribution sketched here is not likely to exceed the suppressed one discussed in the text.

order to the $SU(4)$ identity representation and the operator $\sum \sigma_i \tau_i$ has nonvanishing matrix elements within the appropriate ground-state irreducible representations. We will soon dwell in more detail on questions relating to possible partial suppression of the Σ channel in these hypernuclei. Since such suppression is far from being complete and since the triplet YN interaction is not as important there as for $^5_\Lambda$He, the latter is singled out as *the* s-shell hypernucleus where Σ suppression is bound to yield a major effect on a calculated B_Λ value. The connection with the overbinding problem is then clear.

We would like to stress that the effect considered above, namely that of a $\Lambda N \rightarrow \Sigma N$ suppression within a particular hypernucleus, bears not only on the question of modification of the free ΛN interaction in a nuclear environment, but also on the three-body ΛNN forces used in a phenomenological calculation. These forces have been introduced to compensate for elimination of the Σ (as well as Σ^*) channel from hypernuclear calculations. Once this coupling is suppressed in a serious manner, the related ΛNN interaction should also be modified from that derived in Section 2.3. Whether or not the modification of the three-body interaction enhances or reduces binding is difficult to predict, since the ΛNN force is only a partial byproduct of Σ elimination; the ΛN two-body forces are also being modified under these circumstances. It is the *total* modification of $\Lambda \rightarrow \Sigma \rightarrow \Lambda$ transition in $^5_\Lambda$He that is shown by the above arguments to be suppressed. In the (coherent) limit of a complete Σ suppression no ΛNN forces of the TPE type are to be applied in a calculation at all.

The effect of the Σ channel on s-shell B_Λ calculations was recently examined by Dabrowski and Fedorynska (DF 73), who performed an exact model calculation on $^3_\Lambda$H, using separable s-wave YN potentials fitted by Wycech (Wyc 72) to available low-energy YN cross sections. These potentials do not contain hard cores, nor are they motivated by meson-theoretic reasoning. In particular, no tensor coupling is assumed for the triplet $\Lambda N \rightarrow \Sigma N$ transition. The calculated B_Λ value for the observed $J = \frac{1}{2}$ $^3_\Lambda$H ground state is too high, about 0.6 MeV, which could result from the absence of hard core and/or from adjusting the ΛN potential to almost equal ΛN scattering lengths which somewhat exceed in magnitude those of the fit, Eq. (63). Whereas for a one-channel calculation, using ΛN potentials of the same form as before with a strength adjusted to the same ΛN low-energy parameters, the $J = \frac{3}{2}$ state comes lower than the $J = \frac{1}{2}$ state by about 0.5 MeV, in the two-channel calculation the former state appears as an excited state above the $J = \frac{1}{2}$ ground state, the latter moving slightly to a lower energy.

It is not surprising in itself that in the one-channel calculation the ground state turns out to have $J = \frac{3}{2}$. This follows from a fit to ΛN scattering lengths where the triplet interaction is stronger than the singlet. Within one standard deviation, however, solutions for ΛN scattering lengths may be found which favor the singlet interaction, and would thus lead in a one-channel calculation to a $J = \frac{1}{2}$ ground state for $^3_\Lambda$H. What is striking is that when a $\Lambda N \to \Sigma N$ transition is invoked [in the triplet channel alone, as suggested by polarization data (Yam+ 69) from the reaction $\Sigma^- p \to \Lambda n$ at low energy], the order of levels is appreciably reversed, indicating that in their calculation the triplet $\Lambda N \to \Sigma N$ coupling is strongly suppressed for the $J = \frac{3}{2}$ state (The $J = \frac{1}{2}$ state obtains most of its potential energy from the singlet interaction and thus remains scarcely affected by the triplet $\Lambda \Sigma$ coupling, whereas the $J = \frac{3}{2}$ state obtains *all* of its potential energy from the triplet interaction.) These authors find that the most obvious factor in determining this trend is the effective three-body force, which turns out in their calculation to be attractive for $J = \frac{1}{2}$ and repulsive for $J = \frac{3}{2}$. This Λ-spin dependence of the three-body force, which should vanish for the double OPE three-body force as discussed in Section 2.3, results from insisting that the $\Lambda N \to \Sigma N$ transition is of a $^3S_1$ nature.

Actually, the experimental data mentioned above cannot discriminate between a pure $^3S_1$ transition and a $^3S_1$–$^3D_1$ transition induced by a tensor component. If the latter possibility holds, then the $\Lambda N \to \Sigma N$ transition is proportional to $\boldsymbol{\sigma}_\Lambda$ and upon applying closure and combining both $(\Lambda 1 \to \Sigma 1, \Sigma 2 \to \Lambda 2)$ and $(\Lambda 2 \to \Sigma 2, \Sigma 1 \to \Lambda 1)$ terms, the $\boldsymbol{\sigma}_\Lambda$ dependence drops out. The only difference between modifications (due to the Σ channel) of B_Λ values for $J = \frac{1}{2}$ and $J = \frac{3}{2}$ could then arise from the somewhat different closure excitation energie $\bar{\varepsilon}$ valid for $J = \frac{1}{2}$ and $J = \frac{3}{2}$. Such a difference is in fact clear from the following simple argument: Since $I = 0$ holds for $^3_\Lambda$H and hence for the intermediate ΣNN states, the NN intermediate states must have $I = 1$. For the $J = \frac{3}{2}$ $^3_\Lambda$H state the total spin is $\frac{3}{2}$, implying a spin value of 1 for any of the baryon pairs. Thus for a $\Lambda N \to \Sigma N$ interaction of an arbitrary ratio of triplet to singlet strengths, only the triplet part will contribute, leaving $S = \frac{3}{2}$ unchanged. Since for two nucleons in the intermediate ΣNN state $I = S = 1$ holds, the space part of this nuclear excitation must be antisymmetric, which tends very probably to increase the average excitation energy $\bar{\varepsilon}_{3/2}$ appropriate to such nuclear excitations. On the other hand, for $J = \frac{1}{2}$ and $\Lambda N \to \Sigma N$ transition with a nonvanishing spin dependence the final nuclear spin function necessarily contains an $S = 0$ component and thus symmetric spatial excited nuclear configurations are allowed, which tends very probably to result in $\bar{\varepsilon}_{1/2} < \bar{\varepsilon}_{3/2}$. In this context it may be more appropriate to treat separately the relatively low nuclear excitation ($\sim$2.3 MeV) of the pn virtual $^{31}S_0$ state. Apart from the arguments given above, this nuclear state cannot form an intermediate ΣNN state to be connected to $J = \frac{3}{2}$ $^3_\Lambda$H by s-wave $\Lambda N \to \Sigma N$ transition, simply because $0 + \frac{1}{2} \neq \frac{3}{2}$. The contribution from this particular intermediate nuclear excitation for $J = \frac{1}{2}$ $^3_\Lambda$H, however, is nonzero provided the s-wave $\Lambda N \to \Sigma N$ transition is not spin independent. The $J = \frac{1}{2}$ level gains, with respect to the $J = \frac{3}{2}$ level, by coupling in the Σ channel.

To sum up this discussion on $^3_\Lambda$H, we point out that although the effects considered above are moderately small, typically of the order of a few tenths of an MeV, they may prove extremely significant in determining the spin sequence of levels in $^3_\Lambda$H. In the calculation of DHT the calculated excitation energy $E^*(J = \frac{3}{2})$ turns out to be rather small, 0.09 MeV [0.13 MeV if the latest B_Λ values (Jur+ 73) are used]. The Σ suppression effects, when taken into account properly, may actually increase this calculated excitation, raising the $J = \frac{3}{2}$ level to become unbound and slightly lowering the position of the $J = \frac{1}{2}$ ground state. The observation that for $J = \frac{1}{2}$ the presence of an explicit Σ channel tends to slightly increase the binding points to the possibility that in some special cases the Σ channel is not bound to be suppressed in hypernuclei, but rather enhanced.[†] In the calculation described above (DF 73), the $\Lambda N \to \Sigma N$ hypernuclear transition receives significant contributions by exciting the nuclear core from a $^{13}S_1$ deuteron-like state to a $^{31}S_0$ virtual-like state, which are close in energy to each other. It is not clear whether or not the introduction of more realistic NN and YN interactions would change these results by providing more types of intermediate states, for some of which Σ suppression could be strong. For the OPE $\Lambda N \to \Sigma N$ transition, proceeding mostly through a tensor component, the introduction of a realistic nuclear $^{13}D_1$ component in $^3_\Lambda$H would then allow a strong contribution for $J = \frac{1}{2}$ by coupling to a $^{31}S_0$ virtual-like nuclear state. As before, the analogous contribution for $J = \frac{3}{2}$ vanishes.

A realistic nuclear $^3D_1$ component for $^3_\Lambda$H (giving rise to a $^4D_{1/2}$ configuration for ground state $^3_\Lambda$H) would also allow for an assessment of the role played by an ordinary ΛN tensor force. The latter is expected to have a short range which may tend to decrease its contribution. A smaller effect

---

[†] Recall that the philosophy behind the one-channel calculation is that the presence of the Σ channel acts in a similar manner for free ΛN scattering and for ΛN nuclear interaction; the actual difference between effectiveness of the Σ channel for free ΛN scattering and for ΛN nuclear interaction is manifested by explicitly introducing the Σ channel into a nuclear calculation.

is expected from the spin dependence of the central ΛN force, which gives rise in ${}_\Lambda^3$H to the appearance of an S' state (Bod 66), antisymmetric in the spatial coordinates of the nucleons and necessarily involving odd angular momentum for the Λ with respect to the nuclear center of mass. The probability of such an S' state turns out (BD 67) to be extremely small, reflecting partly the weak spin dependence of the fitted ΛN central s-wave interaction.

We turn now to the $A = 4$ hypernuclei. Experimentally it is well established [for a detailed exposition consult (Dal 69a)] that ${}_\Lambda^4$H has $J = 0$ for its ground state and with less certainty that the same holds for ${}_\Lambda^4$He. The difference in binding energies for these two hypernuclear species has been discussed earlier where the explicit assumption $J = 0$ was made. As discussed by DHT, $J = 1$ excited states are expected to be particle-stable for both ${}_\Lambda^4$H and ${}_\Lambda^4$He. The observation of any of these excited states has an immediate bearing on the spin dependence of the central ΛN interaction. With K^--mesons stopped in ^{6}Li and ^{7}Li a γ line of nonnuclear origin at 1.09 MeV was observed by Bamberger *et al.* (Bam+ 71, 73) and was assigned to a hypernuclear transition in either ^{4}H* or ^{4}He*. The production rate for this 1.09-MeV, $A = 4$ state is $0.14 \pm 0.02\%$ in ^{6}Li and $0.37 \pm 0.04\%$ in ^{7}Li. A second line, the assignment of which is uncertain, at 1.42 MeV was also observed with intensity $0.10 \pm 0.02\%$ in ^{6}Li and $0.04 \pm 0.04\%$ in ^{7}Li. The possibility of a hypernuclear line at about 0.5 MeV or less cannot be ruled out since the huge background for positron annihilation renders any determination of low-intensity hypernuclear lines almost impossible at this energy.

Bamberger *et al.* included the above 1.09 MeV line[†] in a B_Λ analysis of the type performed by DHT, using this γ line as an additional piece of information. For the assignment of this excitation to either ${}_\Lambda^4$H* or ${}_\Lambda^4$He*, the other excitation turns out to be in the range 1.2–1.5 MeV for ΛN potential parameters deduced by the B_Λ analysis and which give rise to Λp potentials that favorably reproduce Λp low-energy cross sections. This does not imply that one should expect the other $A = 4$ excited state to correspond to the experimentally suspected γ line at 1.42 MeV, since small variations in the input value of $B_\Lambda({}_\Lambda^3$H) could change this deduced excitation considerably. The Λp potentials that are in agreement with the B_Λ analysis and fit Λp low-energy cross sections with low values of χ^2 turn out to have an intrinsic range of 1.5 fm and a hard core radius of 0.45 fm. This is not too far from results reached by DHT and is certainly compatible

---

[†] *Note added in proof*: It now appears established that the 1.09 MeV line belongs to ${}_\Lambda^4$H (Bed+ 75).

with theirs within one or two standard deviations. The spin dependence of the ΛN force, however, is appreciably larger in this calculation than that found in the analysis of DHT, who did not include the 1.09-MeV state in their input. The $J = 1$ excitation energy provides a direct measure of the spin–spin term of the ΛN interaction. Thus, if we consider $_\Lambda^4$H, for example, and write down the central ΛN interaction in the form $U + \Delta \mathbf{s}_\Lambda \cdot \mathbf{s}_N$, where $U = \frac{3}{4}U_t + \frac{1}{4}U_s$ is the spin-independent component and $\Delta = U_t - U_s$ measures the spin dependence directly, we obtain for $_\Lambda^4$H the corresponding contributions to the energy

$$2\bar{U}_n + \bar{U}_p + (\tfrac{1}{4})\bar{\Delta}_p, \qquad J = 1 \tag{67a}$$

$$2\bar{U}_n + \bar{U}_p - (\tfrac{3}{4})\bar{\Delta}_p, \qquad J = 0 \tag{67b}$$

where a bar means an expectation value with respect to the appropriate radial wave functions and the subscript p (n) stands for an Λp (Λn) interaction, so that CSB is automatically included. The spin dependence parameter of the Λp interaction appears in (67a) and (67b) since the two neutrons are paired to a spin of zero. Subtracting Eq. (67b) from Eq. (67a) yields the $_\Lambda^4$H* excitation energy

$$E^*(_\Lambda^4\text{H*}, J = 1) = \bar{\Delta}_p \tag{68a}$$

and similarly for $_\Lambda^4$He (interchange $p \leftrightarrow n$):

$$E^*(_\Lambda^4\text{He*}, J = 1) = \bar{\Delta}_n \tag{68b}$$

A schematic excitation diagram is shown in Fig. 8.

It is clear now why most of the various fits reported by Bamberger *et al.* on the assumption that the 1.09-MeV excited state belongs to $_\Lambda^4$H give for $-a_s^{(p)} - (-a_t^{(p)})$ the same value of about 1 fm as required by the constraint $\bar{\Delta}_p = 1.09$ MeV. Depending on whether the 1.09-MeV excitation belongs to $_\Lambda^4$H* or to $_\Lambda^4$He*, the CSB ΛN potential required in their analysis is radically different. In fact, a further identification of *both*

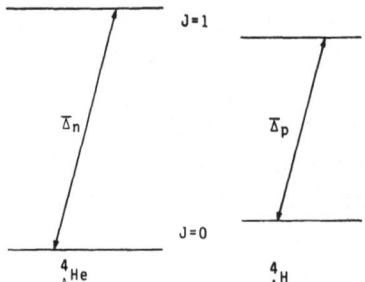

Fig. 8. A schematic level diagram of $A = 4$ hypernuclear states, stable against Λ emission. The excitation energy of the $J = 1$ state with respect to the $J = 0$ ground state is primarily given by the spin–spin Λn (Λp) parameter $\bar{\Delta}_n$ ($\bar{\Delta}_p$) in the case of $_\Lambda^4$He ($_\Lambda^4$H).

$^4_\Lambda$H*$(J = 1)$ and $^4_\Lambda$He*$(J = 1)$ would, according to Eqs. (68a) and (68b), determine $\bar{\Delta}_p$ and $\bar{\Delta}_n$, respectively, so that the spin dependence of the CSB ΛN potential, $\bar{\Delta}_p - \bar{\Delta}_n$, would be fixed. To determine the spin-independent part of the CSB potential, the difference in binding energies of both ground states has to be invoked. At present, only the latter datum is well established and assigned, which results in constraints on CSB parameters, short of a complete determination. Since in the difference

$$[E(^4_\Lambda \text{H*}) - E(^4_\Lambda \text{He*})] - [E(^4_\Lambda \text{H}) - E(^4_\Lambda \text{He})]$$

$$= E^*(^4_\Lambda \text{H*}) - E^*(^4_\Lambda \text{He*}) = \bar{\Delta}_p - \bar{\Delta}_n \qquad (69)$$

the Coulomb effects due to the Λ polarization of the nuclear core are expected to cancel out to a large extent, $\bar{\Delta}_p - \bar{\Delta}_n$ would be more readily and unambiguously determined than the analogous spin-independent CSB term $\bar{U}_p - \bar{U}_n$.

Additional factors might affect the position of the $J = 1$ excitations in $^4_\Lambda$H and $^4_\Lambda$He. Here we mention two such effects. The first is due to the presence of a relatively strong tensor component in the ΛN CSB interaction induced by OPE (see Section 2.4). For the dominant spatial S wave function of $^4_\Lambda$H and $^4_\Lambda$He, the tensor CSB contribution vanishes. However, calculations on the bound $A = 3$ nuclear system reveal that the probability of a D state may be as high as 9%. The tensor ΛN CSB interaction readily connects this appreciable D nuclear component with the dominant S nuclear wave function of the core. This CSB tensor contribution has never been evaluated. It could be quite different for the ground state $J = 0$ ($^4_\Lambda$H, $^4_\Lambda$He) than for the excited state $J = 1$ ($^4_\Lambda$H*, $^4_\Lambda$He*), since the latter state can be based on both ^4D and ^2D nuclear components, whereas the ground state $J = 0$ cannot be built on a ^2D nuclear component as long as the Λ is kept in a $1s$ state with respect to the nuclear core. Consider the CSB contribution arising from an admixture of a nuclear ^4D component. Since the total nuclear spin is $\frac{3}{2}$, the two protons in $^4_\Lambda$He, based on this ^4D nuclear configuration, have a total spin $S_{pp} = 1$, whereas in the leading configuration (^2S for the core) they are coupled to $S_{pp} = 0$. Hence this particular CSB tensor contribution in $^4_\Lambda$He derives from Λp bonds; a Λn contribution necessarily vanishes because of orthogonality of the protons' spin wave functions. This is in contrast to the contribution in $^4_\Lambda$He from the spin–spin component of OPE CSB, which derives from a Λn bond.

The second effect on the $J = 1$ excitations for $A = 4$ could come from a partial suppression of the Σ channel in one of the $J = 0$ and $J = 1$ states, similar to what has been discussed for $^3_\Lambda$H. This would provide a charge-symmetric shift of the excitation energies calculated in the absence of the considered effect. Dabrowski and Fedorynska (DF 73) argue for an increase in the $A = 4$ $E^*(J = 1)$ excitation for the $\Lambda N \to \Sigma N$ ^3S transition, an increase provided mainly by a spin dependence of ΛNN forces. We stress again that if the OPE $\Lambda N \to \Sigma N$ transition potential is employed and closure is applied to the Σ-hypernuclear intermediate states, with the same average excitation energy $\bar{\varepsilon}$ for both $J = 0$ and $J = 1$ states, then the resulting three-body force is independent of the spin of the Λ and hence its contributions in the $J = 0$ and $J = 1$ states are equal. Whether or not the two-body ΛN force is modified differently for $J = 0$ and $J = 1$ is not clear from their argument (DF 73); one cannot repeat here the considerations sketched in the case of $^3_\Lambda$H for increasing hypernuclear excitation energy by inclusion of the Σ channel.

As we shall now show (Dab 73b), a definite conclusion is reached when the $\Lambda N \to \Sigma N$ transition is considered with the nuclear $A = 3$ ground state being the only intermediate nuclear core state involved (coherent approximation or rigid core approximation).

Consider the $\Lambda N \to \Sigma N$ transition represented by the s-wave interaction

$$(1/\sqrt{3})\boldsymbol{\Phi} \cdot \sum_i \boldsymbol{\tau}_i(V_i + \delta_i \mathbf{s}_Y \cdot \mathbf{s}_i) \tag{70}$$

where

$$V_i = \tfrac{3}{4}V_t(r_{Yi}) + \tfrac{1}{4}V_s(r_{Yi}), \qquad \delta_i = V_t(r_{Yi}) - V_s(r_{Yi}) \tag{71}$$

are the spin-independent and spin-dependent potentials, respectively, between the hyperon and the ith nucleon. The $\boldsymbol{\tau}_i$ are the Pauli isotopic matrices for nucleon i and $\boldsymbol{\Phi}$ is an isotopic normalized vector which transforms a Λ into a Σ in the same spin-space state:

$$\Phi_+ = -\frac{1}{\sqrt{2}}\,(1, i, 0), \quad \Phi_0 = (0, 0, 1), \quad \Phi_- = \frac{1}{\sqrt{2}}\,(1, -i, 0) \tag{72}$$

Let us calculate the expectation value of the transition interaction [Eq. (70)] in an $A = 4$ hypernuclear state (Λ on the right and Σ on the left) ($J, I = \tfrac{1}{2}$) based on the dominant $^{22}S_{1/2}$ configuration for the $A = 3$ nuclear core. $\bar{V}$ and $\bar{\delta}$ stand for the radial matrix elements appropriate to Eq. (71). The coefficient of $\bar{V}$ is given by $(1/\sqrt{3})\,\langle\sum^3 \boldsymbol{\Phi} \cdot \boldsymbol{\tau}_i\rangle = (2/\sqrt{3})\,\langle\boldsymbol{\Phi} \cdot \mathbf{T}\rangle = 1$, since formally the isotopic labels of a nucleon and an $A = 3$ ground-state nuclear core are identical, and $(2/\sqrt{3})\boldsymbol{\Phi} \cdot \mathbf{T}$ is the normalized operator for transforming an $N\Lambda$ state into an $N\Sigma$ state. For evaluation of the coefficient of $\bar{\delta}$, tensor algebra methods are used to relate it to a reduced matrix element of $\sum_{i=1}^{3} s_i \tau_i$:

$$\frac{1}{\sqrt{3}} \left\langle \sum_i s_i \cdot s_\Lambda \Phi \cdot \tau_i \right\rangle = \tfrac{1}{2}(-1)^{1+J} \begin{Bmatrix} \tfrac{1}{2} & \tfrac{1}{2} & J \\ \tfrac{1}{2} & \tfrac{1}{2} & 1 \end{Bmatrix} \left\langle \left\| \sum_i s_i \tau_i \right\| \right\rangle \quad (73)$$

Each one of the nine operators $\sum_i s_i \tau_i$ is a generator of the spin–isospin $SU(4)$ group and the above reduced matrix element for the ground-state $SU(4)$ irreducible representation (2, 1) is easily evaluated to equal -3. Hence the coefficient of $\bar{\delta}$ is $+\tfrac{3}{4}$ for $J = 0$ and $-\tfrac{1}{4}$ for $J = 1$ and altogether the $\Lambda N \to \Sigma N$ matrix element (appropriate to $A = 4$) is given by

$$\bar{V} + \tfrac{3}{4}\bar{\delta} = \tfrac{3}{2}\bar{V}_t - \tfrac{1}{2}\bar{V}_s, \qquad J = 0 \qquad (74a)$$

$$\bar{V} - \tfrac{1}{4}\bar{\delta} = \tfrac{1}{2}\bar{V}_t + \tfrac{1}{2}\bar{V}_s, \qquad J = 1 \qquad (74b)$$

or per nucleon

$$\tfrac{1}{2}\bar{V}_t - \tfrac{1}{6}\bar{V}_s, \qquad J = 0 \qquad (75a)$$

$$\tfrac{1}{6}\bar{V}_t + \tfrac{1}{6}\bar{V}_s, \qquad J = 1 \qquad (75b)$$

as obtained by Dabrowski (Dab 73b). If the $\Lambda N \to \Sigma N$ transition occurs predominantly in the triplet configuration, a stronger coupling for $J = 0$ than for $J = 1$ results, the ratio of strengths being 3:1. The $J = 0$ level will then benefit more than the $J = 1$ level from the coupling to the Σ channel and $E^*(J = 1)$ will increase with respect to $E^*(J = 1)$ (assumed still positive) in the absence of the Σ channel. The same occurs for a spin–spin $\Lambda\Sigma$ conversion, $\bar{V}_s = -3\bar{V}_t$.

The above example suggests that in some instances it may be misleading to talk about "suppression" of the Σ channel in hypernuclei. Suppose, for the sake of argument, that the nuclear forces were spin and isospin independent. $SU(4)$ would then become an exact symmetry for the nuclear core and the wave function of the $A = 3$ nuclear core in its ground state would correspond to a pure configuration $^{22}S_{1/2}[3]$, where [3] stands for the Young tableau of highest possible permutation symmetry. We now form Λ-hypernuclear and Σ-hypernuclear $J = 0$ and $J = 1$ states built on this $A = 3$ nuclear ground state. For the $\Lambda N \to \Sigma N$ transition operator [Eq. (70)], according to Eqs. (74a) and (74b) obtain for the contribution of the lowest intermediate state:

$$-\frac{\langle J | V_{\Sigma,\Lambda} | J \rangle \langle J | V_{\Lambda,\Sigma} | J \rangle}{\Delta} = \begin{cases} -\dfrac{1}{\Delta}\left(\dfrac{3}{2}\bar{V}_t - \dfrac{1}{2}\bar{V}_s\right)^2, & J = 0 \quad (76a) \\[2ex] -\dfrac{1}{\Delta}\left(\dfrac{1}{2}\bar{V}_t + \dfrac{1}{2}\bar{V}_s\right)^2, & J = 1 \quad (76b) \end{cases}$$

where $\Delta \sim 80$ MeV as in Section 2.3. This term alone may sometimes exceed the contribution expected from an unquenched $\Sigma\Lambda$ coupling in $A = 4$. In fact, the relevant unsuppressed quantity for the J states in $A = 4$ is given by

$$-\frac{1}{\bar{\varepsilon}} \left\langle J \left| \sum_i \left[\frac{1}{\sqrt{3}} \boldsymbol{\Phi} \cdot \boldsymbol{\tau}_i (V_i + \delta_i \mathbf{s}_\Lambda \cdot \mathbf{s}_i) \right]^2 \right| J \right\rangle \qquad (77)$$

where we have explicitly assumed that the very same nucleon is involved in both the $\Lambda N \to \Sigma N$ and $\Sigma N \to \Lambda N$ transitions, as is in the free case. Terms of the form $(\Lambda N \to \Sigma N) \cdot (\Sigma N' \to \Lambda N')$ with $N \neq N'$ arise because of the Pauli principle and on the average are expected to modify the "unsuppressed" contribution given by Eq. (77). Since $\langle [(1/\sqrt{3})\boldsymbol{\Phi} \cdot \boldsymbol{\tau}_i]^2 \rangle = 1$ and $(\mathbf{s}_\Lambda \cdot \mathbf{s}_i)^2 = (3/16) - \frac{1}{2}\mathbf{s}_\Lambda \cdot \mathbf{s}_i$, we obtain for Eq. (77) the results

$$-(1/\bar{\varepsilon})(\tfrac{3}{2}\overline{V_t^2} + \tfrac{3}{2}\overline{V_s^2}), \qquad J = 0, \quad \bar{\varepsilon} > \Delta \qquad (78a)$$

$$-(1/\bar{\varepsilon})(\tfrac{5}{2}\overline{V_t^2} + \tfrac{1}{2}\overline{V_s^2}), \qquad J = 1, \quad \bar{\varepsilon} > \Delta \qquad (78b)$$

In the "coherent" limit discussed above, the $\Lambda\Sigma$ transition operator is considered as having an infinite range with respect to nuclear sizes and is therefore roughly a constant over nuclear dimensions. Hence the only relevant intermediate hypernuclear state is indeed the one considered in Eqs. (76a) and (76b), $\overline{V_{t,s}^2} = (\bar{V}_{t,s})^2$ and the closure energy $\bar{\varepsilon}$ reduces to the (Σ, Λ) mass difference Δ, $\bar{\varepsilon} = \Delta$. Comparing Eqs. (76a) and (76b) with Eqs. (78a) and (78b), we see that for a pure triplet $\Lambda\Sigma$ transition, enhancement of the $\Lambda\Sigma$ coupling for $J = 0$ and suppression for $J = 1$ occur within this limit. On the other hand, for a spin–spin type of $\Lambda\Sigma$ transition ($\bar{V}_s = -3\bar{V}_t$), both the $J = 0$ and $J = 1$ levels acquire suppression. This example clearly shows that structure effects in light hypernuclei may be crucial for gauging the contribution of the Σ channel with respect to the free hyperon–nucleon system. The Pauli principle, strangely enough, may sometimes enhance the contribution of the Σ channel.[†] On the *average*, however, the Σ channel is always suppressed in nuclei, as can be seen from the following argument. If the contributions to the various hypernuclear levels in Eqs. (76a, b) and (78a, b) are weighted according to $(2J + 1)$,

---

[†] This is reminiscent of enhancement in the π^- decay rates for some light hypernuclear species with respect to the free decay $\Lambda \to p\pi^-$ (DL 59). The origin of enhancement in both cases is the same.

we obtain in the coherent limit

$$-(1/\Delta)(\tfrac{3}{4}\bar{V}_t{}^2 + \tfrac{1}{4}\bar{V}_s{}^2) \tag{76c}$$

$$-(3/\Delta)(\tfrac{3}{4}\bar{V}_t{}^2 + \tfrac{1}{4}\bar{V}_s{}^2) \tag{78c}$$

so that on the average the contribution of the Σ channel to binding in the $A = 4$ hypernuclear systems is suppressed by a factor of 3 in this limit. It can easily be shown that a consideration of terms of the form $(\Lambda N \rightarrow \Sigma N)$ $\cdot (\Sigma N' \rightarrow \Lambda N')$ with $N' \neq N$ leads in the coherent limit to the additional repulsive contribution

$$(2/\Delta)(\tfrac{3}{4}\bar{V}_t{}^2 + \tfrac{1}{4}\bar{V}_s{}^2) \tag{79}$$

which, added to Eq. (78c), reproduces the necessary contribution for $A = 4$ given by Eq. (76c). This simple exercise illustrates Nogami's (Nog 69) observation, originally made for Λ in nuclear matter, that the elimination of the Σ channel in hypernuclei leads to the appearance of (repulsive) $\Lambda NN'$ forces and that these, as well as the "free" ΛN interaction, should be evaluated disregarding the Pauli principle. The above discussion tells us that such $\Lambda NN'$ forces may well prove to be state dependent, in some cases yielding attraction, and that an explicit introduction of the Σ channel in some hypernuclei, at least for discussion of certain phenomena, is unavoidable. It is difficult at present to give more quantitative arguments for the necessity of an explicit introduction of the Σ channel in light hypernuclei. Further calculations would be extremely desirable.

4.3. *p*-Shell Hypernuclei

The information which may be obtained experimentally for hypernuclear level characteristics in the p shell is potentially abundant, in particular when we recall that for the s shell, in addition to spin assignments, only a few binding energies and even fewer excitation energies are partly measured and are even on the experimentalists' agenda. In the p shell we have at present (Jur+ 73) (see Table I) 12 charge-symmetric ground-state binding energies, four binding energy differences within hypernuclear isomultiplets, and two nontrivial spin assignments, for ${}^{8}_{\Lambda}\text{Li}$ (Dal 63, DLR 63) and ${}^{12}_{\Lambda}\text{B}$ (Kie+ 75, ZD 75). In addition, many hypernuclear excited states are generally expected to prove particle-stable and their detection is expected in the next few years through the application of (K^-, π^-) reactions on specific nuclear targets. These excited states will provide valuable information on the effective ΛN interaction, particularly the spin dependence of this interaction.

In Section 4.1 we briefly discussed several models for B_A calculations in the p shell. The first extensive calculations for the hypernuclear p shell were carried out by Bodmer and Murphy (BM 65), who solved a two body Λ–nucleus Schrödinger equation, the potential term of which was obtained by folding an assumed ΛN interaction potential into an observed nuclear density. This method essentially amounts to the "rigid core" approximation considered in Section 4.1. It is reasonable to expect its validity to hold as long as the nuclear-core excited states coupled by the ΛN interaction to the nuclear-core ground state lie at a high excitation energy, say above 10 MeV. Such would be the situation for a central, spin-independent, ΛN interaction. However, the spin–spin central term, operative already in the s shell, as well as ΛN spin–orbit and tensor forces may prove very effective in coupling the nuclear ground state to low-lying nuclear states which basically differ from the ground state by reorienting spins and angular momenta of individual nucleons without involving excitations to a higher shell. This is one of the new features that one meets in passing from s-shell to p-shell calculations. As will become clear below, even for spin-zero hypernuclei in the p shell ($^{7}_{\Lambda}$Be, $^{9}_{\Lambda}$Be, $^{13}_{\Lambda}$C) not all spin-dependent ΛN interaction terms vanish when operating on a nuclear-core ground-state wave function. A spin–orbit ΛN force *induces* a nuclear spin–orbit force which has appreciable nonvanishing matrix elements between the nuclear-core ground state and several (including the ground state) low-lying nuclear-core states. For this reason, even if the "rigid core" approximation were to hold for one (spin-zero core) p-shell hypernuclear species, it would prove almost impossible to extract systematic information on the ΛN interaction parameters by separating their contributions from each other.

Other methods of calculation, such as hypernuclear HF calculations and cluster calculations, are briefly mentioned in connection with some distinct fitting problems encountered below in the following discussion of shell model (SM) calculations. The SM approach (GSD 71, 72; LHC 70; LR 60) provides at present the most systematic approach to B_A values and excited state calculations in the p shell, in spite of its imperfections. In the following discussion we review in detail the SM analysis of B_A values for hypernuclei in the p shell as given by Gal, Soper, and Dalitz (GSD, 71, 72), based on observed B_A values in the $p_{3/2}$ shell.

Two basic assumptions are involved in shell model calculations for hypernuclei in the range $6 \leq A \leq 17$. The first assumption is that the nuclear SM provides a physically sufficient description of energy level patterns, β and γ transition rates, and static moments throughout the so-called "p shell." This assumption has been tested in the past by various calculations, with the result that most of the low-lying (a few MeV) levels for nuclei in the mass range $6 \leq A \leq 16$ are well described by calculated intermediate coupling wave functions of the configuration $(1p)^{4-4}$ (Bar 66, CK 65). The word "intermediate" means that neither LS nor jj coupling is found appropriate for the nuclear $1p$ shell but rather an intermediate coupling scheme which is quite close to LS coupling in the beginning of the shell and to jj at the end of the $1p$ shell. These nuclear SM calculations specify the appropriate combinations of angular, spin, and isospin functions within the $1p$ shell required by the data to yield a wave function with given J^{π} values. The radial form of a $1p$ nucleon wave function, however, is not specified, except for some radial expectation values, with respect to such a form, e.g., $B(E2)$ values. For most purposes a harmonic oscillator wave function gives a satisfactory fit to these radial matrix elements,

$$\psi_N(1p) = (2\nu)^{1/2}(\nu/\pi)^{3/4}\mathbf{r}\exp(-\tfrac{1}{2}\nu r^2) \tag{80}$$

where ν, the oscillator frequency, is to a good approximation constant throughout the p shell. Measured charge radii in electron scattering experiments yield an approximately constant value of about $\nu = 0.41$ fm^{-2}, although somewhat smaller values are required to fit nuclei, such as ^{6}Li and ^{8}Be, which are lightly bound or are slightly unbound, respectively. In the presence of a Λ-particle, however, these nuclear cores acquire additional binding: the p nucleons in $^7_\Lambda$Li are bound by 2.5 MeV more than their counterparts in ^{6}Li and the lowest particle stability threshold for $^9_\Lambda$Be is at about 3.5 MeV (^{4}He $+ ^5_\Lambda$He), whereas ^{8}Be is unstable by 0.1 MeV against a decay into ^{4}He $+ ^4$He. It is reasonable then to expect that the presence of a Λ will make the approximation of one and the same $1p$ wave function through the p shell into more valid.[†]

The second basic assumption of the SM, as appropriate for the present analysis, is that of a uniform Λ $1s$ wave function through the p shell. In Fig. 9, Λ–nucleus wave functions calculated numerically for the hypernuclei $^7_\Lambda$Li, $^8_\Lambda$Li, $^{10}_\Lambda$Be, and $^{13}_\Lambda$C have been plotted and show that this assumption of uniformity is well justified, particularly from $A = 8$ to $A = 13$. It is not so good for $^7_\Lambda$Li due to its relatively low B_Λ value. The 1s harmonic oscillator Λ wave function

$$\psi_\Lambda(1s) = (\lambda/\pi)^{3/4}\exp(-\tfrac{1}{2}\lambda r^2) \tag{81}$$

[†] An exception is $^6_\Lambda$He, in which the p neutron acquires more than 1 MeV binding with respect to its unbound $p_{3/2}$ state in ^{5}He, but is still very lightly bound, $B_n \sim 0.2$ MeV. The situation is somewhat better for $^8_\Lambda$He, where $B_n \sim 1.6 \pm 0.7$ MeV. These two p-shell hypernuclei were not included by GSD in most of their fits.

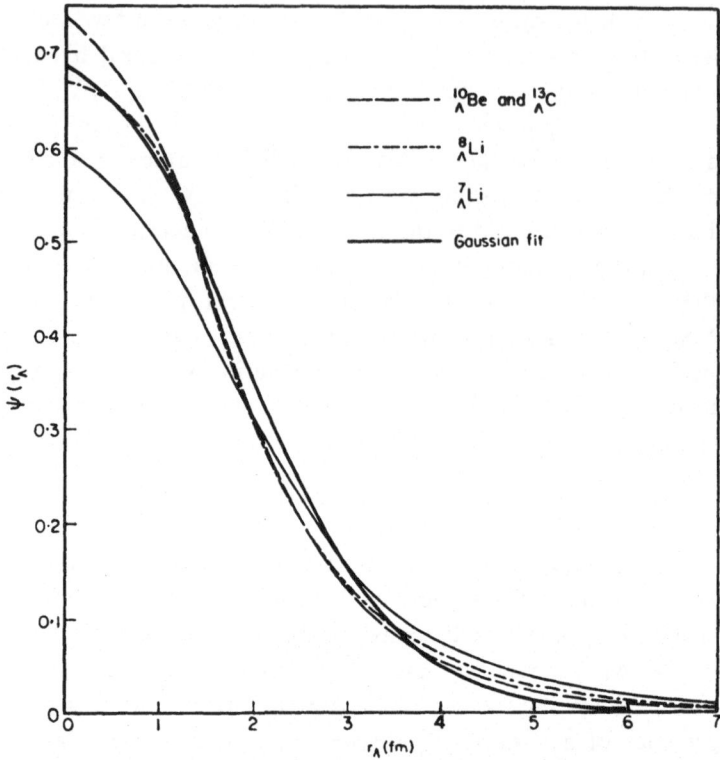

Fig. 9. Λ–nucleus wave functions calculated for a Λ-particle attached to a p-shell nucleus of mass number $A - 1$. The nucleon distribution is of the form $\{1 + (A - 5)\nu r^2/6\}$ $\exp(-\nu r^2)$, with $\nu^{-1/2} = 1.56$ fm and with a Λ–nucleon potential of Gaussian form and intrinsic range corresponding to that for a Yukawa potential with range parameter $1/2m_\pi$. The curves for $^{10}_{\Lambda}$Be and $^{13}_{\Lambda}$C differ very little. The Gaussian form $\exp(-\tfrac{1}{2}\lambda r^2)$, with $\lambda = 0.33$ fm^{-2}, gives quite a good fit to these wave functions for $A = 8$–13. A two-term Gaussian form mentioned in the text provides a really excellent fit; it always lies between the curves shown for $^{8}_{\Lambda}$Li and $^{13}_{\Lambda}$C, and we have not shown it here in order to avoid confusion. [From (GSD 71).]

with $\lambda = 0.33$ fm^{-2} gives quite a good fit to these wave functions for $A = 8$–13. An excellent fit in this range is provided by the double Gaussian form

$$\psi_\Lambda = N^{-1/2}[y \exp(-\tfrac{1}{2}\lambda_1 r^2) + \exp(-\tfrac{1}{2}\lambda_2 r^2)] \qquad (82\text{a})$$

with

$$N = y^2\left(\frac{\pi}{\lambda_1}\right)^{3/2} + 2y\left(\frac{2\pi}{\lambda_1 + \lambda_2}\right)^{3/2} + \left(\frac{\pi}{\lambda_2}\right)^{3/2} \qquad (82\text{b})$$

and

$$\lambda_1 = 0.495 \text{ fm}^{-2}, \qquad \lambda_2 = 0.165 \text{ fm}^{-2}, \qquad y = 3.0 \qquad (82\text{c})$$

Evaluations of radial matrix elements, where necessary, have been performed with ψ_Λ, Eq. (82a). As explained in Section 4.1, the uniform approximation for the Λ wave function results from two opposing trends in the p shell. The increase of B_Λ with A leads to a sharper falloff of ψ_Λ outside the Λ–nuclear interaction region, but the increase of the core radius with A pushes ψ_Λ out before this asymptotic falloff sets in. It is interesting to note that the consistency of the uniform approximation for the Λ wave function may be tested in *double* $\Lambda\Lambda$ hypernuclei. In fact, the additional binding energy $\Delta B_{\Lambda\Lambda}$ (relative to $2B_\Lambda$, where B_Λ as usual denotes the binding energy for the first Λ-particle attached to the nuclear core) appropriate to $_{\Lambda\Lambda}^6$He and $_{\Lambda\Lambda}^{10}$Be is roughly about 4.6 MeV for both these species (Dan+ 63, Pro 66). In the SM analysis this number is given by the expectation value, for two Λ-particles in the s shell, of the effective $\Lambda\Lambda$ interaction and inasmuch as the Λ wave functions are independent of A, this expectation value is expected to be the same for these two species.

We turn now to an enumeration of the ΛN and ΛNN effective interaction matrix elements that are relevant for our discussion. Since the Λ is in a s state, the total orbital angular momentum of the ΛN system for p shell hypernuclei is $L = 1$, whereas the total ΛN spin receives two values $S = 0, 1$. Five matrix elements of the ΛN interaction thus appear as parameters in a SM analysis of B_Λ values[†]:

$$^3P_J = -\langle {}^3P_J \mid V_{\Lambda N} \mid {}^3P_J \rangle, \qquad J = 0, 1, 2 \tag{83a}$$

$$^1P_1 = -\langle {}^1P_1 \mid V_{\Lambda N} \mid {}^1P_1 \rangle \tag{83b}$$

$$P = -\langle {}^3P_1 \mid V_{\Lambda N} \mid {}^1P_1 \rangle \tag{83c}$$

These matrix elements can conveniently be related to matrix elements of particular components of the ΛN interaction. Since only five matrix elements are required in the p shell, these can be related to spin-independent, spin–spin, tensor force, and spin–orbit symmetric and antisymmetric interactions. In the following we write down such interactions and evaluate their matrix elements. The above discussion does not imply that other ΛN interactions, such as a quadratic spin–orbit force, do not exist. In the p shell it is impossible to isolate more than five components of the ΛN interaction; the others simply modify the values expected for the matrix

---

[†] The phase convention adopted here is the standard Condon and Shortley convention. $\mathbf{S} = \mathbf{s}_N + \mathbf{s}_\Lambda$ and $\mathbf{S} + \mathbf{L} = \mathbf{J}$ in *that* order. This agrees with the numerical fitting (GSD 72) given by GSD but differs from their (GSD 71) formulas (A.14) (and following two lines) (A.15e) and (A.16e).

elements of the former. We note that for a SM hypernuclear analysis in the s–d shell, where no data are at present available, more than five effective matrix elements are required, which might in the future shed light on additional components of the ΛN interaction.

The ΛN interaction used by GSD is therefore of the general form

$$V(r_{\Lambda N}) + \Delta(r_{\Lambda N})\mathbf{s}_\Lambda \cdot \mathbf{s}_N + f(r_{\Lambda N})[\mathbf{s}_\Lambda \otimes \mathbf{s}_N]^{(2)} \cdot [\mathbf{r}_{\Lambda N} \otimes \mathbf{r}_{\Lambda N}]^{(2)}$$
$$+ g(r_{\Lambda N})(\mathbf{s}_\Lambda + \mathbf{s}_N) \cdot \mathbf{l}_{\Lambda N} + h(r_{\Lambda N})(\mathbf{s}_\Lambda - \mathbf{s}_N) \cdot \mathbf{l}_{\Lambda N} \qquad (84)$$

The first two terms give the central ΛN interaction, as used in the s shell,

$$V = \tfrac{3}{4}V_t + \tfrac{1}{4}V_s, \qquad \Delta = V_t - V_s \qquad (85)$$

with the appropriate LS matrix elements for their contribution to B_Λ values:

$$\left\langle (p^n, \alpha, S, L)J_N, \tfrac{1}{2}; J \left| - \sum_i^n V(r_{\Lambda i}) \right| (p^n, \alpha', S', L')J_N', \tfrac{1}{2}; J \right\rangle$$
$$= \delta_{\alpha\alpha'}\delta_{SS'}\delta_{LL'}\delta_{J_N J_N'}\, n\bar{V} \qquad (86)$$

$$\left\langle (p^n, \alpha, S, L)J_N, \tfrac{1}{2}; J \left| - \sum_i^n \Delta(r_{\Lambda i})\mathbf{s}_\Lambda \cdot \mathbf{s}_i \right| (p^n, \alpha', S', L')J_N', \tfrac{1}{2}; J \right\rangle$$
$$= \delta_{\alpha\alpha'}\delta_{SS'}\delta_{LL'}(-1)^{L-S+\frac{1}{2}+J}\,\frac{\sqrt{6}}{2}$$
$$\times [S(S+1)(2S+1)(2J_N+1)(2J_N'+1)]^{1/2}$$
$$\times \begin{Bmatrix} S & L & J_N \\ J_N' & 1 & S \end{Bmatrix} \begin{Bmatrix} J_N & \tfrac{1}{2} & J \\ \tfrac{1}{2} & J_N' & 1 \end{Bmatrix} \bar{\Delta} \qquad (87)$$

where $\bar{V}$ and $\bar{\Delta}$ are radial matrix elements of $V(r_{\Lambda N})$ and $\Delta(r_{\Lambda N})$, respectively, for a $1s\,\Lambda$ and a $1p$ nucleon, analogous to the quantities $\bar{U}$ and $\bar{\Delta}$ in the s shell. The other three interaction terms in Eq. (84) do not contribute to the s-shell calculations described in Section 4.2, although the inclusion of realistic nuclear D-waves in those calculations will probably imply a nonnegligible contribution of the ΛN tensor interaction. In the p shell, however, all these three interaction terms contribute in *first order* to B_Λ values. The expressions derived for their matrix elements are more complicated than Eq. (87), involving coefficients of fractional parentage, and will not be given here. Three radial integrals, treated as parameters in the SM analysis, are associated with these interactions: T for the tensor force and S_+ for the symmetric and S_- for the antisymmetric spin–orbit forces.

The symbol α appearing in the LS states of Eqs. (86) and (87) stands for any additional nuclear quantum number, beyond those (S and L) required to specify the nuclear level $J_N{}^\pi$. The nuclear spatial permutation symmetry provides an example, and the nuclear isospin is also implicitly hidden in α. In (86) and (87), $n = A - 5$ is the number of p nucleons. The interactions between the Λ and the $1s$ nucleons, as well as the Λ kinetic energy, are lumped in one constant, $B(5)$, which for a reasonable SM analysis should come as close as possible to the measured value of $B_\Lambda({}^5_\Lambda\text{He})$.

The relations between the phenomenological ΛN matrix elements (83a)–(83c) and the ΛN interaction parameters introduced above are given by

$$ {}^3P_0 = \bar{V} - \tfrac{1}{4}\Delta + 2S_+ + 6T \tag{88a} $$

$$ {}^3P_1 = \bar{V} - \tfrac{1}{4}\Delta + S_+ - 3T \tag{88b} $$

$$ {}^3P_2 = \bar{V} - \tfrac{1}{4}\Delta - S_+ + \tfrac{3}{5}T \tag{88c} $$

$$ {}^1P_1 = \bar{V} + \tfrac{3}{4}\Delta \tag{88d} $$

$$ P = -\sqrt{2}\,S_- \tag{88e} $$

As expected, the symmetric spin–orbit and the tensor force operate only within triplet ΛN states, whereas the antisymmetric spin–orbit force can only connect triplet with singlet $J = 1$ states.

In dealing with ΛN spin–orbit interactions it is useful to recall that since the Λ is in a $1s$ state the operator $l_{\Lambda N}$ effectively reduces to l_N, the angular momentum of a p nucleon. The ΛN spin–orbit forces therefore involve a part effectively looking like $\sum_i \zeta(r_\Lambda, r_i)\mathbf{s}_i \cdot l_i$, i.e., an induced nuclear spin–orbit potential. As we will soon see, this potential contributes in first order, even for hypernuclei whose core nucleus is spinless. Such a contribution, as alluded to before, has not been taken into account by rigid core approximations and hypernuclear cluster calculations. For example, all these calculation methods for ${}^9_\Lambda\text{Be}$ use as an input a central, spin-independent, ΛN interaction derived from ${}^5_\Lambda\text{He}$, where no spin–orbit force could be made operative for the simple wave functions used. However, for ${}^9_\Lambda\text{Be}$ the contribution of the ΛN spin–orbit force, due to the nuclear spin–orbit force induced by it, is far from being negligible. The SM provides a natural framework for incorporating effects due to spin–orbit interactions.

Generalizing the above discussion, we can divide the ΛN effective interaction parameters into two groups. In the first group are the Λ-spin-independent terms, such as the central $\sum_i V(r_{\Lambda i})$ and the induced nuclear spin–orbit $\sum_i \zeta(r_\Lambda, r_i)\mathbf{s}_i \cdot l_i$, terms. These receive contributions roughly proportional to n, the number of p-shell nucleons. The statement is exact, of course, for the coefficient of $\bar{V}$ [Eq. (86)] and holds in the jj coupling limit (appropriate only toward the end of the p shell) for the parameter $(S_+ - S_-)$. Indeed, the contribution of the induced nuclear spin–orbit force is given by

$$-\left\langle \sum_i^n \zeta(r_A, r_i)\mathbf{s}_i \cdot \mathbf{l}_i \right\rangle = -n\langle \zeta(r_A, r_1)\rangle\langle \mathbf{s}_1 \cdot \mathbf{l}_1\rangle$$

$$= -n(S_+ - S_-)\tfrac{1}{2}[j(j+1) - l(l+1) - \tfrac{3}{4}]$$

$$= \begin{cases} -(S_+ - S_-)n/2, & p_{3/2} \text{ shell} & \text{(89a)} \\ (S_+ - S_-)n, & p_{1/2} \text{ shell} & \text{(89b)} \end{cases}$$

where $(S_+ - S_-)$ is the appropriate radial matrix element. Equations (89a) and (89b) hold for all values of J_N, including $J_N = 0$.

On the other hand, the spin–spin, tensor, and part of the spin–orbit interaction [the part having the form $\sum_i \xi(r_A, r_i)\mathbf{s}_A \cdot \mathbf{l}_i \to \xi \mathbf{s}_A \cdot \mathbf{L}$] depend linearly on the spin of the A-particle and form a second group. They give rise to contributions to B_A values that depend in a sensitive way on the spin couplings inherent in the nuclear wave function, but which essentially are due to one (or few) paired-off nucleon(s) and hence no n factor appears in their expressions. These interactions are necessarily of the form $\sum_i^n \mathbf{s}_A \cdot \mathbf{A}_i$, where $\mathbf{A}$ is a nuclear vector. In the jj limit we have for the case $J_N = J_N' \neq 0$

$$\sum_{i=1}^n \langle \mathbf{s}_A \cdot \mathbf{A}_i \rangle = \frac{1}{j(j+1)} \langle \mathbf{A} \cdot \mathbf{j} \rangle \sum_{i=1}^n \langle \mathbf{s}_A \cdot \mathbf{j}_i \rangle = \alpha_j(\mathbf{A})\langle \mathbf{J}_N \cdot \mathbf{s}_A \rangle$$

$$= \alpha_j(\mathbf{A})\tfrac{1}{2}[J(J+1) - J_N(J_N+1) - \tfrac{3}{4}]$$

$$= \begin{cases} \tfrac{1}{4}\alpha_j(\mathbf{A})(2J - 1), & J = J_N + \tfrac{1}{2} & \text{(90a)} \\ -\tfrac{1}{4}\alpha_j(\mathbf{A})(2J + 3), & J = J_N - \tfrac{1}{2} & \text{(90b)} \end{cases}$$

where

$$\alpha_j(\mathbf{A}) = \frac{\langle \mathbf{A} \cdot \mathbf{j}\rangle}{j(j+1)} = \frac{\langle j \| \mathbf{A} \| j\rangle}{\langle j \| \mathbf{j} \| j\rangle} \tag{90c}$$

The vectors $\mathbf{A}$, corresponding to contributions to B_A values from spin–spin, spin–orbit, and tensor AN interactions, are given by $\Delta\mathbf{s}$, $-(S_+ + S_-)\mathbf{l}$, and $6(8\pi)^{1/2}T[\mathbf{s} \otimes Y_2(\hat{\mathbf{r}})]^{(1)}$, respectively, and the coefficients $\alpha_i(\mathbf{A})$ are evaluated to be

$$j = \tfrac{3}{2}: \qquad -\tfrac{1}{3}\Delta, \quad -\tfrac{1}{3}(S_+ + S_-), \quad +\tfrac{4}{5}T \tag{91a}$$

$$j = \tfrac{1}{2}: \qquad +\tfrac{1}{3}\Delta, \quad -\tfrac{4}{3}(S_+ + S_-), \quad -8T \tag{91b}$$

respectively.

The number of ΛNN effective matrix elements involved in a shell model in the p shell is considerably larger than five. GSD restricted their analysis to the TPE ΛNN interaction discussed in Section 2.3. The outstanding features of this interaction from the point of view of the quantum numbers involved are (i) a factor $\tau_1 \cdot \tau_2$ for the two nucleons; (ii) independence of the Λ spin, and (iii) linearity in both spins of the two nucleons. Hence we can decompose the ΛNN force between a $1s$ Λ and two $1p$ nucleons as follows (assuming the radial integration over the Λ wave function has been carried out):

$$\tau_1 \cdot \tau_2 \sum_{k,l,m} Q_{lm}^k (r_1, r_2)[\sigma_1 \otimes \sigma_2]^{(k)} \cdot [C_l(\hat{r}_1) \otimes C_m(\hat{r}_2)]^{(k)}, \qquad k = 0, 1, 2 \tag{92}$$

where

$$C_l(\hat{r}) = [4\pi/(2l + 1)]^{1/2} Y_l(\hat{r}) \tag{93}$$

and $(l, m) = (0, 0)$ and $(2, 2)$ for $k = 0$; $(2, 2)$ for $k = 1$; and $(0, 2)$, $(2, 0)$, and $(2, 2)$ for $k = 2$. Since the nuclear wave functions are antisymmetric in the nucleon labels, the terms Q_{02}^2 and Q_{20}^2 will always appear in the form $(Q_{02}^2 + Q_{20}^2)$ and altogether we have five radial matrix elements to consider. In order to reduce the total number of parameters, GSD evaluated the radial matrix elements $Q_{lm}^k = \langle Q_{lm}^k(r_1, r_2) \rangle$ for the wave function (80), applying a regularization scheme which gets rid of the δ-functions contained in the TPE ΛNN interaction. The ratio of these five parameters is held fixed, as given by their evaluation, through the B_Λ fitting procedure, so that they are all specified by the value Q^* fitted for Q_{22}^0 and by the following fixed ratios:

$$Q_{00}^0 : Q_{22}^0 : Q_{22}^1 : Q_{02}^2 = Q_{20}^2 : Q_{22}^2$$
$$= 0.0259 : 1 : -0.4823 : -0.0446 : 0.2131 \tag{94}$$

The most important terms are Q_{22}^0 and Q_{22}^1 and for the $1p$ nuclear shell they correspond to repulsion and attraction, respectively. The value Q^* is left, together with the ΛN parameters, for the SM fitting procedure; the introduction of all five TPE ΛNN parameters as free parameters would leave almost no room for a "goodness" measure of the fit. It is interesting to note that for the TPE ΛNN force discussed by Bhaduri *et al.* (BLN 67),

$$Q_{22}^0 = -0.73 C_p \text{ (MeV)} \tag{95}$$

and since C_p is somewhat greater than 1 MeV, we also expect this to be the order of magnitude for the value Q^* obtained in the SM fit.

One should not neglect the ΛNN contributions arising from one p nucleon and all the s-shell nucleons. Since the s-shell nucleons are summed over, these contributions modify some of the two-body ΛN parameters discussed above. Even though $\boldsymbol{\sigma}$ and $\boldsymbol{\tau}$ average to zero in the sum over the $1s$ shell for the nucleon, there are still terms which survive for both $k = 0$ and 1 [Eq. (92)] due to exchange effects. The $k = 0$ term is proportional to n and hence modifies $\bar{V}$, whereas the $k = 1$ term gives rise to an additional induced nuclear spin–orbit force $\zeta' \sum_i \mathbf{s}_i \cdot \boldsymbol{l}_i$. No new matrix elements are thus added to the B_Λ analysis, but the theoretical interpretation of the phenomenological values obtained for $\bar{V}$ and $(S_+ - S_-)$ is affected. The ΛNN interactions involving two $1s$ nucleons are assumed the same throughout the p shell and hence are included in the constant $B(5)$.

Before turning to the SM fitting procedure, we list in Table III for the purpose of orientation the coefficients of the contributions of the potential terms nQ^*, $n(S_+ - S_-)$, $(S_+ + S_-)$, Δ, and T to the total B_Λ value for a series of p-shell hypernuclei labeled by their mass number A and isospin I. These coefficients simply give the expectation values of the Λ–nuclear interactions for the intermediate coupling wave function of the appropriate nuclear ground state. Where two entries appear, the upper one corresponds to the case $J = J_N - \frac{1}{2}$ for the hypernuclear spin, and the lower one to the spin state $J = J_N + \frac{1}{2}$. The dash denotes situations $(J_N = 0)$ in which the coefficient is automatically zero. From this table we see that the coefficient of $(S_+ - S_-)$ is by no means proportional to n throughout the $p_{3/2}$ shell, as implied by jj coupling, but in the upper part of this shell it is roughly given by $-0.4n$. The coefficient of Q^* rises through the $p_{3/2}$ shell faster than linearly but in the upper part of this shell the coefficient of Q^* is roughly given by $0.2n$. Thus in the approximation that the core nucleus wave function is simply that for the ground state in the intermediate coupling approximation, the slope of the B_Λ curve in the upper part of the $p_{3/2}$ shell can be roughly approximated per unit step in A by

$$\bar{V} - 0.4(S_+ - S_-) + 0.2Q^* \tag{96}$$

Recall that both $\bar{V}$ and $(S_+ - S_-)$ obtain contributions from ΛNN interactions where one nucleon is in the p shell and the other in the s shell. The coefficients of the Λ-spin-dependent terms $(S_+ + S_-)$, Δ, and T vanish for spinless core nuclei and are of the order of unity in the $p_{3/2}$ shell. In the $p_{1/2}$ shell, however, the coefficient of T assumes large values, in accordance with the large value of $\alpha_{j=1/2}(\mathbf{A})$ for the tensor force as given by Eq. (91b). If we ignore T, on the expectation that it has a small value due to

TABLE III

The Coefficients of the Contributions of the Potential Terms Q^*, S_+, S_-, Δ, and T to the Total B_Λ Value as a Function of the p-Shell Hypernuclear Species (A, I)

These coefficients are appropriate to the situation in which the Λ is coupled only to the ground state of the core nucleus $(A - 1, I)$ through the intermediate coupling wave function of the ground state (GSD 71, 72). Here $n = A - 5$ denotes the number of p-shell nucleons. The dash denotes situations in which the coefficient is automatically zero $(J_N = 0)$. The first set of values is for the spin value $J = |J_N - \frac{1}{2}|$, the values in parentheses corresponding to the spin value $J = J_N + \frac{1}{2}$.

(A, I)	nQ^*	$n(S_+ - S_-)$	$(S_+ + S_-)$	Δ	T
$(7, 0)$	0.130	−0.121	0.023 (−0.012)	0.977 (−0.489)	0.137 (−0.068)
$(7, 1)$	0.099	−0.410	—	—	—
$(8, \frac{1}{2})$	0.204	−0.170	0.825 (−0.495)	0.425 (−0.255)	−0.750 (0.450)
$(9, 0)$	0.276	−0.186	—	—	—
$(9, 1)$	0.135	−0.348	0.832 (−0.555)	0.668 (−0.445)	−0.230 (0.153)
$(10, \frac{1}{2})$	0.222	−0.280	0.914 (−0.498)	0.336 (−0.202)	−0.754 (0.452)
$(11, 0)$	0.155	−0.421	1.441 (−1.081)	0.559 (−0.420)	−1.754 (1.316)
$(12, \frac{1}{2})$	0.189	−0.406	0.997 (−0.598)	0.253 (−0.152)	−1.234 (0.741)
$(13, 0)$	0.252	−0.370	—	—	—
$(14, \frac{1}{2})$	0.248	−0.296	1.004 (−0.335)	−0.254 (0.085)	5.347 (−1.782)
$(15, 0)$	0.308	−0.191	1.371 (−0.686)	−0.371 (0.186)	7.504 (−3.752)
$(16, \frac{1}{2})$	0.413	−0.091	1.000 (−0.333)	−0.250 (0.083)	6.000 (−2.000)
$(17, 0)$	0.532	—	—	—	—

the ineffectiveness of K and η couplings, then the spin dependence of the ΛN hypernuclear interaction is given by the spin–spin (Δ) and spin–orbit ($S_+ + S_-$) terms. As seen from Table III, the coefficients of Δ and ($S_+ + S_-$) have the same sign throughout the $p_{3/2}$ shell but are of opposite sign in the $p_{1/2}$ shell. From the deduced spins of $^8_\Lambda\text{Li}$ and $^{12}_\Lambda\text{B}$, which follow the $J = |J_N - \frac{1}{2}|$ rule, we may argue that at least one [of Δ and ($S_+ + S_-$)] spin-dependent parameter is positive, probably ($S_+ + S_-$) in view of its larger coefficients. If this is the case, then for the $p_{1/2}$ shell we may expect the same pattern, $J = |J_N - \frac{1}{2}|$, to hold. These arguments, however, obtain substantial support only from $^{12}_\Lambda\text{B}$, since for $^8_\Lambda\text{Li}$ the effect of admixing the first low-lying nuclear excited state (0.48 MeV) into the hypernuclear ground state is far from small. We will come back later to this question.

It would be misleading at this point to use Table III for determining the values of $\bar{V}$, ($S_+ - S_-$), and Q^* from the three B_Λ values (with respect to $^5_\Lambda\text{He}$) available for spinless core nuclei, i.e., $^7_\Lambda\text{Be}$, $^9_\Lambda\text{Be}$, and $^{13}_\Lambda\text{C}$. As a result of the Λ distortion (within the $1p$ shell) of the nuclear core, by coupling-in excited nuclear states, the appropriate coefficients depicted in Table III change somewhat. But since these changes are multiplied by n for the terms now being discussed, their effect on the B_Λ fitting may prove nonnegligible.

We now describe the calculational procedure applied by GSD in their SM analysis of B_Λ values in the p shell. First, the expressions for the matrix element of the ΛN and ΛNN interactions, for every possible spin for every hypernuclear species in the p shell, are transformed from their LS basis to the intermediate coupling basis, using the calculated core nuclei eigenstates (Sop 64). This is done once and for all in terms of the parameters Δ, etc. Note that the nuclear interactions are diagonal in this basis and the appropriate eigenvalues can be, and are, taken directly from the observed nuclear spectrum. Next, starting values are assumed for the parameters, which we call $\{X_\alpha\}$ ($\bar{V}$, Δ, S_+, S_-, T, Q^*). Upon diagonalization of the numerical energy matrices, this initial choice yields a spectrum in which the hypernuclear ground states possess certain spin values and these are assumed for the moment to be the self-consistent spin values. The calculated "ground state" energies are then expressed in terms of the set $\{X_\alpha\}$ and a least squares fit is performed to minimize the weighted sum of squares of differences between calculated $[E(k)]$ and observed binding energies

$$\sum_k [E(k) - B_\Lambda(k)]^2/[\sigma(k)]^2 \tag{97}$$

where $\sigma(k)$ denotes the experimental uncertainty in the B_Λ value of the species labeled k. New values for the set $\{X_\alpha\}$ are thus determined which should normally give a better fit to the data than that given by the original starting values. With these new values the energy matrices are again diagonalized and the whole procedure is iterated until a minimum for Eq. (97) is found. It may turn out that the starting values of spin assumed do not actually correspond to ground states derived by the minimization program. This is easily revealed by examining the spectrum of every hyper-nucleus used in the fit. The whole procedure is then repeated for the ground-state spin values serving as starting values. In this way stable states are almost always reached where the final J values are the same as the starting ones. This is a consistent solution. The worst that happens in a few cases is the occurrence of a bistable solution where a recurring cycle ($l = 2$ at most in the calculation of GSD) of sets of J values is reached. This is a rather technical detail and we refer the reader to GSD.

The possibility exists in the SM analysis described above to impose constraints on the values of the parameters considered. Thus it was consid-ered useful to constrain some of the parameters to a value of zero so that the effect of the remaining parameters is made more transparent. For any given set of constraints two solutions are generally found. For example, if only a central ΛN interaction is used, a solution is found both for positive Δ and for negative Δ. As long as we do not require $J(^8_\Lambda\text{Li}) = J(^{12}_\Lambda\text{B}) = 1$, as deduced from experiment and fitted with $\Delta > 0$, there is no general way of preferring one solution to the other. [†] With more free spin param-eters to be varied, such as Δ, T, and $(S_+ + S_-)$, there are quite a few different χ^2 minima for which the ground-state spins mentioned above are correctly reproduced. In other cases, although these spin values do not correspond to ground states, they appear with such a small excitation energy (~ 0.1 MeV) that the branching ratio for their decay is overwhelm-ingly in favor of a weak hypernuclear decay, by means of which these states have actually been observed.

For a consistent solution the following are printed out: the wave function, namely the amplitude with which each of the parent nuclear states enters the hypernuclear state; the value found in the fit for each one of the potential parameters; the energy expressed as a linear combination of the potential parameters. These enable an assessment of the extent to which the core nucleus is polarized and which of the parameters has a significant role in reproducing the B_Λ value. In general, the hypernuclear wave func-

---

[†] Such would have been the case in the s shell if the ground-state spin values for $^3_\Lambda\text{H}$ and $^4_\Lambda\text{H}$ were unknown.

tions are rather pure, that is, they consist mostly of a Λ attached to one particular state of the core nucleus. This statement holds quite well for hypernuclear ground states, with the exception of $^{8}_{\Lambda}$Li, where the $J = 1$ state is built on both $J_N = \frac{3}{2}$ and $J_N = \frac{1}{2}$ (0.48 MeV excitation) with an appreciable mixing. The following comments should be made on the results of GSD:

(i) The sign of Q^* is generally negative, corresponding to a net repulsion. This repulsion is necessary mainly for fitting the relatively low B_Λ value of $^{9}_{\Lambda}$Be (see Table I). In fact, the difference between $B_\Lambda(^{9}_{\Lambda}$Li) and $B_\Lambda(^{9}_{\Lambda}$Be) is about 1.8 MeV, considerably larger than the ~ 1 MeV binding per nucleon observed on the average for p-shell hypernuclei. It is almost impossible to reproduce this large $A = 9$ B_Λ difference with two-body ΛN parameters alone, since for such ΛN interactions there is no special status whatsoever attached to $^{9}_{\Lambda}$Be. To some extent a relatively low binding also appears for $^{13}_{\Lambda}$C, whose B_Λ value is somewhat lower than that of $^{12}_{\Lambda}$B, with one less nucleon for the latter. The TPE three-body force contains terms (in particular the strong Q^0_{22}) which acquire an extra repulsion for core nuclei of the α-group type, such as $^{9}_{\Lambda}$Be and $^{13}_{\Lambda}$C, due to the relative high degree of spatial symmetry in their wave function (Gal 67). From Table III we see that the three-body force contributes $0.55Q^*$ in excess of $^{9}_{\Lambda}$Li to $B_\Lambda(^{9}_{\Lambda}$Be). If $Q^* \doteq -2$ MeV [about twice the value given by Bhaduri *et al.* (BLN 67)], then most of the B_Λ difference for $A = 9$ is thereby accounted for; the rest of it may be attributed to Λ-spin-dependent terms which vanish for $^{9}_{\Lambda}$Be and necessarily provide attraction to $^{9}_{\Lambda}$Li. However, for $A = 10$ the three-body contribution is seen from Table III to be almost identical to that for $^{9}_{\Lambda}$Be, again exceeding by $0.55Q^*$ (repulsion) the analogous contribution to $^{9}_{\Lambda}$Li, whereas $^{10}_{\Lambda}$B does not seem to possess exceptionally low binding. Here the spin-dependent terms cannot be invoked for explanation since they are pretty much the same for both $^{9}_{\Lambda}$Li and $^{10}_{\Lambda}$B and one definitely needs a large value of $\bar{V}$, in the vicinity of 1.5 MeV, to yield a B_Λ difference of about $\bar{V} + 0.55Q^* = 1.5 - 1.1 = 0.4$ MeV, roughly as observed.[†]

We note that the observed slope of the B_Λ curve as function of A is about 1 MeV per nucleon. According to Eq. (96), for $S_+ \approx S_-$ this impies $\bar{V} + 0.2Q^* \sim 1$ MeV, so that for $Q^* = -2$ MeV we obtain $\bar{V} = 1.4$ MeV. On the other hand, fits constrained to have $Q^* = 0$ require very

---

[†] We deliberately refer to $B_\Lambda(^{10}_{\Lambda}B) = 8.89 \pm 0.12$ MeV, exceeding $B_\Lambda(^{9}_{\Lambda}$Li) by 0.36 MeV, not taking into account a CS average of $B_\Lambda(^{10}_{\Lambda}B)$ and $B_\Lambda(^{10}_{\Lambda}Be) = 9.30 \pm 0.26$, since the latter number is based on one established event only.

strong, spin-dependent ΛN terms, in particular spin–orbit and/or tensor terms. From Table III we may again deduce that the low binding of $^9_\Lambda$Be prefers negative values of the induced spin–orbit parameter $S_+ - S_-$ and hence this time $\bar{V}$ is considerably lower than the slope of 1 MeV mentioned above. For $^{13}_\Lambda$C the situation is less acute than for $^9_\Lambda$Be; at least it seems so since no other hypernucleus with the same mass number or with $A = 14$ has been observed to compare with.

(ii) S_+ and S_- almost always are of the same sign for the minima found. Hence it is enough to specify the sign of $S = (S_+ + S_-)$.

(iii) For every solution with a positive Δ, there occurs a corresponding solution with a negative Δ, all the remaining parameters having the same signs and roughly the same magnitudes for the pair of solutions. Such parallel minima correspond to the choice $|J_N - \frac{1}{2}|$ for $\Delta > 0$ and $(J_N + \frac{1}{2})$ for $\Delta < 0$ in the $p_{3/2}$ shell.

(iv) The minima also occur in pairs with respect to S, which again roughly corresponds to two possible sets of spin values. For $S > 0$, $|J_N - \frac{1}{2}|$ holds and for $S < 0$, $(J_N + \frac{1}{2})$ holds for ground-state spin values throughout the p shell.

In Table IV several representative fits are shown, labeled by their χ^2 value, for which the values of the parameters are reasonable from a theoretical standpoint. For example, from the beginning we discard ratios $\Delta/\bar{V} \gtrsim 1$ as unphysical. We insist on the appearance of Q^* with a large

TABLE IV

The ΛN and ΛNN Potential Parameters in MeV (and the χ^2 Corresponding to Them) for Four Representative Least Square Fits of 11 B_Λ Values in the $p_{3/2}$ Shell (GSD 72)

The ΛN tensor parameter was constrained to zero. In two of the fits Δ was also constrained to zero. The values of χ^2 given in parentheses correspond to omission of $B_\Lambda(^8_\Lambda$He) due to the low neutron separation energy in $^8_\Lambda$He. The "slope" of the B_Λ curve in the $p_{3/2}$ region is defined as $\bar{V} - 0.4(S_+ - S_-) + 0.2Q^*$.

χ^2	$\bar{V}$	Δ	S_+	S_-	Q^*	"Slope"	J
42.8 (22.8)	1.48	—	−0.49	−0.73	−1.99	0.98	$J_N + \frac{1}{2}$
36.1 (14.6)	1.44	—	0.39	0.29	−1.99	1.00	$\mid J_N - \frac{1}{2} \mid$
32.8 (17.2)	1.07	0.65	−0.96	−0.95	−0.73	0.92	$J_N + \frac{1}{2}$ [a]
33.7 (9.7)	1.15	0.36	0.33	0.46	−1.08	0.98	$\mid J_N - \frac{1}{2} \mid$

[a] Except ^{7,8}Li.

contribution since otherwise the value of $\bar{V}$ obtained is significantly lower than 1 MeV, which, as will be explained below, appears rather unreasonable. For the first two fits, Q^* is very large, about twice the value obtained by Bhaduri *et al.* (BLN 67) for the TPE ΛNN force. These two fits differ from each other essentially by the signs of the spin–orbit terms. The first of these fits has a minus sign for S_+, in accordance with meson-theoretic expectations and parallel to the NN case (GSD 71, LD 71). Its drawback is that it gives $J = 2$ as ground state for both $^8_\Lambda$Li and $^{12}_\Lambda$B. For $^8_\Lambda$Li, however, an excited state at $E^* = 0.19$ MeV with $J = 1$ exists which would, in view of the small excitation energy, decay preferably by weak hypernuclear decay and could conform with the experimental evidence for a weakly decaying $J = 1$ state. For $^{12}_\Lambda$B, on the other hand, such a $J = 1$ excited state appears at an excitation energy of 1.73 MeV, sufficiently high to force its rapid electromagnetic deexcitation to the calculated $J = 2$ ground state, in disagreement with observation. The second fit gives a reasonable magnitude for the spin–orbit parameters but their sign is opposite to that prefered by meson theory. It reproduces the observed $J = 1$ spin values for ground-state $^8_\Lambda$Li and $^{12}_\Lambda$B. The introduction of another free parameter, Δ, into these two fits gives rise to $\Delta > 0$ for the value at the minimum, decreasing the magnitude of $\bar{V}$ and Q^* and increasing the effectiveness of the Λ-spin-dependent spin–orbit term. Both these fits, the last two fits in Table IV, give $J = 1$ for ground-state $^8_\Lambda$Li, but only the last fit gives $J = 1$ for ground-state $^{12}_\Lambda$B. The value assumed by Q^* is now in conformity with TPE derivations (BLN 67). The ratio $\Delta/\bar{V}$ is not small but this is not in gross contradiction with the analogous ratio required for smooth ΛN central interactions in the s shell to reproduce an excitation of about 1 MeV for $A = 4$, as recently observed (Bam+ 71). In the last column of Table IV the value of the B_Λ slope as defined for the $p_{3/2}$ shell [Eq. (96)] is given for the selected fits. Recall that GSD used in their input B_Λ values only observed quantities of the $p_{3/2}$ shell. It is clear that the slope is the *only* combination well determined by the SM analysis, since even these selected fits differ by almost everything else from each other.

We address ourselves now to an examination of the value expected for $\bar{V}$ and $\Delta/\bar{V}$ on the basis of Λp s-wave cross sections and B_Λ analysis of s-shell hypernuclei. If we assume the Λ–^{4}He interaction to be characteristic of $\bar{V}$ and assume the spin-independent ΛN potential to have a Gaussian shape with an intrinsic range $b = 2.0$ fm, then, knowing the size and shape of ^{4}He from electron scattering data, we can deduce a corresponding estimate for $\bar{V}$: $\bar{V} = 1.76$ MeV. However, if we adopt the viewpoint that such a value of $\bar{V}$ for $^5_\Lambda$He is strongly affected by suppression of the Σ

channel, then we can only estimate it from the measured Λp cross sections. Assuming, again, the same intrinsic range for both triplet and singlet ΛN interactions, we obtain the estimate $\bar{V} = 2.5$ MeV. However, as mentioned above, all the TPE ΛNN interactions that involve one of the s-shell nucleons also contribute to $\bar{V}$. This contribution was estimated by GSD, with the net estimate for $\bar{V}$ of

$$\bar{V} = 2.5 - 0.325C_p = 2.5 + 0.44Q^* \quad \text{(in MeV)} \tag{98}$$

Hence for the $(\bar{V}, S, Q^*)$ fits of Table IV this relation between $\bar{V}$ and Q^* is almost satisfied, whereas for the $(\bar{V}, \Delta, S, Q^*)$ fits this relation is badly violated. It is clear now why we prefer B_Λ fits with strongly repulsive ΛNN forces: These fits usually give values of $\bar{V}$ considerably higher than 1 MeV, thus improving the agreement with Eq. (98) over those fits that give, in the absence of assumed ΛNN parameters, values remarkably smaller than 1 MeV. Such small values of $\bar{V}$ are also inconsistent with the estimate made above for $\bar{V}$ on the basis of a phenomenological fit of $B_\Lambda({}^5_\Lambda\text{He})$, being too low by a factor of two.

Similar to Eq. (98), GSD find the following estimate for the induced nuclear spin–orbit parameter $(S_+ - S_-)$:

$$(S_+ - S_-) = -0.4 - 0.26C_p = -0.4 + 0.36Q^* \quad \text{(in MeV)} \tag{99}$$

Of the fits shown in Table IV, only the last one gives a nonvanishing, negative value $(S_+ - S_-) = -0.13$, which, however, is too small in magnitude to satisfy Eq. (99). We do not try to bring similar arguments for the spin–orbit parameter $(S_+ + S_-)$ since the fitted values for this Λ-spin-dependent parameter differ greatly from each other, as can be deduced from Table IV, where the range of variation of $(S_+ + S_-)$ is from -1.91 to 0.79. On the basis of the measured B_Λ values alone it would be too ambitious at present to discuss whether or not a fitted value of $(S_+ + S_-)$ agrees with our theoretical expectation, which is about -0.2 MeV. The same remark also applies to the spin–spin parameter Δ. Nevertheless, in the fits shown in Table IV we ignored cases where $\Delta < 0$ or $\Delta/\bar{V} \gtrsim 1$ and it seems quite interesting to relate this requirement to the B_Λ analysis of the s shell hypernuclei. Assuming that Σ suppression effects are minimal for $A = 4$ (see discussion in Section 4.2) and CSB can be ignored, we follow DHT in associating a potential strength $U_4 = \frac{1}{2}U_{0t} + \frac{1}{2}U_{0s}$ per nucleon for $B_\Lambda(J = 0) = 2.31$ MeV and $U_4^* = \frac{5}{6}U_{0t} + \frac{1}{6}U_{0s}$ per nucleon for $B_\Lambda(J = 1) = 1.22$ MeV, so that this corresponds to the CS binding energy of ${}^4_\Lambda\text{H}$ and the observed 1.09-MeV excitation for one of the hypernuclei having $A = 4$. Using extrapolation formulas given by DHT, we find $U_4 = 423.88$ MeV and $U_4^* = 401.60$ MeV. From these we deduce

$$\Delta_S = 3(U_4 - U_4^*) = 66.84 \text{ MeV} \tag{100a}$$

$$\bar{V}_S = \tfrac{1}{4}U_4 + \tfrac{3}{4}U_4^* = 407.17 \text{ MeV} \tag{100b}$$

with the following ratio:

$$\Delta_S/\bar{V}_S = 0.16 \tag{100c}$$

This particularly small value for the ratio $\Delta/\bar{V}$ in the s shell, in spite of the relatively large excitation value of about 1 MeV related directly to Δ, results from the singular character of the potential adopted, giving a very deep attraction over a short distance just outside a region of extremely strong repulsion. With such a potential, the B_Λ value changes especially rapidly with increasing potential depth U. Thus, to give a change $\Delta B_\Lambda \sim 1$ MeV involves only a relatively small change in U, relative to the critical value of U_c needed for binding to occur at all. Since in the SM analysis described here for p-shell hypernuclei, smooth ΛN interactions are employed, we should relate our observation to an s-shell calculation which also employs smooth ΛN interactions. Such is the oldest rigid core calculation, by Dalitz and Downs (DD 58), where a Gaussian ΛN potential with intrinsic range 1.5 fm is considered. Extrapolating the results of their paper to the presently acceptable size of the $A = 3$ core nuclei gives the relation

$$U_4 = 209(1 + 0.46\sqrt{B_\Lambda}) \text{ (in MeV-fm}^3) \tag{101a}$$

whereas retaining in DHT the $\sqrt{B_\Lambda}$ term only with the coefficient enlarged, so that this expression fits their exact U_4 value for the physical B_Λ value, leads to

$$U_4 = 350(1 + 0.14\sqrt{B_\Lambda}) \text{ (in MeV)} \tag{101b}$$

We see that the coefficient in front of $\sqrt{B_\Lambda}$, which is the quantity governing the effect we are discussing, is about three times larger for the Gaussian potential than for the hard core potential. For $B_\Lambda = 2.31$ MeV, Eq. (101a) gives $U_4 = 355.3$ MeV-fm³ and for $B_\Lambda = 1.22$ MeV, $U_4^* = 315.6$ MeV-fm³. Hence[†]

$$\frac{\Delta_S}{\bar{V}_S} = \frac{3(U_4 - U_4^*)}{\tfrac{1}{4}U_4 + \tfrac{3}{4}U_4^*} = \frac{119}{326} = 0.37 \tag{102}$$

[†] I owe this observation to R. H. Dalitz, private communication (1972).

We see that of the two last fits ($\Delta \neq 0$) in Table IV, the last one has a similar value of $\Delta/\bar{V}$ in the p shell to that given by Eq. (102).

To sum up this discussion, we stress that the SM analysis presented here appears inconclusive with regard to the values appropriate for the Λ-spin-dependent terms Δ, $(S_+ + S_-)$, and T. In due time when low-lying hypernuclear excitation energies are observed and identified, it will become possible to isolate these spin-dependent terms more reliably. From Table III we see that for the two hypernuclear $^7_\Lambda\text{Li}$ states, one of which is the ground state, based on ground-state $^6\text{Li}(J_N = 1)$, the coefficients of $(S_+ + S_-)$ and T are negligible (as a result of the core nucleus wave function being rather well approximated by the dominant LS configuration $^3S_1$, i.e., $L = 0$). Hence, to a good approximation, $E(J = \frac{3}{2}) - E(J = \frac{1}{2}) = \frac{3}{2}\Delta$ and this excitation, when observed, could determine the appropriate value for Δ. Since the coefficient of T is still small in magnitude for the two hypernuclear $^9_\Lambda\text{Li}$ states based on ground-state $^8\text{Li}(J_N = 2)$ it would become possible, knowing the appropriate value for Δ, to determine from $E(J = \frac{5}{2}) - E(J = \frac{3}{2})$ the appropriate value for $(S_+ + S_-)$. To determine the value of T, it would be best to observe hypernuclear excitation energies in the $p_{1/2}$ shell, since the coefficients of T, evaluated in Table III or given in the jj limit [Eq. (91b)], are particularly large there. Thus for $^{16}_\Lambda\text{O}$, the excitation energy of the Λ doublet based on the nuclear ground state ($J_N = \frac{1}{2}$) is given by

$$E(J = 1) - E(J = 0) = 8T + \tfrac{4}{3}(S_+ + S_-) - \tfrac{1}{3}\Delta$$

heavily emphasizing the T term.

In the SM analysis of GSD, B_Λ values of the $p_{3/2}$ shell were used; the observation of $^{15}_\Lambda\text{N}$ with a B_Λ value of 13.59 ± 0.15 MeV was reported after the completion of their work. It is therefore instructive to report their predictions for B_Λ values and spins in the $p_{1/2}$ shell and confront these with this one available number. In Table V we show these predictions for the four fits specified in Table IV. Since the coefficient of $(S_+ + S_-)$ in the difference between the B_Λ values for the states $(J_N + \frac{1}{2})$ and $|J_N - \frac{1}{2}|$ is much larger than the coefficient of Δ, it is the sign of $S \equiv (S_+ + S_-)$ which determines the values of ground-state spins: $S > 0$ leads to $J = |J_N - \frac{1}{2}|$ and $S < 0$ to $J_N + \frac{1}{2}$.

Since the coefficients of T are found to be considerably larger in magnitude in the $p_{1/2}$ shell than in the $p_{3/2}$ shell, it is important to check the effect of a small nonzero value for T, say $|T| = 0.1$ MeV. We prefer not to present here the results of fits that are obtained by using the last

TABLE V

The B_Λ Values in MeV for the $p_{1/2}$-Shell Hypernuclei as a Function of (A, I) for Each of the Four Fits Specified in Table IV (GSD 72)

In the next to the last column the expected ground-state spin values are shown and in the last column the calculated B_Λ values for $^{13}_\Lambda$C, the heaviest hypernucleus in the p shell whose measured B_Λ value was used as input, are given in MeV. These B_Λ values were supposed to fit an old B_Λ value of 10.51 ± 0.51 MeV (Boh+ 68) instead of the recent value of 11.22 ± 0.08 MeV (Jur+ 73). In parentheses next to each value of χ^2 the corresponding parameters that differ from zero among Δ, S_+, S_-, T, Q^* are given with a superscript to indicate their sign (since S_+ and S_- appear with the same sign, S stands for $S_+ + S_-$). Q^* is always negative when it is significant. For $J_N \neq 0$, in the columns under the (A, I) values, the values (in MeV) given in parentheses correspond to adding the contribution due to a small tensor force, $|T| = 0.1$ MeV and sgn $T =$ sgn S (the latter relation is practically satisfied by all the fits for which T is also varied), to the otherwise predicted B_Λ values.

(A, I) χ^2	$(14, \frac{1}{2})$	$(15, 0)$	$(15, 1)$	$(16, \frac{1}{2})$	$(17, 0)$	J	$(13, 0)$
42.8 (S^-Q^*)	11.78 (11.91)	12.21 (12.54)	11.08 (11.06)	10.36 (10.54)	7.98	$J_N + \frac{1}{2}$	10.26
36.1 (S^+Q^*)	12.10 (12.60)	12.19 (12.89)	11.01 (11.00)	10.40 (11.00)	7.60	$|J_N - \frac{1}{2}|$	10.44
32.8 ($\Delta^+S^-Q^*$)	12.14 (12.24)	13.29 (13.59)	11.63 (11.59)	12.32 (12.48)	11.18	$J_N + \frac{1}{2}$	10.37
33.7 ($\Delta^+S^+Q^*$)	12.02 (12.54)	12.41 (13.13)	11.36 (13.35)	11.54 (12.14)	9.87	$|J_N - \frac{1}{2}|$	10.50

two fits in Table V as "starting fits" and relaxing the constraint $T = 0$, since the values thus found for T are rather large and involve significant changes in the values of the other nonzero parameters (e.g., decreasing $\bar{V}$ to 0.7 and 0.8 MeV, respectively). Since physically we expect a small value for T and the fits generally give a sign of T identical to that of S, we included in Table V (in parentheses) the B_Λ values that would have been obtained by the addition of $|T| = 0.1$ MeV, sgn $T =$ sgn S, to the depicted fits. With this choice of sign, the effect of this small value of T is to increase the otherwise predicted B_Λ value by up to 0.7 MeV, except for the species for which $J_N = 0$, where a negligible decrease due to core polarization or no change occurs. With other choices of sign for T, however, a decrease in B_Λ values could be expected sometimes. We note that only for one fit shown in Table V, $\chi^2 = 32.8$, does the calculated B_Λ value of $^{15}_\Lambda$N come close to the observed value. This should not suggest the failure of the other fits, giving B_Λ values too low by about 1 MeV and more (although the small tensor component discussed above would partially alleviate this problem). Thus the SM analysis of GSD was based on older B_Λ data (Boh+ 68), which gave for $B_\Lambda(^{13}_\Lambda$C) a value of 10.51 ± 0.51 MeV. The fitted values of $B_\Lambda(^{13}_\Lambda$C) for the four fits discussed here were then 10.26, 10.44, 10.37, and 10.50 MeV, considerably lower than the up-dated value (Jur+ 73) of 11.22 ± 0.08 MeV.[†] Since $^{13}_\Lambda$C was the heaviest hypernuclear species, on the border of the $p_{1/2}$-shell region, whose B_Λ value was used as an input, it is reasonable to expect that with the use of the new $B_\Lambda(^{13}_\Lambda$C) value, the B_Λ values predicted for the $p_{1/2}$ shell will also increase with respect to the values given in Table V by about 0.6 MeV for $^{15}_\Lambda$N and 0.8 MeV for $^{17}_\Lambda$O. Even though the B_Λ value for $^{15}_\Lambda$N predicted with this artifice comes close to the observed value of 13.59 ± 0.15 MeV for most of the fits, the B_Λ value predicted for $^{17}_\Lambda$O still remains exceptionally low.

This trend of predicted B_Λ values in the $p_{1/2}$ shell to decrease with increasing A and reach an almost ridiculously low value for $^{17}_\Lambda$O, at the end of the $p_{1/2}$ shell, results from the ΛNN force [see also (Bha+ 68)]. Whereas in the middle of the p shell and in the beginning of the $p_{1/2}$ shell the repulsion provided by the ΛNN force is due to a partial cancellation between an even larger repulsion given by Q^0_{22} and a strong attraction given by Q^1_{22}, the coefficient of Q^1_{22} decreases in magnitude to zero at the end of the p shell, thus leaving only the repulsive, strong-Q^0_{22} term, which increases with A. The quadratic dependence on n of the coefficient of Q^* cannot be neglected in the $p_{1/2}$ shell. As we have pointed out, the last fit shown in Table V corresponds to a "realistic" TPE ΛNN strength and even here the predicted B_Λ value is about 10 MeV (less than 11 MeV, probably, with the new B_Λ value deduced for $^{13}_\Lambda$C), still exceptionally low.

The values predicted (not shown in Table V) for $B_\Lambda(^{17}_\Lambda$O) by fits with $Q^* = 0$ are generally higher than those for $Q^* \neq 0$, but not considerably higher than 13 MeV. The reason for this is that with $Q^* = 0$ a strong induced nuclear spin–orbit force is required to fit the data and consequently a value of $\bar{V} \sim 0.8$ MeV or less is generally determined, the spin–orbit term then bringing up the average slope of the B_Λ curve in the $p_{3/2}$ shell to about 1 MeV. However, this induced spin–orbit force reverses its sign for contributions from $p_{1/2}$ nucleons [Eqs. (89a) and (89b)] and for $^{17}_\Lambda$O at the end of the p shell its total contribution is zero. The B_Λ excess of $^{17}_\Lambda$O

[†] A revised examination of $^{13}_\Lambda$C events suggest $B_\Lambda = 11.69 \pm 0.12$; also $B_\Lambda(^{14}_\Lambda$C) = 12.17 ± 0.33 MeV (Can+ 74). A revised GSD is planned for 1976.

with respect to $^5_\Lambda$He is then simply given by $12\bar{V} \sim 9.6$ MeV, so that $B_\Lambda(^{17}_\Lambda O) = 3.1 + 9.6 = 12.7$ MeV. We note that *all* fits with $Q^* = 0$ are found to have a negative value of $(S_+ - S_-)$, which, by the "slope" argument [Eq. (96)], leads to $\bar{V} < 1$ MeV as elaborated above. This negative sign for $(S_+ - S_-)$ arises mainly from the need to explain a large B_Λ difference for $A = 9$ (see Tables I and III) without invoking an overwhelmingly strong Λ-spin dependence in the fit. If $^9_\Lambda$Be is eliminated, for reasons which are discussed below, from the B_Λ fitting procedure, we can envisage a situation where either the ΛNN interaction required for a satisfactory fit would be moderate, not as strong as obtained by GSD, or, with $Q^* = 0$, the value of $(S_+ - S_-)$ would occasionally be positive. In the latter case $\bar{V} > 1$ MeV is implied by the "slope" argument. Assume for the sake of illustration that $\bar{V} = 1.2$ MeV [leading to $B_\Lambda(^{17}_\Lambda O) = 17.5$ MeV] and $(S_+ - S_-) = 0.5$ MeV. The "slope" of the B_Λ curve in the $p_{3/2}$ shell is then roughly given by $\bar{V} - 0.4(S_+ - S_-) = 1$ MeV as required by the data. However, the "slope" in the $p_{1/2}$ shell (using $^{13}_\Lambda$C as an origin) will be roughly given by [Eq. (89b)] $\bar{V} + (S_+ - S_-) = 1.7$ MeV. Hence for $^{15}_\Lambda$N, even before any Λ-spin dependence is invoked, a B_Λ value of $2 \times 1.7 = 3.4$ MeV in excess of $B_\Lambda(^{13}_\Lambda$C) is derived, which gives for $B_\Lambda(^{15}_\Lambda$N) a value of $11.2 + 3.4 = 14.6$ MeV, much too high compared to the observed 13.6 MeV. This seems to exclude fits with a significant positive value for $(S_+ - S_-)$, once $B_\Lambda(^{15}_\Lambda$N) is considered. Also, the possibility of obtaining by such a SM analysis B_Λ values for $^{17}_\Lambda$O considerably in excess of 13 MeV appears remote. Nevertheless, it is somewhat dangerous to predict B_Λ values for the $p_{1/2}$ shell on the basis of fitted B_Λ values of the $p_{3/2}$ shell alone. It may turn out, in a combined analysis, that the argument that the "slope" of 1 MeV is the only combination of well fitted parameters does not hold and a different combination, related probably to the combined curvature of the B_Λ curve, emerges from the various fits.

As we pointed out above, application of SM analysis to hypernuclei of the p shell particularly suits data pertaining to excitation energies and transition probabilities in *one* hypernucleus. By this procedure, uncertainties due to small variations in nuclear sizes and deviations form uniformity of the Λ wave function are largely eliminated and the radial integrals for $\bar{V}$, Δ, etc. may safely be assumed constant through the hypernuclear analysis. At present such additional data on excited hypernuclear states are scarce and do not yet allow for a meaningful analysis. In Section 5 we will discuss some of our predictions and expectations for excited states, but since these are based exclusively on a shell model fit to ground-state data, these predictions should be taken with a grain of salt.

Finally we would like to refer to other criticisms of the SM analysis of GSD. At the beginning of this section we discussed the criteria for the validity of SM calculations in the p shell. We noted that the presence of a Λ provides the nuclear core with extra binding, so that even-core nuclei such as ^{6}Li and ^{8}Be that are either lightly bound or unbound acquire a normal binding; the lowest threshold for particle instability is found at a few MeV excitation in the presence of the Λ-particle. The SM procedure is therefore expected to work better for p-shell hypernuclei than for their corresponding nuclear cores. Bodmer and Murphy (BM 65) have empha-sized that irregular variations in the nuclear radii with increasing mass number are not excluded by the data and that these affect B_Λ values (cal-culated in a rigid core approximation) by the presence of a term $-\delta a \times (\partial B_\Lambda/\partial a)_{\bar{V}}$, where a denotes the radius of the nuclear core. These authors give values in the range 13–20 MeV fm^{-1} for $(\partial B_\Lambda/\partial a)_{\bar{V}}$. Hence for $\delta a = 0.05$ fm this term contributes 0.65–1.00 MeV to the B_Λ difference be-tween neighboring nuclei. This is quite a large number when compared to the Λ-spin-dependent contributions to B_Λ values in the p shell. It may happen, in a SM approach, that such variations in the size of nuclear cores, provided they are not a fluke, completely mask the intrinsic ΛN spin dependence as deduced from a difference in B_Λ values between two neigh-boring hypernuclei, the core nucleus of one of which is spinless. Thus, according to this criticism, the only reliable hypernuclear quantity derivable in a SM analysis would be the "slope" of the B_Λ curve as function of A. In the present section we have discussed this quantity, which indeed is the same in most reasonable fits: in the range 0.95–1.00 MeV. However, the slope receives contributions from the spin-independent ΛN interaction, from the ΛN spin–orbit interaction, and from ΛNN forces, and a clear separation between these factors is not provided by the SM fits. The Λ-spin-dependent parameters, on the other hand, are radically different from each other in the various fits and this could partly reflect the size effect discussed above.

Since the relatively low value of $B_\Lambda(^9_\Lambda\text{Be})$ is mainly responsible in the SM analysis for the appearance of strongly repulsive ΛNN forces and/or a considerable ΛN spin–orbit component, it is natural to examine in some detail whether or not other special effects could single out $^9_\Lambda\text{Be}$ from the rest of the known p-shell hypernuclei. The question posed here has mo-tivated at least one HF calculation (BG 70) employing a two-body ΛN central interaction. If there were something special in the coupling of Λ to the ^{8}Be core, such as an excitation of a particular deformation mode or symmetry, it would be reflected in the dynamical HF calculation. On the contrary, Bassichis and Gal do not find a significant difference in B_Λ values for $A = 9$ in this model calculation. Instead, the calculated B_Λ values for $A = 9$ systems show the same minor dynamical differences as found, for example, in the $A = 7$ hypernuclei, mainly due to small differences in size, in agreement with Bodmer and Murphy.

Rayet (Ray 73) has recently reported calculations on hypernuclear binding energies in the α-cluster model for hypernuclei with baryon number $A = 4n + 1$ up to $^{25}_{\Lambda}$Mg. He finds a significant nuclear clustering effect only for $^9_\Lambda$Be and $^{13}_\Lambda$C, where the Λ wave function providing the greater binding is of a molecular type, namely a linear combination of $1s$ functions centered around each of the α-particle centers. These calculations do not allow a comparison with neighboring hypernuclei and it is therefore impossible to check the special role of $^9_\Lambda$Be in the p shell. We point out that with this α-cluster model constrained to the observed $B_\Lambda(^5_\Lambda$He), the observed B_Λ value of $^9_\Lambda$Be is reproduced but the results for $^{13}_\Lambda$C and $^{17}_\Lambda$O are 12.01 and 18.75 MeV, respectively. Since the nuclear interaction used by Rayet yields too large a radius for the ^{12}C core, the calculated $B_\Lambda(^{13}_\Lambda$C) should be higher than the above value if the effective nuclear force is to saturate correctly. We conclude that in an α-cluster model and for a simple attractive central ΛN interaction as used by Rayet, no meaningful consistency is reached between calculated B_Λ values of $^5_\Lambda$He (input), $^9_\Lambda$Be, and $^{13}_\Lambda$C. These calculations can in principle be extended to more realistic interactions.

It is possible, however, that $^9_\Lambda$Be plays a similar role in the p shell to that played by $^5_\Lambda$He in the s shell. In Section 4.2 we discussed the suppression of the Σ channel in $^5_\Lambda$He. We first mentioned that the high nuclear excitation energy required for furnishing $T = 1$ intermediate nuclear states detracts from the effectiveness of the Σ channel in $^5_\Lambda$He, i.e., this channel is suppressed. We then showed that in the coherent limit, where the range of the $\Lambda\Sigma$ transition is infinite with respect to nuclear sizes, the Σ channel is completely suppressed due to the spin–isospin structure of the leading ^{4}He configuration. This does not happen in other s-shell hypernuclei, where on the average the Σ channel is not suppressed or enhanced (provided three-body forces are added to the Λ–nuclear interaction). An analogous situation may hold in the p shell, though to a lesser extent, for $^9_\Lambda$Be, $^{13}_\Lambda$C, and $^{17}_\Lambda$O. Although other p-shell core nuclei have $T = 0$ in their ground state (e.g., ^{6}Li, ^{10}B, ^{14}N), their first $T = 1$ level is low lying (3.56, 1.74, and 2.31 MeV, respectively). In ^{8}Be the first $T = 1$ levels are found around 17 MeV excitation, in ^{12}C around 15 MeV, and in ^{16}O around 12.5 MeV. In the supermultiplet representation $[SU(4)]$ these $T = 1$ states in the $A = 4n$ nuclei belong to irreducible representations which are different from the unit representation appropriate to the ground state and in the "coherent" limit cannot be reached from the ground state via pion coupling. However, the supermultiplet representation gradually loses its validity through the p shell, where a passage from LS coupling (essentially the supermultiplet representation) at the beginning of the shell to the jj coupling at the end of the shell occurs. Thus in ^{8}Be the unit $SU(4)$ representation $^{11}S[4]$ is still a dominant ground-state configuration, whereas in ^{12}C such dominance ceases to hold for $^{11}S[4, 4]$. It is reasonable to expect, then, that a similar suppression mechanism holds for both $^5_\Lambda$He and $^9_\Lambda$Be. For heavier nuclear cores of the type $A = 4n$, the first $T = 1$ level is still high relative to that of $A = 4m + 2$ but its absolute excitation goes down to about 7.7 MeV in ^{40}Ca. For still heavier nuclear cores a neutron excess builds up; $T \neq 0$ necessarily holds for all ground states and for $T \gg 1$ about half of the strength of coupling a pion, $T = 1$, remains in states with the same T. Hence, for *real* heavy nuclei, as opposed to hypothetical nuclear matter, Σ suppression is certainly less effective than for $^5_\Lambda$He.

5. HYPERNUCLEAR EXCITATIONS

In the previous section we have mainly discussed ground-state hypernuclear properties such as binding energies and spin assignments. From the discussion in Section 4.2 on B_Λ analysis in the s shell it is clear that a precise assignment of the hypernuclear 1.09-MeV γ line observed by Bamberger *et al.* (Bam+ 71, 73) to one of the $A = 4$ hypernuclei will yield very valuable information on the spin–spin ΛN interaction in the s shell.[†] It is true that CSB contributions, as well as specific enhancement or suppression effects connected with the Σ-hypernuclear channel, are generally expected in this case. Their partial elimination, however, can be accomplished by observing the γ-ray for the $M1$ transition $J = 1 \rightarrow J = 0$ in the other $A = 4$ hypernucleus. Whether or not this corresponds to the uncertain 1.42-MeV line is not clear at the moment. From the discussion in Section 4.3 on B_Λ analysis in the p shell, we concluded that unless several low-lying hypernuclear excited states are observed and identified, there appears to be a substantial uncertainty as to the value appropriate to the Λ-spin-dependent matrix elements of the ΛN interaction in the p shell. We have pointed out that the observation and identification of particular hypernuclear excited states by means of the (K^-, π^-) reaction, or by (K^-, γ) with or without coincidence with the energetic pion, will provide direct information on the properties of these effective interactions.

[†] See *Note added in proof*, p. 51.

In this section we consider excitation mechanisms expected to hold in hypernuclei. The most obvious model of low-lying excited hypernuclear states is the nuclear core-excitation model (Des 61, Wal 71), according to which $s_{1/2}$ hypernuclear doublets are formed as a result of coupling a spin-$\frac{1}{2}$ lambda to well-separated low-lying nuclear levels having nonzero spin value. The doublet splitting gives direct evidence of the spin dependence of the ΛN interaction. This splitting is generally expected to decrease with increasing mass number (Wal 71) because of the poor overlap between a $1s$ Λ wave function and valence single-nucleon wave functions characterized by an appropriate combination of centrifugal barrier factors and several nodes. As will be seen from a consideration of predicted excitation patterns in the hypernuclear p shell, the assumption of $s_{1/2}$ doublets gives a satisfactory first-order approximation in most cases, except where the corresponding nuclear core states lie close in energy to each other ($^{8}_{\Lambda}$Li) or where specific effects which crucially depend on small admixtures of high nuclear excitations are sought ($^{7}_{\Lambda}$He).

The situation with regard to low-lying hypernuclear excitations in deformed nuclei could be radically different from that considered above. We expect the appearance of hypernuclear rotational bands based, for a Λ coupled to an even–even, strongly deformed core nucleus, on intrinsically deformed self-consistent Λ states. Here the Λ-particle couples not to separate core-excited states, but rather to a whole group of nuclear states comprising the members of a rotational band. The emerging hypernuclear rotational band could sometimes appear substantially different from the excitation pattern expected from a core-excitation type of coupling. Thus a strongly deformed nuclear core could produce $\frac{3}{2}^{+}$ or $\frac{5}{2}^{+}$ intrinsic Λ states (Nilsson type) lower in energy than the normal $\frac{1}{2}^{+}$ state, as a result of mixing between major hypernuclear shells in an axially symmetric self-consistent field. A $\frac{1}{2}^{+}$ level will then not appear among the low-lying hypernuclear levels, contrary to naive expectations. In due time, when such information becomes available, it would be extremely instructive to compare such quantities as moment of inertia, $B(E2)$ values, and g-factors for both nuclei and hypernuclei.

Coupling of the Λ to other collective nuclear excitation modes is feasible in principle. One could, however, envisage a dynamical participation of the Λ in such hypernuclear collective excitation. This naturally brings us to the rapidly developing field of (K^{-}, π^{-}) strangeness exchange reactions. Due to the exothermic nature of such reactions, the momentum of the incoming beam can be appropriately chosen so that in the forward direction the momentum transfer to the produced Λ is zero. Under these conditions the (K^-, π^-) reaction in flight provides an ingenious experimental procedure for creating excited hypernuclear states which are obtained from the corresponding nuclear ground state by replacing a neutron by a Λ in the same space–spin state. Coherent combinations of such Λ-particle–neutron hole states are expected to be favorably formed and to decay by characteristic modes. These states, observed to date in $^{12}_{\Lambda}C$ and $^{16}_{\Lambda}O$, are highly excited, in the range of 10–20 MeV excitation. The substitution of a neutron by a Λ creates a state which is totally antisymmetric in the labels of the neutrons and the Λ together. Since the Λ is distinct from nucleons, this Pauli correlation between the Λ and the neutrons in the coherent hypernuclear state is not really required by any fundamental law and in general only hinders the Λ from minimizing its interaction energy with the nuclear core.

In this section we will present various versions of theoretical expectations and predictions for states favorably produced in the (K^-, π^-) reaction in flight. Other K^--initiated reactions have been proposed (Lip 65) which allow a zero baryonic momentum transfer, thereby enhancing the production of a coherent state. Ignoring the (K^-, π^0) reaction, in which a proton is substituted by a Λ, but where the π^0 is not easily detectable, such reactions are

$$K^-\,^AZ \to \pi^+\,^A_{\Sigma}(Z-2) \tag{103}$$

$$K^-\,^AZ \to \pi^+\pi^-\,^A_{\Lambda}(Z-1) \tag{104}$$

Reaction (103) provides a unique signature for producing Σ hypernuclear coherent states. It is not known at present whether such states have reasonable widths. Reaction (104) is similar in character, as pointed out by Lipkin, to the nuclear $(p, 2p)$ reaction, since a proton hole is formed, accompanied by a Λ in the same state, with the energy and angles of the outgoing pions chosen to correspond to zero momentum transfer to the nucleus.

Recently, very energetic oxygen ions (2.1 GeV/c per nucleon) have been observed to produce hypernuclei on a hydrogen target (Hec 73). Though the claim was made for the two-body reaction

$$^{16}O\,p \to\,^{17}_{\Lambda}O\ K^+ \tag{105}$$

the production cross section of the order of 2 μb suggests the formation of other fast hypernuclei, necessarily accompanied by a slow K^+ (T. Bowen, private communication through R. Dalitz, July 1974):

$$^{16}O\,p \to \begin{cases} ^{16}_{\Lambda}N\ p\ K^+ \\ ^{16}_{\Lambda}O\ n\ K^+ \end{cases} \tag{106}$$

Since relativistic hypernuclei travel many centimeters in the laboratory before decaying, these reactions and similar ones allow in principle the deduction of hypernuclear lifetimes as well as ratios of (π^- mesic)/(non-mesic) for the decay rate of these hypernuclei.

5.1. Excited Hypernuclear States in the p Shell

5.1.1. $^7_\Lambda He$

Pniewski and Danysz (PD 62) suggested that $^7_\Lambda$He may have an isomeric excited state which prefers to decay by weak hypernuclear modes rather than, as usual, by γ-emission to its ground state. This suggestion was motivated, on the experimental side, by the broad B_Λ distribution, extending over a few MeV, observed for $^7_\Lambda$He. With the few events surviving the stringent acceptance criteria of the European K^- Collaboration (Jur+ 73), there exists at present no strong evidence for an isomeric $^7_\Lambda$He*, although the individual B_Λ values still remain too widely spread.

The ground state of ^{6}He has $J_N = 0$ (predominantly $^1S_0$) and the first excited state lies at $E^* = 1.71$ MeV, is particle unstable, and has $J_N = 2$ (predominantly $^1D_2$). There are no other nuclear excited states below 4 MeV excitation, so that the approximation of attaching the Λ to these two nuclear-core states is expected to be justified for the first three hypernuclear states. Under these conditions the ground state of $^7_\Lambda$He has (trivially) $J = \frac{1}{2}$, and a core-excited doublet, $J = \frac{3}{2}$ and $J = \frac{5}{2}$, is formed by attaching the Λ to the first ^{6}He excited state. For a γ-deexcitation these hypernuclear doublet states must then decay to the hypernuclear ground state by $E2$ radiation since, in principle, this is the only mode connecting both nuclear core states. For a normal core nucleus, $E2$ transition rates for γ energies between 1 and 2 MeV are of the order of 10^{12} sec^{-1}, much faster than the order of 10^{10} sec^{-1} expected for hypernuclear weak decay. However, in $^7_\Lambda$He all the particles outside the α-particle (inert) core are neutral. An account of the distortion of the α-particle core by the outer nucleons (Law 65) as well as of the recoil of the α-particle (DG 67) leads to the estimate $| q_{\mathrm{eff}} | \lesssim 0.02e$ for the effective charge associated with each of the outer nucleons. The $E2$ rates are then expected not to exceed the rate of 10^8 sec^{-1}, which is negligible relative to the hypernuclear weak decay rate. However, as pointed out by Dalitz and Gal (DG 67), the Λ-spin-dependent interactions may admix a nuclear core state $J_N = 1$ whose LS assignment is $^3P_1$ into the wave function of the $J = \frac{1}{2}$ and $J = \frac{3}{2}$ hypernuclear states. In addition, intermediate coupling calculations give appreciable amplitudes of $^3P_0$ and $^3P_2$ configurations for the states $J_N = 0$ and $J_N = 2$, respectively, of the nuclear core. An $M1$ hypernuclear transition $\frac{3}{2} \rightarrow \frac{1}{2}$ is then allowed by the nuclear matrix elements connecting $^3P_2$ with $^3P_1$, and $^3P_1$ with $^3P_0$. Since a normal nuclear $M1$ rate for a 1-MeV γ-ray is about 10^{13} sec^{-1}, only small admixtures are really required for this. The same Λ-spin-dependent interactions split the $(\frac{3}{2}, \frac{5}{2})$ doublet so that for a splitting energy greater than about 0.1 MeV, the $M1$ transition rate for either $\frac{5}{2} \rightarrow \frac{3}{2}$ or $\frac{3}{2} \rightarrow \frac{5}{2}$ ceases to be small compared with the hypernuclear weak decay rate of about 0.4×10^{10} sec.$^{-1}$

The spin dependence of the ΛN interaction, as fitted by GSD (GSD 71, 72) in their SM analysis of B_Λ values in the p shell, is strong enough to split the $(\frac{3}{2}, \frac{5}{2})$ hypernuclear doublet by more than 1 MeV for most of the SM fits, leading to a rapid $M1$ deexcitation of either $\frac{5}{2} \rightarrow \frac{3}{2}$ or $\frac{3}{2} \rightarrow \frac{5}{2}$, whichever is energetically possible. For all parameter sets with $S > 0$ the low-lying $^7_\Lambda$He* state has $J = \frac{3}{2}$ and for the two appropriate fits depicted in Table IV the admixture of $J_N = 1$ into the wave function of the lowest two hypernuclear states is small enough (not exceeding 0.05 in magnitude of the amplitude) for the $J = \frac{3}{2}$ level to show up as isomeric to some extent. For all parameter sets with $S < 0$ the low-lying $^7_\Lambda$He* state has $J = \frac{5}{2}$ and is automatically isomeric. However, in all these fits, where $^7_\Lambda$He* isomerism is allowed, the excitation energy for the isomer state does not exceed about 0.5 MeV, quite small with respect to the B_Λ spread observed experimentally for $^7_\Lambda$He.

5.1.2. $^7_\Lambda$Li

The predicted excitation spectrum of $^7_\Lambda$Li for the four selected fits depicted in Table IV is given in Table VI, together with the appropriate hypernuclear spin assignments (in parentheses). The hypernuclear levels displayed are the only ones found in the SM calculation below 4 MeV excitation, which is about the lowest threshold in $^7_\Lambda$Li for particle instability (against decay to $^5_\Lambda$He $+ d$). We note that the first two states in $^7_\Lambda$Li correspond to a doublet predominantly built on the nuclear ground state ($J_N = 1$). Since this $J_N = 1$ ground state is well described by a $^3S_1$ configuration, the Λ-spin-dependent contribution which splits this doublet arises overwhelmingly from the parameter Δ, not from $(S_+ + S_-)$. Hence, for the first two fits, where $\Delta = 0$, the doublet splitting is almost zero; both members are expected to decay via weak hypernuclear modes. For the last two fits the order of levels $J = \frac{1}{2}, \frac{3}{2}$ is determined by the positive sign of Δ. However, for the second doublet, based on $J_N = 3$ ($^3D_3$), $L \neq 0$

TABLE VI

The Excitation Energy E^* and the Spin Value J for Each State Lying Below $E^* = 4$ MeV for the Hypernucleus $^7_\Lambda$Li and for Each of the Four Fits in Table IV (GSD 72)

In the last row the input ^{6}Li core nucleus levels are specified by their excitation energy and spin values.

$\chi^2\ (\Delta\pm S\pm Q^*)$	$E^*(^7_\Lambda\text{Li})$ with respect to ground state, MeV (J value)				
42.8 (S^-Q^*)	0 $(\frac{3}{2})$	0.06 $(\frac{1}{2})$	0.80 $(\frac{7}{2})$	2.55 $(\frac{5}{2})$	3.48 $(\frac{3}{2}^*)$
36.1 (S^+Q^*)	0 $(\frac{1}{2})$	0.04 $(\frac{3}{2})$	0.97 $(\frac{5}{2})$	2.60 $(\frac{7}{2})$	3.24 $(\frac{3}{2}^*)$
32.8 $(\Delta^+S^-Q^*)$	0 $(\frac{1}{2})$	0.88 $(\frac{3}{2})$	1.04 $(\frac{7}{2})$	2.42 $(\frac{5}{2})$	3.47 $(\frac{3}{2}^*)$
33.7 $(\Delta^+S^+Q^*)$	0 $(\frac{1}{2})$	0.57 $(\frac{3}{2})$	0.96 $(\frac{5}{2})$	3.23 $(\frac{7}{2})$	3.74 $(\frac{3}{2}^*)$
^{6}Li	0 (1)	2.18 (3)	4.52 (2)	5.50 (1)	9.70 (1)

and the spin–orbit parameter $(S_+ + S_-)$ plays the major role in determining the order of levels and their splittings. Thus $(S_+ + S_-) \lesssim 0$ puts the $|\,(J_N \pm \frac{1}{2})\,|$ spin lowest. For other fits, not shown here, the spin–orbit and spin–spin parameters are even larger than those depicted in Table IV, which causes a "mixing" between the first two doublets: the higher member of the first doublet lies higher in energy than the lower member of the second doublet. A fast $M1$ transition (Λ spin flip) will occur within the split doublet in each case, unless the splitting is less than about 0.1 MeV, in which case the two levels coincide for all practical purposes. The fifth and highest hypernuclear excitation shown in Table VI has $J = \frac{3}{2}$, irrespective of the sign of S. This occurs since for such a choice of hypernuclear spin the strong Λ-spin-dependent terms can and do obtain additional significant contributions by strongly admixing in the $J_N = 1$ level at 5.50 MeV, only 1 MeV above the $J_N = 2$ level on which we expect the third doublet to be built.

Isomerism for $^7_\Lambda$Li has been discussed by Pniewski *et al.* (Pni+ 67). On the basis of the observed, somewhat asymmetric, B_Λ distribution for $^7_\Lambda$Li (Jur+ 73), these authors conjectured that strong, noncentral, spin-dependent ΛN interaction terms could bring either the $\frac{5}{2}$ level or the $\frac{7}{2}$

level of the second doublet of $^7_\Lambda\text{Li}^*$ down sufficiently close to the $^7_\Lambda\text{Li}$ ground state to slow down the $E2$ transition rate between the lower of the former levels and the levels belonging to the first doublet to the extent where hypernuclear weak decay favorably competes with γ decay. This turns out to be the case for every one of the fits discussed here. For minima with $S > 0$ the $\frac{5}{2}$ level is brought low in excitation. Using the measured nuclear $B(E2)$ value for the transition $3 \to 1$ in ^6Li, the $\frac{5}{2}$ state of $^7_\Lambda\text{Li}$ is found to be isomer if it lies less than about 0.8 MeV above the ground state. Note that an $M1$ transition $\frac{5}{2} \to \frac{3}{2}$, allowed in principle, is very slow in view of the purity of the $\frac{3}{2}$ state in its coupling to the ground state of the parent nucleus ^6Li. With about 1 MeV excitation of the $\frac{5}{2}$ level, as shown for the second and fourth fits in Table VI, isomerism may show up to some extent. For minima with $S < 0$, the $\frac{7}{2}$ level lies sufficiently low in excitation to be isomeric for the relevant fits shown in Table VI. Of the four fits depicted in Table VI, the third one will show the clearest evidence for isomerism at about 1 MeV, $^7_\Lambda\text{Li}^*$. This holds since the energy difference between the $\frac{7}{2}$ level and $\frac{3}{2}$ level is extremely small for an $E2$ transition to compete favorably with hypernuclear weak decay modes and the transition $\frac{7}{2} \to \frac{1}{2}$ can proceed only through $M3$ or $E4$ multipoles, which give negligible rates.

An attempt to observe hypernuclear γ-rays directly in $^7_\Lambda\text{Li}^*$ has recently failed (Bam+ 73). The production rate for particle-stable $^7_\Lambda\text{Li}^*$ levels in the reaction

$$K^- \ {}^7\text{Li} \to \pi^- \ {}^7_\Lambda\text{Li}^* \tag{107}$$

is accordingly estimated to be lower than 0.05% per stopped K^-. Nevertheless, observation of hypernuclear γ-rays corresponding to $^7_\Lambda\text{Li}^*$ states is highly desirable in view of the limited number expected for such γ lines and the relatively simple structure predicted for $^7_\Lambda\text{Li}^*$ levels.

5.1.3. $^{12}_\Lambda B$

For this hypernucleus a determination of the ground state spin has very recently been published (Kie+ 75, ZD 75). This has been done by studying the final $\alpha\alpha\alpha\pi^-$ angular correlations for the sequential decay

$$^{12}_\Lambda\text{B} \to \pi^- + {}^{12}\text{C}^*, \qquad {}^{12}\text{C}^* \to 3\alpha \tag{108}$$

where this decay process occurs predominantly via the two $^{12}\text{C}^*$ levels at 12.7 MeV (1^+, $I = 0$) and 16.1 MeV (2^+, $I = 1$). The measured branching ratio for these two levels as well as the observed angular correlations have been used to deduce that $J = 1$ holds for the $^{12}_{\Lambda}$B ground state. Among the four SM fits depicted in Table IV and analyzed here, only those with $(S_+ + S_-) > 0$ give $J = 1$ as a ground state for $^{12}_{\Lambda}$B. The $J = 2$ level, being the other member of the hypernuclear doublet built on the $J_N = \frac{3}{2}$ ^{11}B ground state, is predicted by these fits to occur above 1 MeV excitation and hence its $M1$ decay rate (Λ spin flip) is large, of the order of 10^{13} sec^{-1}, by far exceeding the hypernuclear weak decay rate. An observation of a γ-ray appropriate to the transition $^{12}_{\Lambda}$B*$(J = 2) \rightarrow {}^{12}_{\Lambda}B(J = 1)$ will bear directly on the magnitude of the ΛN spin–orbit parameter $(S_+ + S_-)$, which is mainly responsible for the above hypernuclear doublet splitting. The low-lying states of the ^{11}B nuclear core are well separated from each other, giving rise to a quite distinct doublet structure for the low-lying $^{12}_{\Lambda}$B levels. In particular, the next $^{12}_{\Lambda}$B* $J = 2$ level is generally expected to lie at about 5 MeV excitation energy, so that its presence is not supposed to interfere with the conclusions reached above for the location of the

TABLE VII

The Predicted Spectrum of $^{12}_{\Lambda}$B Below an Excitation Energy of 5 MeV for the Two Fits Specified in Table IV that Yield $J = 1$ for Its Ground State

The core-excitation structure of such $^{12}_{\Lambda}$B is shown on the left, where the indicated Λ couplings to the appropriate ^{11}B core states are rather pure.

Core (^{11}B) excited structure of $^{12}_{\Lambda}$B levels	$^{12}_{\Lambda}$B excitation energy E^*, MeV	
	$\chi^2 = 36.1\ (S^+Q^*)$	$\chi^2 = 33.7\ (\Lambda^+S^+Q^*)$
$E_N = 4.46,\ J_N = \frac{5}{2}$		
$\quad\quad J = 2$	4.89	4.75
$\quad\quad J = 1$	3.70	3.84
$E_N = 2.14,\ J_N = \frac{1}{2}$		
$\quad\quad J = 0$	2.78	2.91
$\quad\quad J = 2$	1.09	1.40
$E_N = 0,\quad J_N = \frac{3}{2}$		
$\quad\quad J = 1$	0	0

first excited ($J = 2$) $^{12}_{\Lambda}$B* level and its deexcitation to the ground state. For those fits where $S > 0$, the excited $J = 1$ $^{12}_{\Lambda}$B* level lies at about 4 MeV excitation, sufficiently high so as not to interfere with the ground state. The level pattern for $^{12}_{\Lambda}$B* below 6 MeV excitation is shown in Table VII for the two fits of Table IV having $S > 0$. The similarity of the predicted spectra is quite impressive. All indicated excited hypernuclear levels are expected to decay fast by $M1$ radiation to the ground state, the decay rates involved being roughly of the order of 10^{13} sec^{-1}. Since the spectrum of $^{12}_{\Lambda}$B is charge-symmetric to that of $^{12}_{\Lambda}$C, our predictions can be tested in due time by refining the experimental resolution in the recently reported reaction (Fae+ 73, Bon+ 73)

$$K^- \, ^{12}C \rightarrow \pi^- \, ^{12}_{\Lambda}C* \qquad (109)$$

5.1.4. $^{8}_{\Lambda}Li$

We left the discussion of this hypernucleus to the end of the section because here, significantly more than for any other p-shell species, admixture of nuclear core excited states plays a major role in the predicted low-lying excitation pattern. The core nucleus has two levels close in energy below 6 MeV (which is about the lowest threshold for particle instability in $^{8}_{\Lambda}$Li): the ground state $J_N = \frac{3}{2}$ and an excited $J_N = \frac{1}{2}$ state at 0.48 MeV excitation. Both these levels belong in the LS limit to the configuration $^2P[3]$. As a result of coupling of the Λ to these two core states, four hypernuclear levels are expected: 0, 1, 1, 2. For the two hypernuclear $J = 1$ states a strong mixing between the two nuclear core states is generally expected, whereas the other low-lying levels come out rather pure in the SM fits. As early as 1963 the spin of the $^{8}_{\Lambda}$Li ground state had been determined (Dal 63, DLR 63) by studying the decay sequence

$$^{8}_{\Lambda}Li \rightarrow \pi^- + \, ^8Be*, \qquad ^8Be* \rightarrow 2\alpha \qquad (110)$$

where the transition through the 2.9-MeV (2^+) level of ^{8}Be dominates. The angular distribution for the decay process ^{8}Be*(2.9 MeV) $\rightarrow 2\alpha$, with respect to the π^- emitted in the weak hypernuclear decay, is well fitted by $(1 + 3 \cos^2 \theta)$, strongly suggestive of $J(^{8}_{\Lambda}$Li$) = 1$. Recently (Boh+ 74), the total rate of transition to ^{8}Be*(2^+) states near 17 MeV in the sequence (110) has been found to be 0.09 ± 0.02 of that to the ^{8}Be* (2.9 MeV) level. This ratio and the angular distribution expected for the subsequent decay into two α's determine two ranges of values for the mixing ratio between $J_N = \frac{3}{2}$ and $J_N = \frac{1}{2}$ in the $J = 1$ ground-state wave function for $^{8}_{\Lambda}$Li (ZD

74). Thus, if the representation

$$| \, _{A}^{8}\text{Li}, \, J = 1 \rangle = (\cos \varepsilon) \, | \, ^{7}\text{Li} \, J_N = \tfrac{3}{2}, \, (\tfrac{1}{2})_A; \, J = 1 \rangle$$
$$+ (\sin \varepsilon) \, | \, ^{7}\text{Li} \, J_N = \tfrac{1}{2}, \, (\tfrac{1}{2})_A; \, J = 1 \rangle \qquad (111)$$

is used for the $_{A}^{8}\text{Li}$ ground state, then the allowed values of ε are found to be in the ranges

$$0.1\text{–}0.4 \text{ rad}, \qquad 0.9\text{–}1.2 \text{ rad} \qquad\qquad (112)$$

In Table VIII we give values of the mixing angle ε obtained for the four fits depicted in Table IV and whose predictions for the excitation pattern of $_{A}^{8}\text{Li}$ are tabulated here. We see that these fits can be divided into two groups, those with a positive value of $\varepsilon \sim 1.1$, for $S < 0$, which fit one of the ranges deduced in (112); and those with a negative value of ε, small in magnitude, for $S > 0$. The latter fits do not conform with the range of values for ε allowed by (112). Although one of the "good" fits shown in Table VIII actually gives $J = 2$ for the $_{A}^{8}\text{Li}$ ground state, in accordance with the rule $J = J_N + \tfrac{1}{2}$ which otherwise holds in the p shell for $S < 0$ (when mixing is negligible), the first excited state in this fit has $J = 1$ and lies low at $E^* = 0.19$ MeV. The introduction of $\Delta > 0$ then reverses the order of these two hypernuclear levels, again leading to a small excitation

TABLE VIII

The Predicted Spectrum of $_{A}^{8}\text{Li}$ Below an Excitation Energy of 5 MeV for the Fits Specified in Table IV

The appropriate values (in radians) of the mixing angle ε are given for the first 1^- $_{A}^{8}\text{Li}$ level.

χ^2 $(\Delta^{\pm}S^{\pm}Q^*)$	ε	$_{A}^{8}\text{Li}$ excitation energy E^*, MeV (J^π value)			
42.8 (S^-Q^*)	1.19	0 (2^-)	0.19 (1^-)	1.81 (1^-)	2.02 (0^-)
36.1 (S^+Q^*)	-0.27	0 (1^-)	0.24 (0^-)	0.98 (2^-)	1.23 (1^-)
32.8 $(\Delta^+S^-Q^*)$	1.07	0 (1^-)	0.14 (2^-)	2.84 (1^-)	3.36 (0^-)
33.7 $(\Delta^+S^+Q^*)$	-0.13	0 (1^-)	0.56 (0^-)	1.31 (2^-)	1.50 (1^-)

energy of 0.14 MeV for the $J = 2$ level. A nonnegligible fraction of $^8_\Lambda$Li*(2) events will then decay by hypernuclear weak modes, particularly with a pion emission to the ^{8}Be*(2$^+$) states at about 17 MeV mentioned before. Zieminska and Dalitz (ZD 74) estimate that if this were the case, 0.35 of the observed events identified as $^8_\Lambda$Li decay would have been the result of direct hypernuclear decay from $^8_\Lambda$Li*(2). However, the experimental data (Boh+ 74) are not in favor of a large fraction for decay from $^8_\Lambda$Li*(2). For the other two fits shown in Table VIII, those with $S > 0$, the $J = 2$ $^8_\Lambda$Li* level is separated by 1 MeV or more from the $J = 1$ ground state and a strong $2 \rightarrow 1$ electromagnetic deexcitation is expected to occur. For these fits, $^8_\Lambda$Li*($J = 0$) appears as the first excited state. Again, $M1$ electromagnetic deexcitation to the $^8_\Lambda$Li($J = 1$) ground state is strong and whatever small fraction of $^8_\Lambda$Li*($J = 0$) decays by hypernuclear weak modes, the transition to ^{8}Be* ($\sim$17 MeV) is strongly suppressed and therefore does not affect the analysis of the decay (110).

The choice $S < 0$ for the fit ($\Delta^+S^-Q^*$:32.8), which is the best fit shown here to reproduce both $J(^8_\Lambda$Li$) = 1$ and one of the ranges allowed for ε, is very attractive from the theoretical standpoint, since this sign is predicted by meson theory for the ΛN spin–orbit parameters, similarly to what has been deduced for the NN spin–orbit parameter (LD 71, GSD 71). However, this choice of a negative S will definitely imply $J(^{12}_\Lambda$B$) = 2$, in contradiction with recent observations discussed above.

We can see more clearly why two groups of values of ε are formed as mentioned above. Consider the LS limit in which both relevant nuclear core states are described by the configuration $^2P[3]$. The usual way of coupling the Λ is to couple its spin $\frac{1}{2}$ to the total nuclear angular momentum J_N: $| ^2P_{J_N}, \frac{1}{2}; J\rangle$. However, instead, we form a hypernuclear LS coupling scheme in which *all* spins, including that of the Λ, are coupled to a total **S** which in turn is coupled to the total **L** (the same as for the nuclear core since the Λ is in a $1s$ state) to give **J** directly. With our phase convention the relation between the two schemes for the two $J = 1$ states of $^8_\Lambda$Li is given by

$$| J_N = \tfrac{3}{2}, \tfrac{1}{2}; 1\rangle = \sqrt{\tfrac{2}{3}} \, | ^1P_1\rangle - \sqrt{\tfrac{1}{3}} \, | ^3P_1\rangle \tag{113a}$$

$$| J_N = \tfrac{1}{2}, \tfrac{1}{2}; 1\rangle = \sqrt{\tfrac{1}{3}} \, | ^1P_1\rangle + \sqrt{\tfrac{2}{3}} \, | ^3P_1\rangle \tag{113b}$$

The $J = 0, 2$ states can be derived only from the state ^3P, so that

$$| J_N = \tfrac{3}{2}, \tfrac{1}{2}; 2\rangle = | ^3P_2\rangle, \qquad | J_N = \tfrac{1}{2}, \tfrac{1}{2}; 0\rangle = | ^3P_0\rangle \tag{113c}$$

It is an easy matter to deduce the matrices for the binding energy of the system in this basis. The spin–spin term arises from a contribution $-\Delta \mathbf{s}_A \cdot \mathbf{S}_N$ which is diagonal in the hypernuclear LS representation:

$$-\Delta \langle S_N = \tfrac{1}{2}, s_A = \tfrac{1}{2}; S \mid \mathbf{s}_A \cdot \mathbf{S}_N \mid S_N = \tfrac{1}{2}, s_A = \tfrac{1}{2}; S \rangle$$

$$= -\tfrac{1}{2}\Delta[S(S+1) - \tfrac{3}{2}] = \begin{cases} \tfrac{3}{4}\Delta, & S = 0 \qquad\qquad (114a) \\ -\tfrac{1}{4}\Delta, & S = 1 \qquad\qquad (114b) \end{cases}$$

irrespective of the value of J (for $S = 1$). The spin–orbit term gives rise to two contributions, $-S_+ \sum_i (\mathbf{s}_A + \mathbf{s}_i) \cdot \mathbf{l}_i$ and $-S_- \sum_i (\mathbf{s}_A - \mathbf{s}_i) \cdot \mathbf{l}_i$, which we rewrite as $(\mathbf{L} = \mathbf{L}_N)$

$$-(S_+ + S_-)\sum_i (\mathbf{s}_A + 3\mathbf{s}_i) \cdot \mathbf{l}_i + (2S_+ + 4S_-)\sum_i \mathbf{s}_i \cdot \mathbf{l}_i$$

$$= -(S_+ + S_-)\mathbf{S} \cdot \mathbf{L} + (2S_+ + 4S_-)\tfrac{1}{3}\mathbf{S}_N \cdot \mathbf{L}_N \qquad (115)$$

where use was made of the total symmetry of the spatial wave function $P[3]$ in the nucleons, $\mathbf{l}_i \to \tfrac{1}{3}\mathbf{L}$. The first term on the right-hand side of Eq. (115) is diagonal in the hypernuclear LS scheme with values given by

$$-(S_+ + S_-)\langle J \mid \mathbf{S} \cdot \mathbf{L} \mid J \rangle$$

$$= -\tfrac{1}{2}(S_+ + S_-)[J(J+1) - S(S+1) - L(L+1)]$$

$$= 0, \qquad\qquad {}^1P_1 \qquad\qquad\qquad\qquad (116a)$$

$$= 2(S_+ + S_-), \qquad {}^3P_0 \qquad\qquad\qquad\qquad (116b)$$

$$= S_+ + S_-, \qquad\quad {}^3P_1 \qquad\qquad\qquad\qquad (116c)$$

$$= -(S_+ + S_-), \qquad {}^3P_2 \qquad\qquad\qquad\qquad (116d)$$

The second term on the right-hand side of Eq. (115) is diagonal in the representation (113), where the total nuclear angular momentum is diagonal, with values given by

$$\tfrac{1}{3}(2S_+ + 4S_-) \langle J_N \mid \mathbf{S}_N \cdot \mathbf{L}_N \mid J_N \rangle$$

$$= \tfrac{1}{3}(S_+ + 2S_-)[J_N(J_N + 1) - S_N(S_N + 1) - L_N(L_N + 1)]$$

$$= \tfrac{1}{3}(S_+ + 2S_-)[J_N(J_N + 1) - (11/4)]$$

$$= \tfrac{1}{3}(S_+ + 2S_-), \qquad J_N = \tfrac{3}{2} \qquad\qquad\qquad (117a)$$

$$= -\tfrac{2}{3}(S_+ + 2S_-), \qquad J_N = \tfrac{1}{2} \qquad\qquad\qquad (117b)$$

Denoting by B_0 the spin-independent contributions to the binding energy B_A and by δ the excitation energy (0.48 MeV) of ${}^7\text{Li}^*(J_N = \tfrac{1}{2})$ with respect to the ^{7}Li$(J_N = \tfrac{3}{2})$ ground state, we obtain, using the transformation (113), the binding matrix in the hypernuclear LS basis (we assume $T = 0$)

$$B_0 - \tfrac{1}{4}\varDelta - \tfrac{2}{3}S_+ - \tfrac{1}{3}S_- \qquad\qquad {}^3P_2 \quad (118a)$$

$$\begin{pmatrix} B_0 - \tfrac{1}{4}\varDelta + \tfrac{2}{3}S_+ + \tfrac{1}{3}S_- - \tfrac{2}{3}\delta & -\sqrt{2}\,(\tfrac{1}{3}S_+ + \tfrac{2}{3}S_- + \tfrac{1}{3}\delta) \\ -\sqrt{2}\,(\tfrac{1}{3}S_+ + \tfrac{2}{3}S_- + \tfrac{1}{3}\delta) & B_0 + \tfrac{3}{4}\varDelta - \tfrac{1}{3}\delta \end{pmatrix} \quad \begin{matrix} {}^3P_1 \quad (118b) \\[4pt] {}^1P_1 \quad (118c) \end{matrix}$$

$$B_0 - \tfrac{1}{4}\varDelta + \tfrac{4}{3}S_+ + \tfrac{2}{3}S_- - \delta \qquad\qquad {}^3P_0 \quad (118d)$$

For the fits depicted in our tables $|\,S_+ - S_-\,| \ll |\,S_+ + S_-\,|$ and to a first approximation we can neglect $S_+ - S_-$ and consider only $S \equiv \tfrac{1}{2}(S_+ + S_-)$ to be nonzero. Since the magnitude of S is greater than the magnitudes of $\varDelta$ and δ, we ignore these terms in Eqs. (118b) and (118c) and diagonalize the matrix

$$\begin{pmatrix} S & -\sqrt{2}\,S \\ -\sqrt{2}\,S & 0 \end{pmatrix} \tag{119}$$

The relative binding energies and eigenfunctions are then given by

$$(-\tfrac{1}{4}\varDelta) \qquad -S, \qquad |\,J_N = \tfrac{3}{2}, \tfrac{1}{2}; J = 2\rangle \tag{120a}$$

$$(\tfrac{5}{12}\varDelta - \tfrac{8}{9}\delta) \quad -S, \qquad \tfrac{1}{3}|\,J_N = \tfrac{3}{2}, \tfrac{1}{2}; J = 1\rangle + \tfrac{2}{3}\sqrt{2}\,|\,J_N = \tfrac{1}{2}, \tfrac{1}{2}; J = 1\rangle \tag{120b}$$

$$(\tfrac{1}{12}\varDelta - \tfrac{1}{9}\delta) \quad 2S, \qquad \tfrac{2}{3}\sqrt{2}\,|\,J_N = \tfrac{3}{2}, \tfrac{1}{2}; J = 1\rangle - \tfrac{1}{3}|\,J_N = \tfrac{1}{2}, \tfrac{1}{2}; J = 1\rangle \tag{120c}$$

$$(-\tfrac{1}{4}\varDelta - \delta) \quad 2S, \qquad |\,J_N = \tfrac{1}{2}, \tfrac{1}{2}; J = 0\rangle \tag{120d}$$

where in parentheses we have written the small terms which are to be added in order to remove the apparent degeneracy; the terms in parentheses on the left-hand side of (120b) and (120c) are equal to the expectation values of the small terms in the states given in (120b) and (120c). We now see that in the limit $|\,S\,| \to \infty$ the four $^8_\Lambda$Li levels degenerate in energy to two levels. The mixing angle for the lowest $J = 1$ level is $\varepsilon = -0.34$ for $S > 0$ and $\varepsilon = 1.23$ for $S < 0$, in good agreement with Table VIII. The degeneracy is removed, however, by the small terms so that two well-separated groups of levels are finally formed. The following classification holds:

(i) For $S > 0$ the lowest group consists of $J = 0, 1$ and the higher group consists of $J = 1, 2$ states lying at about a value of $3S$ above the first group. This is indeed the case in Table VIII. According to (120a)–(120d), the small terms remove the degeneracy such that the level order becomes

1 (g.s.), 0, 1, 2 for the values of Δ and δ appropriate here. Comparison with Table VIII shows that the actual order within the second group is reversed, which could be due in part to a repulsion caused by the next $J = 2$ level and in part to other small effects.

(ii) For $S < 0$ the lowest group consists of $J = 1, 2$ and the higher of the $J = 0, 1$ states, opposite to the previous case, but the difference in average excitation between these two groups is given by $3 \mid S \mid$, as before. This is well borne by the corresponding fits in Table VIII (see Table IV for the appropriate values of S). The degeneracy is removed by the small terms, with the result of a level order of 2(g.s.), 1, 1, 0 for $\Delta = 0$ and of 1(g.s.) 2, 1, 0 for $\Delta \neq 0$, as appropriate. This compares well with the results of the fits depicted in Table VIII.

We conclude[†] that the above simple limit, in which only the Λ-spin-dependent part of the ΛN spin–orbit interaction is significant, reproduces quite well the features of fitted $^{8}_{\Lambda}$Li* low-lying excited states. In particular, we are able to explain the two ranges of values obtained in the SM fits for the $J = 1$ mixing angle [Eq. (111)] by noting the two limiting values for ε: $\varepsilon = -0.34$ for $S > 0$ and $\varepsilon = 1.23$ for $S < 0$. We recall that if the excitation properties of $^{8}_{\Lambda}$Li were dependent mainly on Δ, instead of on S, the hypernuclear LS scheme would have automatically become a diagonal one and the appropriate value of ε for $\Delta > 0$ [invert Eqs. (113a) and (113b)] would have been 0.62 rad, far from the two ranges allowed by (112). At present, these equations suggest that $S < 0$ for $^{8}_{\Lambda}$Li, which is in contradiction with the conclusion $S > 0$ drawn on the basis of the observation of $J = 1$ for the $^{12}_{\Lambda}$B ground state.

We note that our results for ε are expected to hold also for the case where $\mid S_{+} - S_{-} \mid$ is nonnegligible with respect to $\mid S_{+} + S_{-} \mid$. In fact, working within a scheme where the total nuclear angular momentum J_{N} labels the base states, the only nondiagonal term (for $T = 0$) connecting $J_{N} = \frac{3}{2}$ with $J_{N} = \frac{1}{2}$ is proportional to $\Delta - (S_{+} + S_{-}) \equiv \Delta - 2S$. This is easy to understand since the induced nuclear spin–orbit force, with a coefficient $(S_{+} - S_{-})$, is diagonal in J_{N} and for $\Delta = S_{+} + S_{-}$ the sum of the spin–spin force and the Λ-spin-dependent part of the ΛN spin–orbit force given by $\Delta \mathbf{s}_{\Lambda} \cdot (\mathbf{S}_{N} + \mathbf{L}_{N}) = \Delta \mathbf{s}_{\Lambda} \cdot \mathbf{J}_{N}$, which again is diagonal in J_{N}. Hence the nondiagonal matrix element in the J_{N} representation is proportional to $\Delta - 2S$. With a large value of $\mid S_{+} - S_{-} \mid$ but small one

---

[†] Correspondence with R. H. Dalitz on this topic is gratefully acknowledged. The above presentation differs somewhat from that given recently by Zieminska and Dalitz (ZD 74).

for $|\Delta - 2S|$, the interaction matrix for $J = 1$ is almost diagonal, so that $\varepsilon \sim 0$. However, the same result obtains if the signs of Δ and S are the same, so that $|\Delta - 2S|$ remains small. This is why for $\Delta > 0$ and $S > 0$ we obtained $\varepsilon = -0.13$ (Table VIII). When the signs of Δ and S are opposite from each other the off-diagonal matrix element is large in magnitude, leading to large values for ε, as confirmed by the appropriate fits shown in Table VIII.

5.2. Strangeness Exchange Reactions

The usefulness of SEX (Strangeness EXchange) reactions on nuclei, in particular (K^-, π^-) two-body hypernuclear production,

$$K^- \,^A Z \to \pi^- \,_\Lambda^A Z^* \tag{121a}$$

was first pointed out by Lipkin (Lip 65), who stressed the collective nature of the expected hypernuclear states in such formation. Feshbach and Kerman (Fes 66) noted that in the

$$K^- n \to \pi^- \Lambda \tag{121b}$$

reaction in-flight, a minimal baryonic momentum transfer can be reached. Thus, for p_K in MeV/c, the following values of momentum transfer in MeV/c are determined in the lab system for π^- emitted in the forward direction:

$$
\begin{array}{lllllllll}
p_K = & 0 & 300 & 350 & 400 & 550 & 700 & 1000 & 2000 \\
q = & 250 & 75 & 55 & 40 & 0 & 40 & 75 & 130
\end{array}
\tag{122}
$$

We see that by choosing the incident momentum of the beam in the range 300 MeV/c–1000 MeV/c the momentum transfer at $0°$ can be made considerably smaller than the nuclear Fermi momentum $p_F \sim 250$ MeV/c. Hence, a neutron is replaced in the nuclear target by a Λ in the same space–spin state.[†] Inasmuch as this formed state may have a considerable overlap with a narrow group of hypernuclear eigenstates, a strong excitation in the residual hypernucleus should be observed.

Yet, it happened for the first time with stopped K^- in emulsion that resonant hypernuclear states were discovered (Boh+ 70) in $_\Lambda^{12}C$ by looking

---

[†] This observation was earlier made by Podgoretsky (Pod 63), who did not, however, discuss the nature of hypernuclear excitations involved in the application of (121a), as other authors did. I am grateful to J. Zakrzewski for pointing out this reference to me.

for reactions of the type $(K^- + \text{nucleus}) \rightarrow (\pi^- + \text{charged particle} + \text{hypernucleus})$. Bohm *et al.* observed the reaction

$$K^- \,{}^{12}C \rightarrow \pi^- p \,{}^{11}_\Lambda B \tag{123}$$

where a clustering of events in the π^- kinetic energy distribution implied a two-step process. Thus, a resonant $^{12}_\Lambda C$ is produced and subsequently decays by emission of a secondary proton:

$$K^- \,{}^{12}C \rightarrow \pi^- \,{}^{12}_\Lambda C^*$$
$$^{12}_\Lambda C^* \rightarrow p \,{}^{11}_\Lambda B \tag{124}$$

Recently, the existence of a particle-unstable state of $^{12}_\Lambda C$ has been confirmed (Jur+ 72) by the same method and also (Fae+ 73) from counter measurements of the π^- spectrum for stopped K^- in ^{12}C. The latter measurement is not sensitive to the decay mode of $^{12}_\Lambda C^*$, and in fact the $^{12}_\Lambda C$ ground state has probably been observed, too. The binding energy of the Λ for this resonant $^{12}_\Lambda C^*$ state was determined to be 0.1 ± 0.2 MeV, which is consistent with the results of the counter experiment: $B_\Lambda(^{12}_\Lambda C^*) = 0 \pm 1$ MeV. The low-lying $^{12}_\Lambda C$ state observed in the counter experiment has a $B_\Lambda(^{12}_\Lambda C)$ $= 11.1 \pm 1$ MeV value, consistent with the charge-symmetric $B_\Lambda(^{12}_\Lambda B)$ $= 11.37 \pm 0.06$ MeV value (Table I).

Recently, two in-flight experiments on ^{12}C have been reported:

$$K^- \,{}^{12}C \rightarrow \pi^- \,{}^{12}_\Lambda C^* \tag{125}$$

with π^- detected in the forward direction. In the Torino–CERN experiment (Bon+ 74), a beam of 390-MeV/c K^- was used and π^- emitted in the forward direction, $\theta_{\text{lab}} \leq 10°$, were detected. This means that baryonic momentum transfers from 40 MeV/c to 80 MeV/c were allowed. The spectrum is shown in Fig. 10. In spite of the poor hypernuclear energy resolution of 6 MeV, a clear peak corresponding to 10-MeV excitation energy in $^{12}_\Lambda C$ is observed, in rough agreement with previous experiments. Higher excitation energies are governed by the $K \rightarrow 2\pi$ background. It is somewhat surprising that a small peak corresponding to low-lying states in $^{12}_\Lambda C$, possibly including the 1^- ground state, is extracted from the raw data. Such states require for their formation the substitution of a $(1p)$ neutron by a $(1s)$ lambda where the momentum transfer involved is considerably larger than that supplied by the kinematical conditions of this experiment, The differential cross sections reported for the production of the ground-state and 10-MeV peaks are 0.39 ± 0.08 mb/sr and 0.87 ± 0.13 mb/sr, respectively.

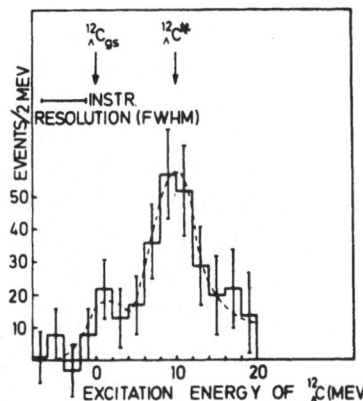

Fig. 10. Excitation energy of $^{12}_{\Lambda}$C as resulting from a preliminary analysis of the data obtained in the reaction $K^- {}^{12}C \rightarrow \pi^- {}^{12}_{\Lambda}C^*$ for K^- in flight ($p_{K^-} = 390$ MeV/c) and π^- emitted in the forward direction. [From (Bon+ 73).]

In the Heidelberg–CERN experiment (Bru+ 75), a K^- beam of 900 MeV/c was employed. The energy resolution was 2 MeV and the acceptance angle for pions emitted in the forward direction was roughly given by $4°$. Hence, the baryonic momentum transfers involved in this experiment are similar to those of the previous one. However, the $K^- \rightarrow 2\pi$ background is particularly strong at the $^{12}_{\Lambda}$C excitation energy roughly corresponding to low-lying states, contrary to the conditions of the other in-flight experiment. The spectrum is shown in Fig. 11, where a strong peak at 11-MeV

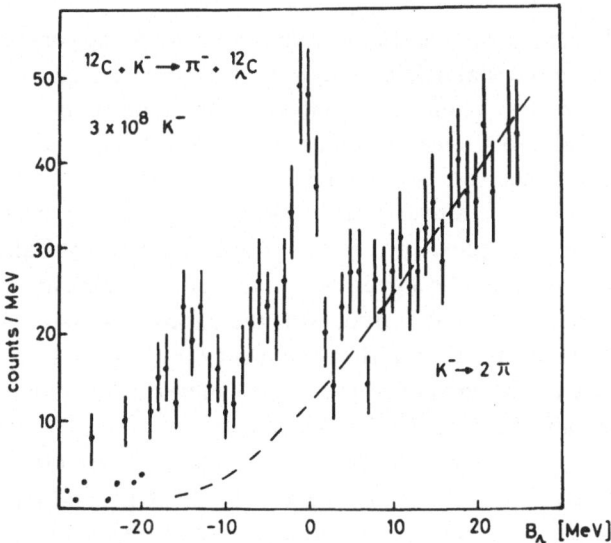

Fig. 11. Excitation energy of $^{12}_{\Lambda}$C resulting from a preliminary analysis of the data obtained in the reaction $K^- {}^{12}C \rightarrow \pi^- {}^{12}_{\Lambda}C^*$ for K^- in flight ($p_K = 900$ MeV/c) and π^- emitted in the forward direction. [From Bru+ 75.]

$^{12}_{\Lambda}$C excitation energy is prominent. There is no indication for low-lying states. However, a second peak, weaker than the first one but considerably broader, at roughly 24-MeV $^{12}_{\Lambda}$C excitation, is noticed. The overall hypernuclear formation rate in this experiment was estimated to be 1.2 ± 0.4 mb/sr.

Similar continuum hypernuclear excitations were reported by these two groups for SEX reactions [Eq. (121a)] on ^{9}Be and ^{16}O (Bru+ 75) and ^{16}O, ^{27}Al (Bon+ 75). The strongest excitations are found in the energy range of 10–20 MeV hypernuclear excitation. The main questions arising from these findings are as follows:

1. What is the structure of the continuum hypernuclear excitations observed? Are these single-particle Λ excitations (shape resonances for the systems $\Lambda + {}^{11}$C, $\Lambda + {}^{15}$O, etc.) or are they collective hypernuclear excitations?

2. In what way does the (K^-, π^-) reaction mechanism enhance observation of these continuum states?

3. Do any predictions follow for observation of similar states in heavier elements, and if so, what excitation energies and production cross sections are expected?

4. What new type of information do these experimental findings bring about? Is the validity of the shell model description for the Λ and the magnitude of residual ΛN interaction thereby implied?

In the following we refer to these questions, invoking general arguments only. Detailed models and calculations are certainly needed, but the experimental data are too poor at present to allow a meaningful comparison.

Dalitz (Dal 69b) suggested that the strongest hypernuclear continuum $^{12}_{\Lambda}$C* excitation observed corresponds to an excitation of the Λ particle into a $1p$ state in the single-particle potential generated by the Λ interactions with the nucleons. In the shell model analysis by Gal, Soper, and Dalitz of B_Λ values in the hypernuclear p shell, the validity of a uniform Λ $1s$ wave function was discussed, with the conclusion that a $1s$ harmonic oscillator Λ wave function with a frequency $\nu_\Lambda = 0.33$ fm^{-2} provides a satisfactory fit to the gross features of B_Λ values in the range $A = 8$–13. This corresponds to $\hbar\omega_\Lambda = 11.5$ MeV, so that if the Λ single-particle potential is only weakly state dependent, the $1p$ Λ excitation should be found at about 11.5 MeV above the hypernuclear ground state, in rough agreement with the strongest excitation reported above for $^{12}_{\Lambda}$C.

It is not obvious why the widths of these hypernuclear excitations are small enough to permit observation of a peak in the π^- energy spectrum. This should not be taken at face value to imply that the residual Λ–nucleus interaction is negligible. It was pointed out by Dalitz for the emulsion experiment that it could well be that the low Q value of about 1 MeV for proton emission [the only available particle emission channel allowed for $^{12}_{\Lambda}C^*(11$ MeV)] is responsible for the phenomenon. Thus, for a formation of $^{12}_{\Lambda}C^*$ component with a spin value $J^{\pi} = 0^+$ (feasible in view of the s-wave nature of the K^- capture from rest), the subsequent decay $^{12}_{\Lambda}C^*(J = 0)$ $\to p + {}^{11}_{\Lambda}B_{\text{g.s.}}.(J = \frac{5}{2}$ or $\frac{7}{2})$ necessarily involves a d-wave centrifugal barrier at least. The low Q value may barely allow a decay into excited $^{11}_{\Lambda}B$ states having lower possible spin values, by which decay the centrifugal barrier would have been eliminated.[†] An example for such a candidate is $^{11}_{\Lambda}B^*(\frac{1}{2})$ based on core-excited $^{10}B^*(J = 1; 0.72$ MeV).

Let us concentrate on the $^{12}_{\Lambda}C$ excitation spectrum expected for (K^-, π^-) in-flight for π^- detected in the forward direction and a vanishing momentum transfer. The kinematical conditions of this reaction ensure, as discussed earlier, that one target neutron is substituted by a Λ in the same spin–space state, thus creating a particle–hole excitation Λn^{-1}. Since the nuclear wave function is antisymmetric with respect to neutron exchanges and the reaction operator is symmetric in the substituted neutrons, being given schematically by a form $\sum_i f(i \to \Lambda)$, the hypernuclear state which is instantaneously formed is also antisymmetric in Λn exchanges as well as nn exchanges. It has a definite isospin value $I = \frac{1}{2}$ (as is the case for $^{16}_{\Lambda}O^*$) as well as spin–parity assignment $J^{\pi} = 0^+$. For this $^{12}_{\Lambda}C^*(t = 0)$ state, two components shown in Fig. 12a are relevant. The first component involves a Λ in the p shell and a p-shell neutron hole, whereas the other component involves a Λ in the s shell and an s-shell neutron hole. These components combine to give a coherent combination (KL 71)

$$^{12}_{\Lambda}C^*(t = 0) = \sqrt{2}\,|\,\Lambda(1s)n^{-1}(1s)\rangle + \sqrt{4}\,|\,\Lambda(1p)n^{-1}(1p)\rangle \quad (126)$$

in which the coefficients squared are equal to the number of neutrons in the appropriate shell, relevant for the nuclear target ^{12}C. The configuration appropriate to $^{12}_{\Lambda}C$ ground state is shown in Fig. 12b; the Λ is in a $1s$ state and the neutrons fill their lowest orbits. The excitation energy of the two components of $^{12}_{\Lambda}C^*(t = 0)$, Eq. (126), with respect to the $^{12}_{\Lambda}C$ ground state is $\hbar\omega_N \sim 17.5$ MeV (the nuclear oscillator parameter being

---

[†] Actually, since the K^- capture occurs from an atomic orbit, values of J^{π} other than 0^+ are allowed for $^{12}_{\Lambda}C^*$. In the p-wave excitation region, as we discuss later, a $^{12}_{\Lambda}C^*$ $J^{\pi} = 2^+$ excitation is also possible for which there is no centrifugal barrier to overcome in its decay to $p + {}^{11}_{\Lambda}B(\frac{3}{2})$. However, this 2^+ excitation is generally expected to lie below the 0^+ excitation and may make the particle stable (HLW 74, DG 75).

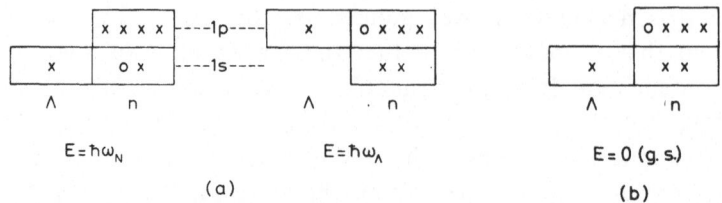

Fig. 12. Λn^{-1} **particle–hole excitations of** ^{12}**C which are expected to be observed as** $^{12}_{\Lambda}$**C states formed in the reaction** $K^-\ ^{12}$**C** $\to \pi^-\ ^{12}_{\Lambda}$**C*.** (a) Particle–hole excitations involving the same shell. The unperturbed shell model excitation energies with respect to the $^{12}_{\Lambda}$C ground state are given. (b) The shell model description of the $^{12}_{\Lambda}$C ground state; note that the Λ is in the $1s$ state, whereas the neutron hole belongs to the $p_{3/2}$ shell. [From (AG 74).]

fixed by the "size" of ^{12}C) for the first component on the right-hand-side of Eq. (126) and is $\hbar\omega_\Lambda \sim 11.5$ MeV for the second component. Since these two single-particle energies differ from each other by more than a reasonable ΛN residual interaction could correct for, the coherence of the $^{12}_{\Lambda}$C*($t = 0$) state will be destroyed with time, and for $t \gtrsim 1/\Delta\omega$ we may look for two separate $^{12}_{\Lambda}$C* excitations corresponding to $E = \hbar\omega_N$ and $E = \hbar\omega_\Lambda$ above the $^{12}_{\Lambda}$C ground state (AG 74). Thus the observed $E \sim 11$ MeV $^{12}_{\Lambda}$C* excitation corresponds in this picture to a $1p\ \Lambda$ excitation relative to the $^{12}_{\Lambda}$C ground state. Another $^{12}_{\Lambda}$C* excitation, corresponding to a $1s$ neutron–hole relative to the $^{12}_{\Lambda}$C ground state is expected a few MeV higher than the observed $^{12}_{\Lambda}$C* excitation. For this predicted excitation there may correspond many modes of decay which contribute to its width. Thus a nuclear Auger effect, accompanied by nuclear breakup, is supposed to be particularly effective in filling the $1s$ neutron hole. The $1s$ hole may actually lie at considerably higher excitation energies ($\gtrsim 20$ MeV) with an appreciable width, as indicated for $1s$ protons by $(p, 2p)$ and $(e, e'p)$ reactions (Wag 73). This could render its experimental detection quite difficult. Nevertheless, it is tempting to identify the observed (Bru+ 75) 24-MeV broad $^{12}_{\Lambda}$C excitation with such an excitation.

In this discussion we have ignored several effects. First, the Λ orbitals differ to some extent from the neutron orbitals, a difference which is already inherent in our use of distinct harmonic oscillator spring constants for Λ and n. Incidentally, in any realistic calculation for ^{12}C, its ground-state wave function contains some $(2s)^2(1p)^{-2}$ two-particle, two-hole components. If one of these $2s$ neutrons is replaced by a Λ in the same orbit, there is a nonnegligible chance for the Λ to wind up in its own $1s$ shell, since the overlap between $(2s)_n$ and $(1s)_\Lambda$ wave functions is nonzero. In this way we obtain a $^{12}_{\Lambda}$C* state with a Λ in its ground state, whereas the

^{11}C nuclear core is in a particle–hole excitation relative to its ground state. A natural candidate for such a resultant $^{12}_{\Lambda}$C* state is the $J = 0$ hypernuclear state obtained by coupling a Λ to $J^{\pi}_N = \frac{1}{2}^+$ in ^{11}C*(6.34 MeV), the latter nuclear state deriving an appreciable strength of a $(2s)(1p)^{-1}$ excitation of ^{11}C. Similarly, for $^{16}_{\Lambda}$O an excitation of a state near 5 MeV relative to the ground state, which is based primarily on ^{15}O*(5.18 MeV) $J^{\pi}_N = \frac{1}{2}^+$, is expected.

The second thing neglected in our previous considerations is the non-zero momentum transfer in the (K^-, π^-) in-flight reactions reported so far. With the π^- emitted in the forward direction, but a momentum transfer $q \neq 0$, an excitation of other than $J^{\pi} = 0^+$ states becomes possible in $^{12}_{\Lambda}$C* (as well as in $^{16}_{\Lambda}$O*). Such states must have a natural parity, i.e., 0^+, 1^-, 2^+, ... (Lip 73), and thus the $^{12}_{\Lambda}$C, $J^{\pi} = 1^-$, ground state (this value is evidently deduced from the charge-symmetric $^{12}_{\Lambda}$B) may, in principle, be excited in the reaction. Its excitation is bound to be considerably weaker than $^{12}_{\Lambda}$C*(11 MeV), since in the experiments reported above the momentum transfer to the Λ is quite small: 40 MeV/$c \lesssim q \lesssim 80$ MeV/c. For capture at rest, on the other hand, this transition, $(1p)_n \rightarrow (1s)_{\Lambda}$, should occur more favorably than in flight since the momentum transfer is about 250 MeV/c. However, even for K^- capture from rest, the transition $(1p)_n \rightarrow (1p)_{\Lambda}$, giving rise probably to $^{12}_{\Lambda}$C*(11 MeV) in the CERN–Heidelberg–Warsaw experiment, is expected to be the strongest one. This transition is generally expected to be split into 0^+ and 2^+ states.

Another neglect in our presentation is that of spin–orbit splittings. We have referred to "the p shell," ignoring its splitting into $p_{3/2}$ and $p_{1/2}$ subshells. The peak corresponding to $^{12}_{\Lambda}$C*(11 MeV) arises probably mainly due to the substitution $(1p_{3/2})_n \rightarrow (1p_{3/2})_{\Lambda}$, with the resulting J^{π} states 0^+ and 2^+, although a weak component $(1p_{1/2})_n \rightarrow (1p_{1/2})_{\Lambda}$, $J^{\pi} = 0^+$ might also be separated out. However, the low-lying $^{12}_{\Lambda}$C peak is expected to include, in addition to the dominant $J^{\pi} = 1^-$ $^{12}_{\Lambda}$C(0 MeV) ground state based roughly on a $(p_{3/2}^{-1})$ ^{11}C ground state, the next $J^{\pi} = 1^-$ $^{12}_{\Lambda}$C*(~ 3.5–4.0 MeV) (see Table VII) based primarily on the first $\frac{1}{2}^-$ ^{11}C state ($p_{1/2}^{-1}$ to a large extent). With the present experimental resolution, or production rates, these secondary $^{12}_{\Lambda}$C* excitations have not yet been observed. In $^{16}_{\Lambda}$O, where the nuclear $p_{3/2}$–$p_{1/2}$ splitting is large, about 6 MeV, we expect in addition to the observed (Bon+ 75) formation of $^{16}_{\Lambda}$O(1^-), primarily based on the ^{15}O(g.s.) $p_{1/2}^{-1}$ configuration, also a comparable formation of $^{16}_{\Lambda}$O*(1^-), which is roughly based on the ^{15}O($\frac{3}{2}^-$, 6.18 MeV) $p_{3/2}^{-1}$ configuration.

Another exciting observation would be a resolution of the main $^{16}_{\Lambda}$O* peak (Bru+ 75) into its two $p_{3/2}$ and $p_{1/2}$ components. The resultant two peaks are not necessarily separated by 6 MeV as predicted by the nuclear spin–orbit splitting observed in this region; the ΛN spin–orbit interaction will also contribute to the splitting. This latter contribution cannot be directly deduced from the SM analysis of p-shell hypernuclear ground states, where the orbital angular momenta of nucleons, not of the Λ, affect the spin–orbit contribution. However, if the ΛN spin–orbit strength is indeed of the same order of magnitude as for NN, and consequently the Λ single-particle $p_{3/2}$–$p_{1/2}$ splitting is similar in magnitude to the nuclear spin–orbit splitting, we face two possibilities. Either both single-particle splittings have the same sign, in which case the corresponding $^{16}_{\Lambda}$O* states are separated by considerably less than 6 MeV (as indicated by the recent Heidelberg–CERN experiment) or they have a different sign, in which case the separation between the two $^{16}_{\Lambda}$O* states is greater than 6 MeV and one of these two states is pushed down to the energy region where the background is appreciable.

Very recently SEX reactions on ^{12}C were discussed by Hüfner, Lee, and Weidenmuller (HLW 74), who considered both capture from rest as well as capture in flight with π^- emitted in the forward direction. These authors allowed for K^- capture from the atomic $3d$ level so that the strangeness exchange is peripheral in their calculation, occurring mostly on a $(1p)$ neutron. The low energy (K^-, π^-) reaction on such a neutron is considered with s-waves and hence a spin flip in the baryonic system is excluded, similar to the same reaction in flight for π^- emitted in the forward direction. Consequently, only natural parity states, $J^\pi = 0^+, 1^-, 2^+, \ldots$, are formed. The simplest ^{12}C configuration is studied, namely closed $(1s_{1/2})(1p_{3/2})$ neutron shells. With this simplification some candidates for $^{12}_{\Lambda}$C* are excluded, notably a $(1p_{1/2})_\Lambda(1p^{-1}_{1/2})_n$ $J^\pi = 0^+$ excitation. A residual ΛN zero-range interaction is adjusted, by folding it into the observed charge density of ^{12}C, to give $B_\Lambda \sim 10$ MeV for $1s$ lambda and $B_\Lambda \sim 0$ MeV for $1p$ lambda as roughly suggested by the experimental data. This residual ΛN interaction, of the order of 1 MeV for the relevant matrix elements, is too weak to appreciably admix the $J^\pi = 0^+$ $(1p_{3/2})_\Lambda(1p^{-1}_{3/2})_n$ excitation with the $(1s_{1/2})_\Lambda(1s^{-1}_{1/2})_n$ excitation since the latter is separated by about 10 MeV from the former. The reaction is treated in the distorted wave impulse approximation, where the distortion is described by eikonal wave functions for the K^- and π^-. Under these assumptions SEX occurs on one baryon with resultant spin parity assignments of $J^\pi = 0^+, 1^-, 2^+$ only. With all these input data, energies and widths of $^{12}_{\Lambda}$C excitations and their formation rates are calculated. The $(1s_{1/2})_\Lambda(1s^{-1}_{1/2})_n$ excitation turns out to be very broad, $\Gamma = 5$–10 MeV, and its calculated formation rate, $\sim 10^{-4}$

per stopped K^-, seems too low at present to allow its observation. On the other hand, the other positive parity $^{12}_{\Lambda}C^*$ states, essentially given by the possible spin couplings of a lambda $1p$ excitation with a $(1p^{-1}_{3/2})_n$, have a small width, considerably less than 1 MeV and their formation rates roughly agree with each other. Their sum, however, exceeds the measured formation rate $[(3 \pm 1) \times 10^{-4}$ per stopped K^- (Fae+ 73)] by a factor of 4–5 and the calculated rate for the $(1s_{1/2})_{\Lambda}(1s^{-1}_{1/2})_n$ excitation by an order of magnitude. For the low-lying 1^- $^{12}_{\Lambda}C$ state, possibly the ground state, the calculated rate lies between 3×10^{-4} and 5×10^{-4} per stopped K^-, in rough agreement with the observed rate $(2 \pm 1) \times 10^{-4}$.

For SEX in-flight on ^{12}C, $p_{K^-} = 400$ MeV/c and π^- emitted in the forward direction, Hufner, Lee, and Weidenmuller find in their calculation of forward cross section the following values for the dominant 0^+ $^{12}_{\Lambda}C^*$ states:

(i) $(p_{3/2})_{\Lambda}(p^{-1}_{3/2})_n$: $d\sigma/d\Omega \sim 1.8$ mb/sr;

(ii) $(s_{1/2})_{\Lambda}(s^{-1}_{1/2})_n$: $d\sigma/d\Omega \sim 0.6$ mb/sr.

Similar values up to a factor of 2–3 should be obtained for $p_{K^-} = 900$ MeV/c, since the basic $K^-n \to \pi^-\Lambda$ forward cross sections as well as the momentum transfers for these two momenta are similar to each other. With the small momentum transfers involved in the reported in-flight experiments, excitation of 2^+ $^{12}_{\Lambda}C^*$ states pertaining to the transition $(1p)_n \to (1p)_{\Lambda}$ is rather weak, by more than an order of magnitude with respect to the 0^+ states.

Similarly, only a weak excitation of the 1^- $^{12}_{\Lambda}C$ ground state may be expected, since a strong transition $(1p)_n \to (1s)_{\Lambda}$ necessarily requires a typical momentum transfer of about 170 MeV/c, whereas the reported experiments provide 40–80 MeV/c. This argument, however, is misleading for the following reason: Within the nucleus the relevant momentum transfer may markedly change from its value in free space. This change occurs since the real parts of the forward K^-N and π^-N elastic scattering amplitude modify the K^- and π^- wave numbers, respectively, in the nuclear medium. Such modifications are typically given in the interesting energy range by a magnitude of roughly 30 MeV/c each, and are negligible for most purposes, particularly in view of the strong nuclear absorption. However, for a "bare" momentum-transfer value of about 60 MeV/c, these modifications can result in a wide range of relevant momentum-transfer values, from zero to about 120 MeV/c, depending on the magnitude and *relative sign* of the K^-N and π^-N real part of the forward elastic scattering amplitude with respect to each other. This effect introduces an uncertainty of one to two orders of magnitude in the above evaluation of formation cross sections of the $^{12}_\Lambda C$ ground state. Even if we knew precise values for those real parts (almost nothing is known currently on the K^-n real part), the mere sensitivity of the eikonal calculation to variations in them would suggest that more reliable K^- and π^- optical wave functions should be used. Unfortunately, experiments relating to the determination of such wave functions are nonexistent at present.

At this point it is useful to record the following two points (DGL 75):

(i) The sizable uncertainty, discussed above, in the theoretical evaluation of the small-momentum-transfer formation of the $1^-\,^{12}_\Lambda C$ state is typical of the transition $(1p)_n \to (1s)_\Lambda$; it does not show up in the formation of 0^+ state resulting from the $(1p)_n \to (1p)_\Lambda$ and $(1s)_n \to (1s)_\Lambda$ transitions. The latter are normally strong for a momentum transfer in the range $q \sim 0\text{--}100$ MeV/c and variations of q in the nuclear environment within such a range of values are therefore not crucial. This point is borne out in Table IX, where calculated forward cross sections for $p_K = 400$ MeV/c and $q = 60$ MeV/c formation of the 0^+ $[(1s)_\Lambda(1s^{-1})_n]$ and 1^- $[(1s)_\Lambda(1p^{-1})_n]$ $^{12}_\Lambda C$ states are grouped as a function of the ratio of real to imaginary part of the forward elastic K^-N and π^-N scattering amplitudes (α_K and α_π, respectively) in the range $[-1, +1]$ for these ratios. For $p_K = 900$ MeV/c and a similar *magnitude* of q, the Λ recoils in the forward direction in the lab system rather than backward as for $p_K = 400$ MeV/c. Since, otherwise, the relevant absorption cross sections as well as the strength of the basic process (121b) are nearly the same for these two momenta, formation cross sections for $p_K = 900$ MeV/c are given by the numbers in Table IX upon the substitution $\alpha \to -\alpha$ (which together with $q \to -q$ amounts to a complex conjugation of the overall result). It is not clear at present whether or not the variations discussed above in the strength of the $(1p)_n \to (1s)_\Lambda$ transition offer a resolution to the puzzling situation where the $1^-\,^{12}_\Lambda C$ ground state has been claimed to be rather strongly excited in one in-flight experiment (Bon+ 74) and unnoticed in another (Bru+ 75).

(ii) A similar situation is expected for SEX in flight in other p-shell nuclei in that the strongest transitions are $(1p)_n \to (1p)_\Lambda$ and $(1s)_n \to (1s)_\Lambda$ to hypernuclear states similar in their space–spin structure to that of the parent nucleus. The generally weaker $(1p)_n \to (1s)_\Lambda$ transition is expected, however, to prefer population of "low-lying" hypernuclear states in the excitation energy range of 5–12 MeV (and even higher for ^{9}Be) to popula-

TABLE IX

Calculated Forward Cross Sections (in mb-sr) for Hypernuclear States Formed in SEX, Eq. (125), as a Function of the Ratios of the Real to Imaginary Parts of π^-N (α_π) and K^-N (α_K) Forward Elastic Scattering Amplitude

The K^- incident momentum is 400 MeV/c and the momentum transfer is 60 MeV/c. To obtain results for $p_{K^-} = 900$ MeV/c, make the substitution $\alpha \to -\alpha$ everywhere. The assumed total cross sections for K-N and π-N are 40 mb. The currently accepted values are $\alpha_\pi \sim -1$, $\alpha_{K-p} \sim 0.2$ at 400 MeV/c and $\alpha_\pi \sim 0$, $\alpha_{K-p} \sim 0.6$ at 900 MeV/c (Bai+ 74).

α_K \ α_π	-1	0	$+1$
(a) $0^+[(1s)_\Lambda(1s^{-1})_n]$ s			
-1	0.7	0.7	0.5
0	1.0	1.0	0.7
$+1$	1.0	1.0	0.7

α_K \ α_π	-1	0	$+1$
(b) $1^-[(1s)_\Lambda(1p^{-1})_n]$			
-1	0.07	0.20	0.27
0	0.012	0.10	0.20
$+1$	0.010	0.012	0.07

tion of ground-state hypernuclear doublets. The reason is that in most cases the $p_{3/2}$ neutron single-particle strength (spectroscopic factor) of the parent nucleus ^AZ is concentrated in high-lying states of the nucleus ^{A-1}Z, the latter providing the nuclear core for the product hypernucleus $^A_\Lambda Z$. The ^{12}C spectrum, discussed above, is exceptionally simple since most of the single-particle p strength of ^{12}C ground state is $p_{3/2}$ and is concentrated in the ^{11}C ground state (CK 67). Moreover, in other p-shell nuclei the location of the unperturbed $(1p)_\Lambda(1p^{-1})_n$ strongest excitation may be, for the same reason, pushed considerably above the hypernuclear excitation energy value $\hbar\omega_\Lambda$, and this may bring it to a region nearly overlapping with the position of the $(1s)_\Lambda(1s^{-1})_n$ excitation, so that a further collectivity might be reached. This mechanism can provide a natural explanation for the wide variation observed in the location of the strongest hypernuclear excitation in the mass region near ^{12}C. In the absence of any detailed calculation, we prefer not to discuss further SEX in the p shell.

For heavier nuclear targets the following crude argument can be used to estimate the Λ single-particle excitations (Gal 68, AG 74). Suppose the Λ self-consistent field is approximated by a harmonic oscillator well with a spring constant ω_Λ. For hypernuclear ground states we obtain by the virial theorem

$$\tfrac{1}{2} m_\Lambda \omega_\Lambda^2 \langle r_\Lambda^2 \rangle = \tfrac{1}{2} \cdot \tfrac{3}{2} \hbar \omega_\Lambda \tag{127}$$

and hence

$$\hbar \omega_\Lambda = \tfrac{3}{2} \hbar^2 / m_\Lambda \langle r_\Lambda^2 \rangle \tag{128}$$

In order for the Λ to derive as much interaction over the nuclear volume, the $1s$ Λ wave function should have an extension roughly measured by

$$\langle r_\Lambda^2 \rangle = \tfrac{3}{5} (r_0 A^{1/3})^2, \qquad r_0 \approx 1.2 \text{ fm} \tag{129}$$

Substitution of this value in Eq. (127) yields

$$\hbar \omega_\Lambda \sim \tfrac{5}{2} (\hbar^2 / m_\Lambda r_0^2) A^{-2/3} = 60 A^{-2/3} \text{ MeV} \tag{130}$$

For $A = 12$ we obtain $\hbar \omega_\Lambda = 11.3$ MeV, in good agreement with the value of 11.5 MeV found in the SM analysis. The A dependence in Eq. (130) can be also inferred for other types of single-particle potential well. For a square well, the coefficient in front of $A^{-2/3}$ turns out to be closer to 80 MeV, which gives a general idea of the uncertainty involved in the estimate (130). We note here that the *same* type of argument, when applied to the nuclear case, gives the well known (and realistic) expression for the analogous nuclear excitation (BM 69)

$$\hbar \omega_N \sim 40 A^{-1/3} \text{ MeV} \tag{131}$$

Equations (130) and (131) for the Λ spring constant and for the nucleon spring constant, respectively, were derived under the assumptions that a single-particle description for nucleons and for Λ provides a satisfactory starting point for level classification in hypernuclei and that the size of the hypernucleus is not much changed from that of a normal nucleus with the same A number. These assumptions make sense for $A \gtrsim 12$.

No dynamical assumptions, such as the strength of the ΛN interaction and its nature relative to the NN case, are involved in deriving Eqs. (130) and (131). The difference in the A dependence expressed in these equations for ω_Λ and ω_N stems from the fact that there are many nucleons but only *one* Λ-particle in $S = -1$ hypernuclei and that this Λ is not constrained by any special requirement such as the Pauli principle. However, the derivation of Eq. (130) breaks down for those Λ hypernuclear states in which the Λ is antisymmetrized with respect to the neutrons. Within a subset of hypernuclear states, namely those in which the total wave function is antisymmetric not only with respect to space–spin exchange of neutrons, but also with respect to such Λn exchanges, a typical single-particle Λ excitation $\hbar\omega_\Lambda$ should have the usual nuclear dependence expressed by Eq. (131). Since the hypernuclear Hamiltonian is not symmetric in the labels of Λ and neutrons, it connects totally antisymmetric hypernuclear wave functions with hypernuclear wave functions having a different type of symmetry, such as the ground state. It follows that the state instantaneously produced in the substitution reaction (121a), which is completely antisymmetric under the exchange of space–spin labels of any Λn pair [such as the state (126) for $A = 12$], may strongly decay to low-lying or nearby hypernuclear states which have different permutation properties. We must therefore be very careful in applying shell-model arguments to Λ-hypernuclear excited states. Many of these states, where basically the Λ is excited, are characterized by a single-Λ excitation given by Eq. (130). More complicated states, in particular the "doorway" Λ-hypernuclear state, antisymmetric in all neutrons and lambda, obtained from the parent-nucleus ground state by substituting $n \rightarrow \Lambda$ [Eq. (121a)], do not obey Eq. (130). Such highly collective states, from the point of view of Λ excitations, are very different in their nature from single-particle excitations.[†] In the following we will describe two distinct views as to the nature of Λ-hypernuclear excitations expected to be obtained prominently in the substitution reaction (121a) in flight. These, to some extent opposing, views will soon be tested in heavy nuclei.

Auerbach and Gal (AG 74) proposed that a band of hypernuclear lambda particle–neutron hole excitations is formed in the reaction (121a) when the baryonic momentum transfer is close to zero. They start from

[†] The general question of the validity of the shell model description for Λ-hypernuclear states has been raised by several "outsiders," such as D. H. Wilkinson at Brookhaven in 1973 and S. G. Cohen at Jerusalem in 1973. For Λ in nuclear matter this question was dealt with by Bodmer (Bod 73). I am grateful to E. F. Redish for stimulating discussions on the topic (1974).

Eqs. (130) and (131) for Λ and n single-particle excitations, noting that their difference

$$f(A) = 40A^{-1/3} - 60A^{-2/3} \tag{132}$$

is larger than 5 MeV between $A = 12$ and $A = 208$ and attains a maximum of about 6.6 MeV for $A \sim 27$. Since to create a neutron hole in the kth inner major neutron shell (measured with respect to $k = 1$ for the valence neutrons) together with a Λ in the corresponding Λ shell involves a hypernuclear excitation of

$$(m - k)\hbar\omega_\Lambda + (k - 1)\hbar\omega_N, \qquad k = 1, \dots m \tag{133}$$

in a nucleus where the neutrons fill m major shells ($m = 2$ for the p-shell cases considered above), the difference between two neighboring excitations is 5–6.5 MeV. The lowest excitation energy, $(m - 1)\hbar\omega_\Lambda$, corresponds to the Λ replacing a valence neutron in the last major neutron shell to be filled. Exciting deeper lying neutrons involves larger excitation energies, $(\hbar\omega_\Lambda - \hbar\omega_N) \gtrsim 5$ MeV per one neutron shell. This argument is a very qualitative one. It assumes the validity of Eqs. (130) and (131) for highly excited Λ orbitals or deeply lying neutron orbitals, whereas these equations are not really expected to hold in such situations; Eq. (131) for the nuclear case is checked in pickup reactions, involving in most cases neutron pickup from a shell close to the valence shell. Nevertheless, two conclusions are still expected to hold[†]:

(i) The various $(k = 1, \dots, m)$ particle–hole excitations greatly differ in their excitation energies from each other. Thus for $k = 1$ we obtain from Eq. (133)

$$(m - 1)\hbar\omega_\Lambda = 10 \pm 1.5 \text{ MeV} \tag{134}$$

over most of the periodic table. This is the lowest possible excitation and consequently has a good chance to be observed. In addition, it is produced with an enhancement factor proportional to the number of neutrons in the major valence shell, which usually contains a larger number of neutrons than in the lower, filled, major shells. On the other hand, for $k = m$ we know that above ^{40}Ca the $1s$ proton separation energy saturates between 50 and 60 MeV (Wag 73). This is also definitely expected for neutrons.

[†] For a parent nucleus with a nonzero value of isospin the particle–hole excitations are further split to $I = I_N \pm \frac{1}{2}$. In this presentation we ignore isospin and will come back to it below.

Even if the estimate (134) is in error, we do not expect it to come close to the other extreme of exciting a 1s neutron hole.

(ii) A band of particle–hole excitations is expected. Normally the lowest and most enhanced member of such a band corresponds to Λ-particle–neutron-hole excitation from the last major valence neutron shell and is located around 10 MeV excitation. The higher members of the band correspond to deeper neutron-hole excitations and are expected to follow each other by several MeV. Fine splittings due to the actual shell structure within a major shell and the residual ΛN interaction within that major shell will certainly cause some splitting of the strength acquired by one member of the band over an interval ΔE somewhat smaller than the interband distance. More detailed arguments based on quantitative estimates are required for an estimate of the widths of these states. At present we see no general mechanism by which the neutron-hole states produced will become either significantly more or less stable in the presence of a Λ than for a corresponding proton-hole produced in the (p, 2p) reaction. The only new ingredient is the ΛN residual interaction, about which our knowledge is rather poor.

As noticed by Dalitz and Lipkin (Dal 73, Lip 73), a somewhat pre-ferable situation of producing the hypernuclear ground state in (K^-, π^-) reactions in flight occurs when the last neutron shell to be filled is an ns shell. Because of the difference in single-particle wave functions, the over-lap between the appropriate neutron ns wave function and the Λ 1s wave function is nonnegligible and the resultant hypernuclear spectrum will include a low-lying state (possibly the ground state) in which the Λ is bound in its 1s orbit and the neutrons orderly fill their inner shells. It would be instructive to test this hypothesis in the s–d shell, just above ¹⁶O.

From the detailed study sketched above of strangeness exchange reactions in the p shell one should not generalize that Λ excitations into the p shell are expected to be observed throughout the periodic table. For a heavy nucleus, exciting the Λ into its p shell is accompanied in the strangeness exchange reaction by a highly excited nuclear hole state. Only in cases where the last neutron shell is of p type is some production of low-lying Λ 1p excitation expected, similar to the considerations given above for the production of hypernuclear ground states. In a heavy hypernucleus the Λ 1p excitation (and similarly for other single-particle states) is expected to appear at low energies. For the lead region $\hbar\omega_\Lambda \lesssim 2$ MeV, so that single-particle 1p, 1d, and 2s Λ states are certainly particle-stable and, where low-lying nuclear levels are not too dense (as in the immediate vicinity of ^{208}Pb), strongly disturb the picture assumed for light and medium elements of core excitation dominance of low-lying hypernuclear states.

Since the isospin of Λ is zero, resultant hypernuclear states reached by the strangeness exchange reaction are in general split into $I = I_N \pm \frac{1}{2}$ for $I_N \neq 0$. However, for heavy nuclei where $I_N \gg 1$ and the neutrons fill shells empty of protons, some simplification occurs. The lowest expected hypernuclear excitation, corresponding to substitution of one of the valence neutrons by the Λ, has $I = I_N - \frac{1}{2}$. This is easy to understand since one can ignore all shells that are filled by both protons and neutrons. The nuclear isospin properties in the shell model description are then completely determined by the excess neutrons, namely $I_N = \frac{1}{2}n_{\text{exc}}$. Hence $I = \frac{1}{2}(n_{\text{exc}} - 1) = I_N - \frac{1}{2}$. For higher hypernuclear excitations where inner neutron holes are excited this ceases to hold, but it can easily be shown that for $I_N \gg 1$ most of the above considerations for hypernuclear excitations remain valid for the $I = I_N - \frac{1}{2}$ component, which obtains most of the $n \rightarrow \Lambda$ strength. To evaluate the position of $I = I_N + \frac{1}{2}$ excitations, one hypothetically invokes the (K^-, π^0) reaction in which a proton is substituted by a Λ, which necessarily leads to an increase in the neutron excess and hence $I = I_N + \frac{1}{2}$.

Kerman and Lipkin (KL 71) suggested that production of hypernuclear states in strangeness exchange reactions in flight is dominated by the formation of a hypernuclear collective state which they called the strangeness analog resonance (SAR). This state is obtained by substituting coherently each one of the target neutrons by a Λ in the same space–spin state, similar to the definition of the isospin analog resonance (IAR) which is excited by (p, n) charge exchange. In the case of IAR, however, we know that isospin is a good nuclear symmetry and is broken mainly by the relatively smooth proton–nucleus electrostatic potential. No such symmetry really exists in the hypernuclear case between Λn and nn interactions in the same states. Nevertheless, what matters in a shell model description is the difference between the Λ–nucleus potential and the n–nucleus potential. These authors make the plausible assumption that the Λ–nucleus potential is similar in shape to the nucleon–nucleus potential, namely that these potentials follow quite well the shape of the nuclear core. Thus in their model the difference between the Λ–nucleus and the n–nucleus potentials is almost constant over the nuclear interior. Therefore the energy to excite a Λ in the nuclear well from one state to another is the same as the energy released when a neutron decays between the corresponding states. Under this assumption all Λn^{-1} particle–hole excitations are degenerate. The residual ΛN interaction will remove this degeneracy to some extent, but most of the (K^-, π^-) reaction strength will remain in the (nearly) hypernuclear eigenstate SAR, which is given by a specific combination ("the coherent combination") of these degenerate particle–hole excitations:

$$| \text{SAR}, {}_{\Lambda}^{A}Z^* \rangle = \frac{1}{\sqrt{N}} \sum_{\{\alpha\}} c_{\alpha}^{+} b_{\alpha} \,|\, \text{ground state } {}^{A}Z \rangle \qquad (135)$$

where c_{α}^{+} creates a Λ in the state α, b_{α} destroys a neutron in the same state, and N is the number of neutrons in the target. This state has been considered earlier for the discussion of ${}_{\Lambda}^{12}C^*$ [see Eq. (126)].

The name SAR is derived from the observation made by Kerman and Lipkin that the operator $U_- = \sum_{\alpha} c_{\alpha}^{+} b_{\alpha}$ which appears in Eq. (135) is nothing else but the lowering U spin operator for a $SU(3)$ algebra, the basic representation of which consists of the *physical* proton, neutron, and lambda. To the extent that this algebra corresponds to a symmetry of the shell model Hamiltonian, the SAR defined in Eq. (135) by the opera- tion of U_-, which is one of the algebra generators, is an eigenstate of the system. The use of this particular $SU(3)$ algebra, related to the old Sakata symmetry, facilitates evaluation of quantities connected with the formation, position, and decay of the SAR. As stressed by Kerman and Lipkin, this $SU(3)$ algebra is radically different than the octet $SU(3)$ algebra, where the Λ and the Σ^0, separated in reality by about 80 MeV, are considered degenerate (Kis 67).

As we have pointed out in the preceding discussion, the SAR defined by Eq. (135) involves deeply lying neutron hole states for which the repre- sentation of the shell model potential by a local state-independent potential is not expected to hold. This does not occur in the case of IAR, since the substitution $n \rightarrow p$ affects only the excess neutrons. Even within the most simplified shell model the Λ single-particle excitations and the Λ single- particle orbitals differ significantly from the corresponding nuclear quan- tities, as exemplified by Eqs. (130) and (131). This difference derives from the distinct roles played by nuclear orbitals and by Λ orbitals. The former are constructed in a self-consistent manner, so that the single-particle nuclear potential formed with them generates the same orbitals. These orbitals are constrained to yield the nuclear shape. On the other hand, the Λ orbitals, at least the low-lying ones, are formed under the requirement that the Λ occupies the lowest possible state in a nuclear well whose shape is *predetermined* by the nuclear calculation. For medium and heavy nuclei, where the distortion capability of the Λ is necessarily negligible, this separa- tion of the hypernuclear problem into a nuclear-core self-consistent calculation followed by a Λ eigenvalue problem in a determined single-particle potential makes sense. Such a distinction between the single-particle properties of nucleons and a Λ holds as long as there is one Λ-particle which is not restricted in any manner by an antisymmetrization postulate. On the other hand, if the Λ had degenerated in its mass and nuclear interactions to a neutron-like particle, we would have found it convenient to work once and for all in a hypernuclear scheme where all baryons are antisymmetrized. This is accomplished by introducing U and V spins which together with I spin form a $SU(3)$ algebra. In this hypothetical case a participation of the Λ in a self-consistent calculation is required; otherwise the low-lying orbitals obtained for the Λ as prescribed above would have a nonzero overlap with orbitals already occupied by nucleons. The resultant single-particle energies are then identical for lambda and nucleons. Nevertheless, no contradiction with our previous discussion arises, since even in this limit there exists a quantity which may split baryonic single-particle excitations according to their unitary-spin contents. This quantity is the U-spin symmetry energy, and there is no *a priori* reason why a typical value for it could not reach 10 to 20 MeV. It may well turn out that the apparent differences in the conclusions of the two approaches discussed above will be ironed out in a careful application (DG 75).

Before closing this section we would like to discuss some of the predictions made by Kerman and Lipkin and compare them with our previous discussion. In the following we ignore neutron and lambda masses, which drop out anyway in the final result. The difference between the ground-state energy of ^{12}C and the energy of $^{12}_{\Lambda}$C*(SAR) is given in this model by $\Delta V_{n\Lambda} = V_n - V_\Lambda$, the difference between the corresponding single-particle potentials. Figure 13a shows how to relate ^{12}C and $^{12}_{\Lambda}$C* through ^{11}C by known quantities: the binding energy of the last neutron in ^{12}C, the binding energy of the Λ in $^{12}_{\Lambda}$C, and the observed $E^* \sim 10$ MeV excitation of the assumed SAR in $^{12}_{\Lambda}$C. A value

$$\Delta V_{n\Lambda} \sim -18 \text{ MeV} \tag{136}$$

is deduced as shown in Fig. 13(a). If we use this fitted value[†] in the vicinity of ^{12}C to predict SAR excitations in ^{16}O we obtain

$$E^*(\text{SAR}) \sim 17 \text{ MeV} \tag{137}$$

[†] It is tempting to "derive" this value from the difference in separation energies of the last neutron in ^{4}He and of the last (and only) Λ in $^4_{\Lambda}$He ground state, both states belonging to a $U = 1$ U-spin multiplet in the limit of $SU(3)$ symmetry. This difference in fact yields $\Delta V_{n\Lambda} \sim -18$ MeV.

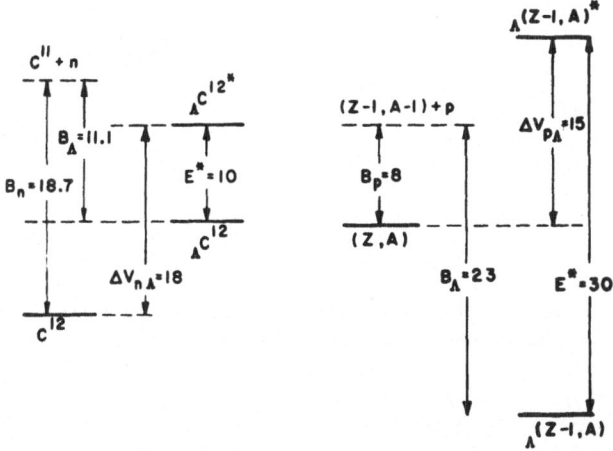

Fig. 13. Energy level diagram for strangeness analog resonances in the model of Kerman and Lipkin: (a) in ^{12}C; (b) in a heavy nucleus, where for the sake of simplicity the hypernuclear $I = I_N + \frac{1}{2}$ state obtained in the (K^-, π^0) reaction is considered. [From (KL 71).]

for $^{16}_\Lambda$O*, where for the $^{16}_\Lambda$O ground state we used a value of about 14 MeV. A broad peak has indeed been observed in the recent CERN experiments (Bru+ 75), on ^{16}O with a $^{16}_\Lambda$O excitation energy whose value is similar to that given by Eq. (137). However, as discussed earlier, such a peak may be attributed in the other approach to a 0^+ $(p_{3/2})_\Lambda(p_{3/2}^{-1})_n$ $^{16}_\Lambda$O* excitation, since the nuclear $p_{3/2}-p_{1/2}$ splitting of 6 MeV adds up to $\hbar\omega_\Lambda \sim 10$ MeV to give $E^* \sim 16$ MeV for this excitation.[†] We must therefore turn to heavier nuclear targets if we want experiment to discriminate between these opposing approaches.

Figure 13(b) shows the Kerman–Lipkin prediction for heavy nuclei. They choose $p \to \Lambda$ substitution since, as explained above, this corresponds to a definite isospin value, $I = I_N + \frac{1}{2}$. By using $\Delta V_{p\Lambda} = -15$ MeV (which allows for $\Delta V_{np} = -3$ MeV due to the Coulomb potential in the mass region where a value of $B_\Lambda \sim 23$ MeV is deduced), they find

$$E^*(\text{SAR}) \sim 30 \text{ MeV} \qquad (138)$$

in a heavy nucleus. This is roughly also the estimate for the $I = I_N - \frac{1}{2}$ (lower) component of SAR obtained in the $n \to \Lambda$ substitution. The reason is that the other, $I = I_N + \frac{1}{2}$, (upper) component is located above its isospin analog $p \to \Lambda$ excitation by an amount given by the Coulomb energy of the substituted proton, and above the $I = I_N - \frac{1}{2}$ (lower) com-

---

[†] For a more careful discussion of $^{16}_\Lambda$O*, see DG 75.

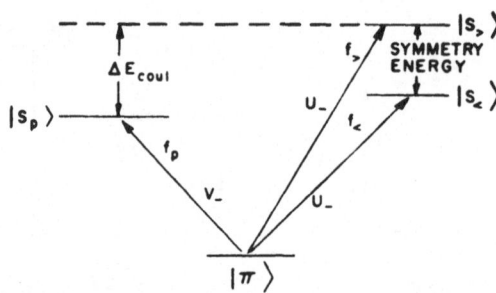

Fig. 14. Schematic representation of the strangeness analog resonances, in the model of Kerman and Lipkin, that are obtained from the nuclear parent state ($| \pi \rangle$) by application of (K^-, π^0) reactions in flight. The state $| S_p \rangle$ is obtained from $| \pi \rangle$ by substituting a proton by Λ in the same space–spin state. The states $| S_> \rangle$ and $| S_< \rangle$, with $I = I_N \pm \frac{1}{2}$, respectively, are obtained by similarly substituting a neutron by a Λ. [From (KL 71).]

ponent by an amount given by the symmetry energy as shown in Fig. 14. Since for heavy nuclei the symmetry energy and the Coulomb energy of one proton are almost equal to each other, the excitation energy of the $I = I_N - \frac{1}{2}$, $n \to \Lambda$, state is almost equal to the excitation energy of the (necessarily $I = I_N + \frac{1}{2}$) $p \to \Lambda$ state, that is, about 30 MeV.

We see that in the SAR model of Kerman and Lipkin the strength of the (K^-, π^-) reaction in flight is expected to be narrowly distributed around a hypernuclear excitation energy of 30 MeV. Calculations by Deloff (Del 73) give cross sections of the order of 1 mb in heavy elements.[†] This prediction markedly differs from the prediction by Auerbach and Gal that the (K^-, π^-) strength is expected to be distributed over a wide range of energy, peaking around specific particle–hole excitations, where the strongest excitation will appear invariably around 10 MeV hypernuclear excitation.

Experiments in heavy elements are needed to discriminate between the two alternative models presented for the distribution of the (K^-, π^-) in-flight strength over hypernuclear states.

6. CONCLUSIONS

In the preceding sections we have discussed various Λ-hypernuclear data accessible through experiments. The current ideas on B_Λ values for s-shell hypernuclei, p-shell hypernuclei, and Λ in nuclear matter have been

---

[†] Unlike previous calculations (BM 72, Esc 73), the calculation by Deloff takes into account in an $SU(3)$ symmetric way the distortion of the K^- and π^- in the entrance and exit channels, respectively.

reviewed, with the conclusion that some prominent problems, such as suppression of the ΛN interaction in the nuclear environment, are only qualitatively understood. Clearly, hypernuclear data of a nature rather different than that of hupernuclear ground-state separation energies are required to resolve these problems. Among the many proposals, the (K^-, π^-) reaction in flight is the most promising one for producing such data. With the near completion of π^- spectrometers, existing hypernuclear data will become considerably richer. The study of excited hypernuclear states over the nuclear periodic table is expected to yield valuable information as to the various coupling schemes for Λ in nuclei: core excitations, single-particle excitations, and collective excitations. The effective ΛN interaction will then probably become well determined. However, it would seem premature to assume that basic properties of the two-body ΛN interaction can be deduced in this way. The nuclear problem has taught us that knowing the NN free interaction to a considerable extent is only the first step in a lengthy and delicate procedure for deducing nuclear effective interaction parameters. One should try therefore, independent of progress in hypernuclear research, to improve on information related to Λ reactions with light nuclear systems. Measurement of Λ polarization, final-state Λn interaction, and spin–orbit splitting are examples of such information.

In some cases, in particular where the core nucleus is particle unstable, purely nuclear properties may be deduced from the corresponding hypernuclear analysis. Thus, if a low-lying isomer $^7_\Lambda$He* exists, an upper limit on the effective neutron charge in ^{6}He is implied, $|q/e| \lesssim 0.1$. Another example relates to Coulomb displacement energies.

We have not touched on properties of hypernuclear weak decay modes, the study of which requires considerable experimental skill. The small branching ratio for Λ leptonic decays at present prevents experimental exploration of these decay modes in hypernuclei. In principle, some of the most exciting properties of strangeness-nonpreserving weak vertices can be deduced from hypernuclear leptonic decays. Pionic decays of hypernuclei have provided valuable information regarding spins of hypernuclear ground states. The dominance of the nonleptonic, nonpionic, $\Lambda N \to NN$ weak decay mode in heavy hypernuclei singles the latter out as a means of investigating weak, strangeness-nonpreserving currents.

Another topic not covered in this review is that of double $(\Lambda\Lambda)$ hypernuclei. The $\Lambda\Lambda$ interactions, as well as some other nuclear properties, are expected to be deduced solely from the study of these exciting species. In particular, a direct verification of the statistics obeyed by Λ particles should emerge from such a study. More than for single-Λ hypernuclei, the presence of two lambdas is expected to stabilize nuclear core configurations which are otherwise considerably unstable. The importance of this observation to the study of nuclear properties is beyond any doubt.

As is true in so many branches of physics, new experiments bearing on exciting ideas or a breakthrough in experimentation will force the creation of new views and theories about hypernuclei. The last significant happening of this nature occured in 1966–68 with the measurement of Λp low-energy cross sections. As a result, the ΛN interaction parameters which had been deduced from hypernuclei proved to be totally wrong. Concepts such as Σ suppression, ΛNN interaction, and tensor ΛN interaction have been consequently suggested and investigated. It is probable that with the recent (K^-, π^-) experiments on nuclear targets a revolution in hypernuclear physics is beginning.

7. REFERENCES

AB 67a   S. Ali and A. R. Bodmer, *Nuovo Cimento* **50A**:511 (1967).
AB 67b   S. Ali and A. R. Bodmer, *Phys. Letters* **24B**:343 (1967).
AG 74   N. Auerbach and A. Gal, *Phys. Letters* **48B**:22 (1974).
AGG 67   G. Alexander, A. Gal, and A. Gersten, *Nucl. Phys.* **B2**:1 (1967).
AGK 67   S. Ali, M. E. Grypeos, and L. P. Kok, *Phys. Letters* **24B**:543 (1967).
Ale+ 68   G. Alexander, U. Karshon, A. Shapira, G. Yekutieli, R. Engelmann, H. Filthuth, and W. Lughofer, *Phys. Rev.* **173**:1452 (1968).
Ale+ 69   G. Alexander, B. H. Hall, N. Jew, G. Kalmus, and A. Kernan, *Phys. Rev. Letters* **22**:483 (1969).
AMB 65   S. Ali, J. W. Murphy, and A. R. Bodmer, *Phys. Rev. Letters* **15**:534 (1965).
BA 64   A. R. Bodmer and S. Ali, *Nucl. Phys.* **56**:657 (1964).
BA 65   A. R. Bodmer and S. Ali, *Phys. Rev.* **138**:B644 (1965).
Bai+ 74   P. Baillon, C. Bricman, M. Ferro-Luzzi, J. M. Perreau, R. D. Tripp, T. Ypsilantis, V. Declais, and J. Seguinot, *Phys. Letters* **50B**:383, 387 (1974).
Bam+ 71   A. Bamberger, M. A. Faessler, U. Lynen, H. Piekarz, J. Piekarz, J. Pniewski, B. Povh, H. G. Ritter, and V. Soerghel, *Phys. Letters* **36B**:412 (1971).
Bam+ 73   A. Bamberger, M. A. Faessler, U. Lynen, H. Piekarz, J. Piekarz, J. Pniewski, B. Povh, H. G. Ritter, and V. Soerghel, *Nucl. Phys.* **B60**:1 (1973).
Bar 66   F. C. Barker, *Nucl. Phys.* **83**:418 (1966).
BD 67   J. Borysowicz and J. Dabrowski, *Phys. Letters* **B24**:549 (1967).
BDI 70   J. T. Brown, B. W. Downs, and C. K. Iddings, *Ann. Phys.* (*N.Y.*) **60**:148 (1970).
BDI 72   J. T. Brown, B. W. Downs, and C. K. Iddings, *Nucl. Phys.* **B47**:138 (1972).
Bed+ 75   M. Bedjidian, A. Filipkowski, I. Y. Grossiord, A. Guichard, M. Gusakow, S. Majewski, H. Piekarz, J. Piekarz, and J. R. Pizzi, Preprint (June 1975).
BG 70   W. H. Bassichis and A. Gal, *Phys. Rev. C* **1**:28 (1970).
Bha+ 68   R. K. Bhaduri, Y. Nogami, P. Friesen, and E. Tomusiak, *Phys. Rev. Letters* **21**:1828 (1968).
BL 70   R. K. Bhaduri and J. Law, *Nucl. Phys.* **A140**:214 (1970).

BLN 67   R. K. Bhaduri, B. A. Loiseau, and Y. Nogami, *Ann. Phys. (N.Y.)* **44**:57 (1967).

Blo+ 63   M. M. Block, R. Gessaroli, J. Koppelman, S. Ratti, M. Schneeberger, L. Grimellini, T. Kikucki, L. Lendinara, L. Monari, W. Becker, and E. Harth, in: *Proceedings International Conference on Hyperfragments* (W. Lock, ed.), CERN, Geneva (1964), p. 63.

BM 65   A. R. Bodmer and J. W. Murphy, *Nucl. Phys.* **64**:593; **73**:664 (1965).

BM 69   A. Bohr and B. R. Mottelson, *Nuclear Structure*, Vol. I, W. A. Benjamin, New York (1969).

BM 72   F. J. Bloore and E. Middleton, *Nucl. Phys.* **A190**:565 (1972).

BNV 67   R. K. Bhaduri, Y. Nogami, and W. Van Dijk, *Phys. Rev.* **155**:1671 (1967).

Bod 66   A. R. Bodmer, *Phys. Rev.* **141**:1387 (1966).

Bod 73   A. R. Bodmer, in: *Proceedings of the Summer Study Meeting on Nuclear and Hypernuclear Physics with Kaon Beams* (H. Palevsky, ed.), BNL report 18335 (1973), p. 64.

Boh+ 68   G. Bohm, J. Klabuhn, U. Krecker, F. Wysotzki, G. Coremans, W. Gajewski, C. Mayeur, J. Sacton, P. Vilain, G. Wilquet, D. O'Sullivan, D. Stanley, D. Davis, E. Fletcher, S. Lovell, N. Roy, J. Wickens, A. Filipkowski, K. Garbowska-Pniewska, T. Pniewski, E. Skrzypezak, T. Sobczak, J. Allen, V. Bull, A. Conway, A. Fishwick, and P. March, *Nucl. Phys.* **B4**:511 (1968).

Boh+ 70   G. Bohm, J. Klabuhn, U. Krecker, F. Wysotzki, G. Coremans-Bertrand, J. Sacton, P. Vilain, J. Wickens, G. Wilquet, D. Stanley, D. H. Davis, J. E. Allen, J. Pniewski, T. Pniewski, and J. Zakrzewski, *Nucl. Phys.* **B24**:248 (1970).

Boh+ 74   G. Bohm, U. Krecker, G. Coremans-Bertrand, D. Kielczewska, J. Sacton, T. Cantwell, A. Montwill, P. Moriarty, D. H. Davis, T. Tymiencka, and O. Adamovic, *Nucl. Phys.* **B74**:237 (1974).

Bon+ 73   G. C. Bonazzola, T. Bressani, R. Cester, E. Chiavassa, G. Dellacasa, A. Fainberg, D. Feschi, N. Mirfakhrai, A. Musso, and G. Rinaudo, in: *Proceedings of the Summer Study Meeting on Nuclear and Hypernuclear Physics with Kaon Beams* (H. Palevsky, ed.), BNL report 18335 (1973), p. 106.

Bon+ 74   G. C. Bonazzola, T. Bressani, R. Cester, E. Chiavassa, G. Dellacasa, A. Fainberg, N. Mirfakhrai, A. Musso, and G. Rinaudo, *Phys. Letters* **53B**:297 (1974).

Bon+ 75   G. C. Bonazzola, T. Bressani, E. Chiavassa, G. Dellacasa, A. Fainberg, M. Gallio, N. Mirfakhrai, A. Musso, and G. Rinaudo, *Phys. Rev. Letters* **34**:683 (1975).

BR 71   A. R. Bodmer and D. M. Rote, *Nucl. Phys.* **A169**:1 (1971).

BR 72   A. R. Bodmer and D. M. Rote, in: *Symposium on the Present Status and Novel Developments in the Nuclear Many-Body Problem*, Rome, September 1972.

BR 73   A. R. Bodmer and D. M. Rote, *Nucl. Phys.* **A201**:145 (1973).

BRM 70   A. R. Bodmer, D. M. Rote, and A. L. Mazza, *Phys. Rev. C* **2**:1623 (1970).

Bru+ 75   W. Brückner, M. A. Faessler, K. Kilian, U. Lynem, B. Pietrzyk, B. Povk, H. G. Ritter, B. Schürlein, H. Schröder, and A. H. Walenta, *Phys. Letters* **55B**:107 (1975).

BS 62   A. R. Bodmer and S. Sampanthar, *Nucl. Phys.* **31**:251 (1962).

Can+ 74   T. Cantwell, D. H. Davis, D. Kielczewska, J. Zakrzewski, M. Juric, U. Krecker, G. Coremanz-Bertrand, J. Sacton, T. Tymiencka, A. Montwill, and P. Moriarty, *Nucl. Phys.* **A236**:445 (1974).

CK 65   S. Cohen and D. Kurath, *Nucl. Phys.* **73**:1 (1965).

CK 67   S. Cohen and D. Kurath, *Nucl. Phys.* **A101**:1 (1967).

CLM 68   D. Cline, R. Laumann, and J. Mapp, *Phys. Rev. Letters* **20**:1452 (1968).

Dab 73a   J. Dabrowski, *Phys. Letters* **47B**:306 (1973).

Dab 73b   J. Dabrowski, *Phys. Rev.* C **8**:835 (1973).

Dal 63   R. H. Dalitz, *Nucl. Phys.* **41**:78 (1963).

Dal 67   R. H. Dalitz, in: *Interactions of High Energy Particles with Nuclei* (T. E. O. Ericson, ed.), Academic Press, New York (1967), p. 89.

Dal 69a   R. H. Dalitz, in: *Nuclear Physics* (C. DeWitt and V. Gillet, eds.), Gordon and Breach, New York (1969), p. 701.

Dal 69b   R. H. Dalitz, in: *Proceedings of International Conference on Hypernuclear Physics* (A. R. Bodmer and L. G. Hyman, eds.) (1969), pp. 728–729.

Dal 73   R. H. Dalitz, in: *Proceedings of the Summer Study Meeting on Nuclear and Hypernuclear Physics with Kaon Beams* (H. Palevsky, ed.), BNL report 18335 (1973), p. 1.

Dan+ 63   M. Danysz, K. Garbowska, J. Pniewski, T. Pniewski, J. Zakrzewski, E. R. Fletcher, J. Lemonne, P. Renard, J. Sacton, W. T. Toner, D. O'Sullivan, T. P. Shuh, A. Thompson, P. Allen, M. Heeran, A. Montwill, J. E. Allen, M. J. Beniston, D. H. Davis, D. A. Garbutt, V. A. Bull, R. C. Kumar, and P. V. March, *Nucl. Phys.* **49**:121 (1963).

DD 58   R. H. Dalitz and B. W. Downs, *Phys. Rev.* **111**:967 (1958).

DD 59   B. W. Downs and R. H. Dalitz, *Phys. Rev.* **114**:593 (1959).

Del 68   A. Deloff, *Nucl. Phys.* **B4**, 585 (1968).

Del 73   A. Deloff, *Nucl. Phys.* **B67**:69 (1973).

Des 61   A. de-Shalit, *Phys. Rev.* **122**:1530 (1961).

DF 73   J. Dabrowski and E. Fedorynska, *Nucl. Phys.* **A210**:509 (1973).

DG 67   R. H. Dalitz and A. Gal, *Nucl. Phys.* **B1**:1 (1967).

DG 75   R. H. Dalitz and A. Gal, submitted for letter publication (August 1975).

DGL 75   R. H. Dalitz, A. Gal, and S. Y. Lee, in preparation.

DHT 72   R. H. Dalitz, R. C. Herndon, and Y. C. Tang, *Nucl. Phys.* **B47**:109 (1972).

DK 69   B. W. Downs and P. D. Kunz, in: *Proceedings of International Conference on Hypernuclear Physics* (A. R. Bodmer and L. G. Hyman, eds.) (1969), p. 796.

DL 59   R. H. Dalitz and L. Liu, *Phys. Rev.* **116**:1312 (1959).

DLR 63   D. H. Davis, R. Levi-Setti, and M. Raymund, *Nucl. Phys.* **41**:73 (1963).

DP 65   B. W. Downs and R. J. N. Phillips, *Nuovo Cimento* **36**:120 (1965).

DP 66   B. W. Downs and R. J. N. Phillips, *Nuovo Cimento* **41A**:374 (1966); **43A**:454 (1966).

DR 64   R. H. Dalitz and G. Rajasekaran, *Nucl. Phys.* **50**:450 (1964).

DV 64   R. H. Dalitz and F. Von Hippel, *Phys. Letters* **10**:153 (1964).

Esc 73   R. J. Esch, *Can. J. Phys.* **51**:1524 (1973).

Fae+ 73   M. A. Faessler, G. Heinzelmann, K. Kilian, U. Lynen, H. Piekarz, J. Piekarz, B. Pietrzyk, B. Povh, H. G. Ritter, B. Schurlein, H. W. Siebert, V. Soergel, A. Wagner, and A. H. Walenta, *Phys. Letters* **46B**:468 (1973).

FHS 69   G. Fast, J. C. Helder, and J. J. de Swart, *Phys. Rev. Letters* **22**, 1453 (1969).

FK 66   H. Feshbach and A. K. Kerman, in: *Preludes in Theoretical Physics* (A. de-Shalit, H. Feshbach, and L. Van Hove, eds.), North Holland, Amsterdam (1966), p. 260.

Gal 66   A. Gal, *Phys. Rev.* **152**:975 (1966).

Gal 67   A. Gal, *Phys. Rev. Letters* **18**:568 (1967).

Gal 68   A. Gal, *Ann. Phys. (N.Y.)* **49**:341 (1968).

Gal 75   A. Gal, in preparation (1975).

GGW 69   B. F. Gibson, A. Goldberg, and M. S. Weiss, *Phys. Rev.* **181**:1486 (1969).

GGW 72   B. F. Gibson, A. Goldberg, and M. S. Weiss, *Phys. Rev. C* **6**:741 (1972).

GGW 73   B. F. Gibson, A. Goldberg, and M. S. Weiss, *Phys. Rev. C* **8**:837 (1973).

GS 67   A. Gal and F. Scheck, *Nucl. Phys.* **B2**:110 (1967).

GSD 71   A. Gal, J. M. Soper, and R. H. Dalitz, *Ann. Phys. (N.Y.)* **63**:53 (1971).

GSD 72   A. Gal, J. M. Soper, and R. H. Dalitz, *Ann. Phys. (N.Y.)* **72**:445 (1972).

Hec 73   H. H. Heckman, in: *Proceedings Uppsalla Conference, June 1973*.

HLW 74   J. Hüfner, S. Y. Lee, and H. A. Weidenmuller, *Phys. Letters* **49B**:409 (1974); *Nucl. Phys.* **A234**:429 (1974).

HS 71   K. Hartt and E. Sullivan, *Phys. Rev. D* **4**:1353, 1366 (1971).

HT 67a   R. C. Herndon and Y. C. Tang, *Phys. Rev.* **153**:1091 (1967).

HT 67b   R. C. Herndon and Y. C. Tang, *Phys. Rev.* **159**:853 (1967).

HT 68   R. C. Herndon and Y. C. Tang, *Phys. Rev.* **165**:1093 (1968).

HV 69   T. H. Ho and A. B. Volkov, *Phys. Letters* **30B**:303 (1969).

HV 70   T. H. Ho and A. B. Volkov, *Phys. Letters* **31B**:259 (1970).

Jur+ 72   M. Jurić, G. Bohm, U. Krecker, G. Coremans-Bertrand, J. Sacton, T. Cantwell, P. Moriarty, A. Montwill, D. H. Davis, D. Kielczewska, T. Tymieniecka, and J. Zakrzewski, *Nucl. Phys.* **B47**:36 (1972).

Jur+ 73   M. Jurić, G. Bohm, J. Klabuhn, U. Krecker, F. Wysotzki, G. Coremans-Bertrand, J. Sacton, G. Wilquet, T. Cantwell, F. Esmael, A. Montwill, D. H. Davis, D. Kielczewska, T. Pniewski, T. Tymieniecka, and J. Zakrzewski, *Nucl. Phys.* **B52**:1 (1973).

Key+ 68   G. Keyes, M. Derrick, T. Fields, L. G. Hyman, J. G. Fetkovich, J. McKenzie, B. Riley, and I.-T. Wang, *Phys. Rev. Letters* **20**:819 (1968).

Kie+ 75   D. Kielczewska, J. Sacton, T. Cantwell, A. Montwill, P. Moriarty, D. H. Davis, T. Tymieniecka, J. Zakrzewski, M. Juric, and U. Krecker, *Nucl. Phys.* **A238**:437 (1975).

Kis 67   L. S. Kisslinger, *Phys. Rev.* **157**:1358 (1967).

KL 71   A. K. Kerman and H. J. Lipkin, *Ann. Phys. (N.Y.)* **66**:738 (1971).

Lag+ 64   J. Lagnaux, T. Lemonne, J. Sacton, E. Fletcher, D. O'Sullivan, T. Shah, A. Thompson, P. Allen, S. Heeran, A. Montwill, J. Allen, D. Davis, D. Garbutt, V. Bull, P. March, M. Yaseen, T. Pniewski, and J. Zakrzewski, *Nucl. Phys.* **60**:97 (1964).

Law 65   J. Law, *Nuovo Cimento* **38**:807 (1965).

LD 71   J. T. Londergan and R. H. Dalitz, *Phys. Rev. C* **4**:747 (1971).

LD 72   J. T. Londergan and R. H. Dalitz, *Phys. Rev. C* **6**:76 (1972).

Lem+ 65   J. Lemonne, C. Mayeur, J. Sacton, P. Vilain, G. Wilquet, D. Stanley, P. Allen, D. Davis, E. Fletcher, D. Garbutt, M. Shaukat, J. Allen, V. Bull, A. Conway, and P. March, *Phys. Letters* **18**:354 (1965).

LGB 69   J. Law, M. R. Gunye, and R. K. Bhaduri, *Phys. Rev.* **188**:1603 (1969).

LHC 70   T. Lee, S. Hsieh, and C. Chen-Tsai, *Phys. Rev. C* **2**:366 (1970).

Lip 65   H. J. Lipkin, *Phys. Rev. Letters* **14**:18 (1965).

Lip 73   H. J. Lipkin, in: *Proceedings of the Summer Study Meeting on Nuclear and Hypernuclear Physics with Kaon Beams* (H. Palevsky, ed.), BNL report 18335 (1973), p. 147.

LR 60   R. D. Lawson and M. Rotenberg, *Nuovo Cimento* **17**:449 (1960).

LR 67   L. Lovitch and S. Rosati, *Nuovo Cimento* **51A**:647 (1967).

MB 66   J. W. Murphy and A. R. Bodmer, *Nucl. Phys.* **83**:673 (1966).

Nog 69   Y. Nogami, in: *Proceedings International Conference on Hypernuclear Physics* (A. R. Bodmer and L. G. Hyman, eds.) (1969), p. 244; Y. Nogami and E. Satoh, *Nucl. Phys.* **B19**:93 (1970).

NRS 73   M. M. Nagels, T. A. Rijken, and J. J. de Swart, *Phys. Rev. Letters* **31**:569 (1973).

PD 62   J. Pniewski and M. Danysz, *Phys. Letters* **1**:142 (1962).

Pni+ 67   J. Pniewski, Z. Szymanski, D. Davis, and J. Sacton *Nucl. Phys.* **B2**:317 (1967).

Pod 63   M. I. Podgoretsky *JETP* **44**:695 (1963).

Pro 66   D. J. Prowse, *Phys. Rev. Letters* **17**:782 (1966).

Ray 72   M. Rayet, *Nucl. Phys.* **B38**:387 (1972).

Ray 73   M. Rayet, *Nucl. Phys.* **B57**:269 (1973).

RB 70   D. M. Rote and A. R. Bodmer, *Nucl. Phys.* **A148**:97 (1970).

Ris 72   D. O. Riska, *Phys. Letters* **40B**:177 (1972).

Sec.+ 68   B. Sechi-Zorn, B. Kehoe, J. Twitty, and B. Burnstein, *Phys. Rev.* **175**:1735 (1968).

SI 62   J. J. de Swart and C. K. Iddings, *Phys. Rev.* **128**:2810 (1962).

SI 63   J. J. de Swart and C. K. Iddings, *Phys. Rev.* **130**:319 (1963).

Sop 64   J. M. Soper, unpublished.

Swa+ 71   J. J. de Swart, M. M. Nagels, T. A. Rijken, and P. A. Verhoeven, in: *Tracts in Modern Physics*, Vol. 60 (G. Hohler, ed.), Springer-Verlag, Berlin (1971), p. 138.

Tan 69   T. H. Tan, *Phys. Rev. Letters* **23**:395 (1969).

Wag 73   G. J. Wagner, in: *Lecture Notes in Physics*, Vol. 23 (U. Smilansky, I. Talmi, and H. A. Weidenmuller, eds.), Springer-Verlag, Berlin (1973), p. 16.

Wal 60   J. D. Walecka, *Nuovo Cimento* **16**:342 (1960).

Wal 71   J. D. Walecka, *Ann. Phys.* (*N.Y.*) **63**:219 (1971).

Wyc 72   S. Wycech, *Acta Physica Polonica* **B3**:307 (1972).

Yam+ 69   S. S. Yamamoto, D. Stephen, G. W. Meisner, R. R. Kofler, S. S. Herlzbach, J. Button-Shafer, P. Yamin, and D. Berley, in: *Proceedings International Conference on Hypernuclear Physics* (A. R. Bodmer and L. G. Hyman, eds.) (1969), p. 939.

ZD 74   D. Zieminska and R. H. Dalitz, *Nucl. Phys.* **B74**:248 (1974).

ZD 75   D. Zieminska and R. H. Dalitz, *Nucl. Phys.* **A238**:453 (1975).

Chapter 2

OFF-SHELL BEHAVIOR OF THE NUCLEON–NUCLEON INTERACTION

M. K. Srivastava

Physics Department
University of Roorkee
Roorkee, India

and

Donald W. L. Sprung

Physics Department
McMaster University
Hamilton, Ontario, Canada

1. INTRODUCTION

1.1. Outline

The nuclear force is conventionally treated in a nonrelativistic framework and represented by a potential model. The potential may be deduced from theory, or be a phenomenological fit to certain nuclear data. In practice, any realistic potential that fits the two-body data well contains some measure of phenomenology. That the resulting model is far from unique is all too obvious from the literature [see, for example, the review by Moravcsik (Mor 72)]. It became evident at an early stage, chiefly through a study of nuclear matter, that these different potential models can predict quite different results in many-body calculations. Even if the fits to data on the two-body problem—so-called on-energy-shell data—were identical,

121

different models can still disagree when compared in many-body calculations. This is because in a many-body situation, although energy and momentum must be conserved overall, they need not be conserved in every two-body interaction. Thus one requires the off-energy-shell matrix elements. This realization has prompted a great body of work aimed at understanding the degree of off-energy-shell liberty in the *NN* force and its importance for various nuclear many-body problems. This chapter is a review and consolidation of this work. We believe that it is opportune to review this work now, while the existing literature, though voluminous, is still manageable. Many facets of the problem are understood, and we have reached a stage where extensions of the nonrelativistic potential model are required.

The relevant features of scattering theory are outlined in Section 2. We then turn to the question of phase shift equivalent potentials (Section 3), ways of avoiding potentials (Section 4), and separable approximations to the *T*-matrix (Section 5). Sections 6–8 are concerned with general features of off-energy-shell matrix elements which have been observed. Practical methods for calculating *T*-matrix elements are outlined in Section 9. Finally, the various physical problems in which off-energy-shell differences are noted are the subject of Section 10.

1.2. Present Status of Off-Shell Behavior

A knowledge of the off-energy-shell behavior of the nucleon–nucleon interaction is basic to nuclear physics. The on-energy-shell information, however complete (which it is not), is not adequate to permit unambiguous calculation of the properties of systems of more than two nucleons. Any reasonable agreement with nuclear binding energy and other properties obtained by using different potential models without knowing the correct off-energy-shell behavior of the interaction is probably fortuitous and therefore not very meaningful. The results obtained with different potentials only rarely agree with each other.

This realization has prompted considerable effort during the past few years directed toward an understanding of the off-energy-shell properties of the nucleon–nucleon force. This study has been in terms of the following model, which is more or less the viewpoint of nuclear structure physics. The nucleons are considered as nonrelativistic point particles interacting by a force that can be described by a nonrelativistic potential. The study is thus restricted to a nonrelativistic, quantum mechanical description of the system. The existence of mesons is ignored except insofar as they are responsible for generating the *NN* force. Naturally, relativistic effects, nucleon form factors, and meson currents can be invoked later on as corrections in specific circumstances where they show up.

Our present information about nuclear forces comes mainly from the analysis of the two-particle elastic scattering data at energies up to about 350 MeV and from the properties of the deuteron. These data are becoming quite detailed (GMW 67, Sig 69a) but can only determine on-energy-shell nuclear reaction matrix elements (phase shifts and coupling constants) up to about 350 MeV. Beyond this, in the high-energy region, the data are incomplete and of lower accuracy, necessitating phenomenological extrapolation of the scattering parameters by imposing model constraints. Even if the data were determined unambiguously at all energies by idealized, high-precision elastic scattering experiments, the interaction can be determined uniquely (in the absence of any bound state) only if it is assumed to be local or to have a specified form of nonlocality. If bound states are present, the binding energies and some properties of the bound state wave functions are required in addition to the phase shifts. There can be infinitely many types of nonlocality of the interaction which correspond to different off-energy-shell extensions of the reaction matrix elements. The principal uncertainties in the off-energy-shell nuclear reaction matrix elements are therefore the following:

(i) Even if the force is local, the lack of elastic scattering data at energies greater than about 350 MeV and experimental uncertainties in its values at all energies. The limit of 350 MeV is simply conventional; fairly good data exist at 425 MeV and some at 650 and 1000 MeV. However, the nonrelativistic model is clearly inadequate and meson production is important at these energies, so we must regard these data above 350 MeV as simply arbitrary extrapolations.

(ii) Lack of knowledge of whether the force is local, or just what is the most important type of nonlocality present, even if complete on-energy-shell information were available.

(iii) Relativistic effects which might be incorporated in a potential model. Their influence on the triton has been discussed by Jackson and Tjon (JT 70) and in nuclear matter by Brown, Jackson, and Kuo (BJK 69), and Weng and Kuo (WK 75).

(iv) Inadequacy of the potential model; dynamic meson effects, possible three-body forces, etc.

Barring a complete inadequacy of the nonrelativistic model, the most interesting uncertainties are (i) and (ii). Some investigations have been made regarding the former with a view to studying the extent of variation in the off-energy-shell reaction matrix elements for arbitrary variations in the high-energy phase shifts. This raises another question. How far within any potential model do the on-energy-shell reaction matrix elements affect the off-energy-shell ones? We shall come to this later in Sections 8 and 10.

Let us now take up (ii), assuming that the on-energy-shell data are known at all energies. We restate the problem: Given on-energy-shell reaction matrix elements for a state of specified l, S, and j, which potential should one use to get the correct extrapolation of the reaction matrix off the energy shell? The solution depends on finding some method of determining the correct off-energy-shell behavior. Meson theory (FK 67, FK 68a, FK 68b), at present, is only of little help in providing an unambiguous extrapolation from on-energy-shell to off-energy-shell matrix elements. We must therefore look at systems whose properties depend on off-energy-shell matrix elements, i.e., systems involving three or more particles. Any of the following two alternatives could be followed in principle.

1. In the conventional approach one chooses a potential (a local one being the simplest) which provides a good description of the available nucleon–nucleon data and then calculates the off-energy-shell matrix elements via the Lippmann–Schwinger equation. This approach can be supplemented by theoretical guidance as to the terms (local or nonlocal) to be included in the potential (BS 64, SW 65, LF 67, KPY 68, SH 68). For example, at large distances it must agree with the one-pion exchange (OPE) potential. This still leaves considerable freedom and many potential models have been proposed (GT 57, HJ 62, Las+ 62, Gre 62a, Gre 62b, TD 66, GS 67, AT 68, Ing 68, KPY 68, Rei 68, BKR 69, EHB 69, MT 69a, SBS 69, SS 69, GPT 70, IP 70, Pet 70, TS 73, TRS 74). They differ partly in the assumption that they make about the on-energy-shell matrix elements at high energies where the data do not exist (e.g., hard core, Yukawa core, finite square core, and super soft core local potentials) and partly in the form of nonlocal interaction assumed (purely local or momentum dependent). Separable potentials (Yam 54, MN 59, Naq 64, Tab 64, Tab 68, Mon 68, Mon 69a, Str 68, Lee 69, Hod 69, HH 69, ACS 71, SS 72, SS 73a, Pie 74) have also been used because of the simplicity they bring to the calculations. They have the additional advantage that the reaction matrix elements can sometimes be written as specific functions of the potential parameters rather than occurring only as numbers in the computer output. The off-energy-shell matrix elements generated by these

potentials can then be used to calculate various nuclear properties and to obtain good agreement with experiment in an attempt to distinguish among different potentials and find criteria for favoring a particular one. But these calculations giving good agreement with experiment are at present very much incomplete.

2. One can look at other experiments (those involving three or more particles) such as electron–deuteron scattering, neutron-deuteron scattering, deuteron photodisintegration, nucleon–nucleon bremsstrahlung, etc., to parameterize the values of the off-energy-shell reaction matrix elements. But these experiments have been performed only at very few energies and are not far off-energy-shell. Besides this, the calculations introduce complications which make the interpretation of the results difficult and ambiguous. For example, the analysis of elastic electron–deuteron scattering suffers from the drawback of introducing poorly known meson-exchange effects (Bec 67), the three-nucleon problem introduces three-body forces (BGG 68), and so on. A careful analysis of various experiments together could lead to a meaningful parameterization of the off-energy-shell matrix elements, and could help in sorting out which effects were due to meson-exchange effects, which due to three-body forces, and so on (LLS 69), but such an analysis remains to be done.

Although it has not yet been possible to pursue either of the above two alternatives to any reasonable degree of success, the problem has been appreciated (GMW 67), a lot of ground work done, and the necessary apparatus developed during the last few years.

Laughlin and Scott (LS 68) have studied some local potentials in the $l = 0$ state to investigate the effect of a hard core. In order to compare potentials which may not agree on-energy-shell, it has been proposed to compare the Kowalski–Noyes (Kow 65, Noy 65) half-off-energy-shell function $f_l(p, k) = t_l(p, k; k^2)/t_l(k, k; k^2)$. The differences in the on-energy-shell behavior are thereby suppressed. Mongan (Mon 69b) did this for several separable potentials, as we (SS 69) did for several local models.

In these comparisons it is difficult to identify a particular feature of the many-body calculation with the differences in the on-shell* behavior of the model assumed or with its off-shell properties. In order to isolate effects due to locality versus nonlocality, several potential models which are precisely equivalent on the energy shell at all energies have been constructed. That infinitely many such potentials exist was pointed out by

---

* For brevity the off-energy-shell, half-off-energy-shell, and on-energy-shell matrix elements shall be called off-shell, half-shell, and on-shell, respectively.

Ekstein (Eks 60). Several practical methods have been proposed to construct such phase shift equivalent (or elastically equivalent) potentials. The details are discussed in Section 3. The off-shell matrix elements obtained from these potentials have been studied and some success has been achieved in correlating various features of the off-shell behavior with the characteristics of the potential (Fie 69a, SS 70, Sri 70b, Sri 71, Spr 70). These matrix elements have also been used in various many-body calculations to investigate how far the differences in the off-shell behavior are reflected in the physical properties and which many-body system is more sensitive to a particular feature. In Section 10 we discuss these calculations.

Attempts have also been made to dispense with potentials altogether. These are described in Sections 4 and 6. T-matrix (or K-matrix) elements are naturally much more closely related to both the elastic scattering experiments and the many-body applications than the potential which is required simply to generate them. The usual procedure of constructing V by a complicated fitting process (fitting certain on-shell T-matrix elements) and then doing almost the inverse procedure in calculating the off-shell T-matrix elements is wasteful.

It is well known that the use of a separable T-matrix makes the many-body calculations much simpler. With this calculational ease in mind, a number of separable approximations to the T-matrix have been proposed. This was also the aim of the separable potentials introduced by Mitra (Mit 62). For example, the Faddeev equations for the three-body problem reduce by this choice to a finite number of coupled one-dimensional integral equations (WN 67, Mit 69). The use of a separable approximation to the T-matrix does not mean that the potential is separable. The aim here is not to have a different extrapolation for the off-shell matrix elements, but to approximate those obtained from any given model. Such an approximation should be able to reproduce reasonably well the given T-matrix over an appreciable region of the variables or at least over the region of interest in any particular problem. In Section 5 we discuss the various proposed separable approximations and point out their special features.

In many cases, such as $(p, 2p)$ reactions, nucleon–nucleon bremsstrahlung, low-energy pion production, etc., one needs half-shell K-matrix elements in the neighborhood of the on-shell point. For these calculations, Fuda (Fud 70) and Redish et al. (RSP 72) have given expansions of the half-shell K-matrix element $K(p, k; k^2)$ in powers of $p - k$. The expansion coefficients are found to be smooth and rapidly decreasing functions of energy and offer a simple way of parameterizing the off-shell behavior. In Section 7 we give details of these methods.

General characteristics of the off-shell matrix elements and their relationship to special features (such as the presence of the hard core or momentum-dependent terms) in the potential have been understood. The effect on the off-shell matrix elements of varying the on-shell ones, i.e., the interplay between them, has also been investigated. Details are given in Section 8.

Section 9 contains methods for calculating the matrix elements $K(p, q; s)$. The choice of the method depends on the nature of the interaction (separable, soft core local, hard core local, or momentum dependent) and on whether s is positive or negative.

In the next section we present the two-body scattering problem, define the transition and reaction matrices, consider their off-shell continuation, and point out their symmetry properties and the extent to which the off-shell properties are arbitrary.

2. THE TWO-BODY SCATTERING PROBLEM

The two-particle Schrödinger equation in the center-of-mass system may be written in the form

$$(H - k^2) \mid \Psi_{\mathbf{k}} \rangle = 0 \tag{1}$$

or

$$(\nabla^2 + k^2) \mid \Psi_{\mathbf{k}} \rangle = (k^2 - H_0) \mid \Psi_{\mathbf{k}} \rangle = V \mid \Psi_{\mathbf{k}} \rangle \tag{2}$$

where $k^2 = E$ is the relative energy measured in fm^{-2} and $H_0 = -\nabla^2$ is the kinetic energy operator; the factor $\hbar^2/m = 41.47$ MeV fm^2 has been removed everywhere. We assume that the potential is of a finite range and supports at most a finite number of bound states.

For the purpose of scattering theory it is convenient to replace Eq. (2) by the Lippmann–Schwinger equation

$$\mid \Psi_{\mathbf{k}}^{(+)} \rangle = \mid \mathbf{k} \rangle + (k^2 - H_0 + i\varepsilon)^{-1} V \mid \Psi_{\mathbf{k}}^{(+)} \rangle \tag{3}$$

$$= \mid \mathbf{k} \rangle + G_0(k^2 + i\varepsilon) V \mid \Psi_{\mathbf{k}}^{(+)} \rangle \tag{4}$$

incorporating the boundary condition that at large distances the wave function consists of a plane wave plus an outgoing spherical wave

$$\Psi_{\mathbf{k}}^{(+)}(\mathbf{r}) \rightarrow \exp(i\mathbf{k} \cdot \mathbf{r}) + f_E(\theta, \phi) e^{ikr}/r \tag{5}$$

where $f_E(\theta, \phi)$ is the elastic scattering amplitude. The ket $\mid \mathbf{k} \rangle$ represents a plane wave and satisfies the free particle equation

$$(H_0 - k^2) \,|\, \mathbf{k}\rangle = 0 \tag{6}$$

and G_0 is the free particle Green's function.

The usual definitions (GW 64) of the scattering matrix S and the transition matrix T are

$$\langle \mathbf{p} \,|\, S \,|\, \mathbf{k}\rangle = \langle \Psi_\mathbf{p}^{(-)} \,|\, \Psi_\mathbf{k}^{(+)}\rangle$$
$$= \langle \mathbf{p} \,|\, \mathbf{k}\rangle - 2\pi i\delta(p^2 - k^2)\langle \mathbf{p} \,|\, T \,|\, \mathbf{k}\rangle; \quad |\mathbf{p}| = |\mathbf{k}| \tag{7}$$

and

$$\langle \mathbf{p} \,|\, T \,|\, \mathbf{k}\rangle = \langle \mathbf{p} \,|\, V \,|\, \Psi_\mathbf{k}^{(+)}\rangle$$
$$= \langle \mathbf{p} \,|\, V \,|\, \mathbf{k}\rangle + \langle \mathbf{p} \,|\, V(k^2 - H_0 + i\varepsilon)^{-1}V \,|\, \Psi_\mathbf{k}^{(+)}\rangle \tag{8}$$

2.1. Off-Shell Continuation of the Schrödinger Equation

In order to discuss the off-shell continuation of Eq. (2) it is convenient to introduce the Möller wave operator Ω defined by

$$|\, \Psi_{\mathbf{k},k^2}^{(+)}\rangle = \Omega(k^2 + i\varepsilon) \,|\, \mathbf{k}\rangle \tag{9}$$

The off-shell continuation of $|\, \Psi_{\mathbf{k},k^2}^{(+)}\rangle$ given by $|\, \Psi_{\mathbf{k},z}^{(+)}\rangle = \Omega(z + i\varepsilon) \,|\, \mathbf{k}\rangle$ satisfies the Bethe–Goldstone equation (BG 57)

$$(\nabla^2 + z) \,|\, \Psi_{\mathbf{k},z}^{(+)}\rangle = V \,|\, \Psi_{\mathbf{k},z}^{(+)}\rangle + (z - k^2) \,|\, \mathbf{k}\rangle \tag{10}$$

The off-shell continuation of Eq. (8) is

$$\langle \mathbf{p} \,|\, T(z) \,|\, \mathbf{k}\rangle = \langle \mathbf{p} \,|\, V \,|\, \mathbf{k}\rangle + \langle \mathbf{p} \,|\, V(z - H_0)^{-1}V \,|\, \Psi_{\mathbf{k},z}^{(+)}\rangle \tag{11}$$

or

$$T(z) = V + V(z - H_0)^{-1}T(z) \tag{12}$$

whose formal solution (GW 64) is

$$T(z) = V + V(z - H)^{-1}V \tag{13}$$

An acceptable T-matrix must satisfy an equation of the form (12). The corresponding relations for the wave operator are

$$\Omega(z) = 1 + (z - H_0)^{-1}V\Omega(z) = 1 + (z - H_0)^{-1}T(z) \tag{14}$$

and

$$\Omega(z) = 1 + (z - H)^{-1}V \tag{15}$$

By taking the potential as composed of two parts V_1 and V_2, Eqs. (14) and (15) lead to a very useful two-potential formula (GW 64, MST 71a)

$$T(z) = (V_1 + V_2)\Omega(z) = T_1(z) + \Omega_1^\dagger(z^*)V_2\Omega(z) \tag{16}$$

where $T_1(z)$ is the transition operator and $\Omega_1^\dagger(z)$ is the Hermitian conjugate of the wave operator for the interaction V_1 alone.

2.2. Partial Wave Decomposition

If the potential is spherically symmetric, it is useful to carry out the partial wave decomposition. In the coordinate space representation, Eq. (3), for any partial wave, reduces to

$$\psi_{k,k^2}^{(+)}(r) = \mathscr{j}_l(kr) + \frac{2}{\pi}\int_0^\infty \frac{\mathscr{j}_l(k'r)t_l(k',k;k^2)}{k^2 - k'^2 + i\varepsilon} k'^2\, dk' \tag{17}$$

and Eq. (10) to

$$\left\{\frac{d^2}{dr^2} - \frac{l(l+1)}{r^2} + z\right\}\psi_{k,z}^{(+)}(r)$$
$$= \int_0^\infty v_l(r,r')\psi_{k,z}^{(+)}(r')\,dr' + (z - k^2)\mathscr{j}_l(kr) \tag{18}$$

where

$$v_l(r,r') = \int \langle \mathbf{r}\,|\,V\,|\,\mathbf{r}'\rangle P_l(\cos\theta_{\mathbf{r},\mathbf{r}'})\,d\Omega \tag{19}$$

is the potential in the lth partial wave and $\mathscr{j}_l(x) = xj_l(x)$ is the Riccati Bessel function (AS 66) of the first kind. The subscript l on ψ has been omitted for clarity. The t_l-matrix element given by

$$t_l(p,q;z) = \frac{\pi}{2}\int d\Omega_\mathbf{q}\int d\Omega_\mathbf{p}\,\langle\mathbf{p}\,|\,T(z)\,|\,\mathbf{q}\rangle Y_l^{m^*}(\Omega_\mathbf{p})Y_l^m(\Omega_\mathbf{q})$$
$$= \frac{1}{pq}\int_0^\infty \mathscr{j}_l(pr)v_l(r,r')\psi_{q,z}^{(+)}(r')\,dr\,dr' \tag{20}$$

satisfies the integral equation

$$t_l(p,q;z) = v_l(p,q) + \frac{2}{\pi}\int_0^\infty \frac{v_l(p,k)t_l(k,q;z)}{z - k^2 + i\varepsilon} k^2\,dk \tag{21}$$

Of course, for a local potential $\langle \mathbf{r} \mid v \mid \mathbf{r}' \rangle = v(r)\, \delta(\mathbf{r} - \mathbf{r}')$, we have $v_l(r, r') = v(r)\, \delta(r - r')$, so the integrals over r' in Eqs. (18)–(20) are trivial.

The scattering wave function $\psi_{k,k^2}^{(+)}(r)$ can be expressed* in terms of Jost solutions,

$$\psi_{k,k^2}^{(+)}(r) = \frac{1}{2if_l(-k)} \{f_l(k)e^{-l\pi i/2}f_l(-k, r) - f_l(-k)e^{l\pi i/2}f_l(k, r)\} \qquad (22)$$

These Jost solutions $f_l(\pm k, r)$ are the two irregular solutions of the radial Schrödinger equation with simple exponential behavior at large r, i.e.,

$$\lim_{r\to\infty} e^{\pm ikr} f_l(\pm k, r) = 1 \qquad (23)$$

and the Jost functions $f_l(\pm k)$ are defined by either (i) the Wronskian $W(f_l(k, r), \phi_l(k, r))$

$$f_l(k) = \frac{k^l e^{l\pi i/2}}{(2l+1)!!}\, W \qquad (24)$$

where $\phi_l(k, r)$ is the regular solution of the radial equation defined by the boundary condition at the origin,

$$\phi_l(k, r) \xrightarrow{r\to 0} r^{l+1} \qquad (25)$$

or (ii) the condition

$$f_l(k) = \frac{k^l e^{l\pi i/2}(2l+1)}{(2l+1)!!} \lim_{r\to 0} r^l f_l(k, r) \qquad (26)$$

The Jost function can be expressed in terms of the phase shift and the bound state energies (JK 52) and can be shown to be related to the Fredholm determinant of the Lippmann–Schwinger equation (17) for the lth partial wave (JP 51, WB 71, SW 71). In general the Jost function, corresponding to a particular partial wave, has two very important properties: Its phase is the phase shift for the partial wave and its zeros in the lower half k-plane correspond to the energies of the bound states occurring in the partial wave (GW 64, New 66). The Jost function for a two-particle system is thus directly related to the observables of the system.

The scattering amplitude $f_E(\theta, \phi)$ for any spherically symmetric potential can be expressed as

$$f_E(\theta, \phi) = (1/k) \sum_l (2l+1)e^{i\delta_l(k)} \sin \delta_l(k)\, P_l(\cos\theta) \qquad (27)$$

* For a nonlocal potential the Jost solutions are not always linearly independent (CAM 70). In that case Eq. (22) breaks down.

where the phase shift $\delta_l(k)$ is determined by the asymptotic behavior of the wave function $\psi_{k,k^2}^{(+)}(r)$:

$$\psi_{k,k^2}^{(+)}(r) \to e^{i\delta_l(k)} \sin\{kr - \tfrac{1}{2}l\pi + \delta_l(k)\} \tag{28}$$

This behavior depends on the singularities of the integrand in Eq. (17). The pole at $k' = k + i\varepsilon$ will dominate the behavior of $\psi(r)$ at large distances. It gives

$$\psi_{k,k^2}^{(+)}(r) \to j_l(kr) - kt_l(k, k; k^2)\hat{h}_l^{(+)}(kr)$$
$$= \sin(kr - l\pi/2) - kt_l(k, k; k^2)e^{i(kr - l\pi/2)} \tag{29}$$

where $\hat{h}_l^{(+)}(x) = xh_l^{(+)}(x)$ is the Riccati Hankel function for outgoing waves. Comparing Eq. (29) with Eq. (28), we get

$$t_l(k, k; k^2) \equiv t_l(k) = -e^{i\delta_l(k)}\{\sin \delta_l(k)\}/k \tag{30}$$

Thus the on-shell t_l-matrix element is simply the usual scattering amplitude and is directly related to the elastic scattering data. The half-shell t_l-matrix elements are equivalent to a complete knowledge of the wave function at energy k^2, through Eq. (17). In NN bremsstrahlung they have direct physical significance, for Cromer and Sobel (CS 66) expressed the bremsstrahlung cross section in terms of them (calling them quasi phase shifts).

The asymptotic form of the solution $\psi_{k,z}^{(+)}(r)$ of Eq. (18) can similarly be obtained:

$$\psi_{k,z}^{(+)}(r) \to \sin(kr - l\pi/2) - kt_l(k, \tau; z)e^{i(\tau r - l\pi/2)} \tag{31}$$

where

$$z = \tau^2 + i\varepsilon \tag{32}$$

2.3. Off-Shell Generalization of the Jost Function

An off-shell generalization of the Jost function has been developed by Fuda and Whiting (FW 73) by considering the behavior, for small r, of the irregular solutions of the equation

$$\left\{\frac{d^2}{dr^2} - \frac{l(l+1)}{r^2} + z - v(r)\right\}f_l(\tau, \pm k, r) = (z - k^2)e^{\mp l\pi i/2}\hat{h}_l^{(\mp)}(kr) \tag{33}$$

with the asymptotic behavior

$$f_l(\tau, \pm k, r) \to e^{\mp ikr} \tag{34}$$

Equation (33) is similar to Eq. (18) except for the right-hand side and goes over to the simple radial Schrödinger equation when $\tau = \pm k$. The solutions $f_l(\tau, \pm k, r)$ in that case become identical to the irregular solutions defined earlier:

$$f_l(\pm \tau, r) = f_l(\tau, \pm \tau, r) \tag{35}$$

Fuda and Whiting define an off-shell Jost function by

$$f_l(\tau, k) = \frac{k^l e^{l\pi i/2}(2l+1)}{(2l+1)!!} \lim_{r \to 0} r^l f_l(\tau, k, r) \tag{36}$$

When $\tau = k$, it becomes the ordinary Jost function [Eq. (26)]. Equations (3), (34), and (36) imply that

$$f_l(\tau, -k) = f_l^*(\tau, k) \qquad (\tau \text{ and } k \text{ real}) \tag{37}$$

Using Eqs. (18), (31), and (33)–(35), the solution $\psi_{k,z}^{(+)}(r)$ can be related to $f_l(\tau, \pm k, r)$:

$$\psi_{k,z}^{(+)}(r) = (1/2i)\{e^{-l\pi i/2}f_l(\tau, -k, r) - e^{l\pi i/2}f_l(\tau, k, r)\}$$
$$- kt_l(\tau, k; z)e^{-l\pi i/2}f_l(-\tau, r) \tag{38}$$

Equation (38) is just a generalization of Eq. (22) for the scattering solution $\psi_{k,k_2}^{(+)}(r)$.

The limit $r \to 0$ in Eq. (38) and Eq. (36) gives

$$t_l(\tau, k; z) = \left(\frac{\tau}{k}\right)^l \frac{f_l(\tau, -k) - f_l(\tau, k)}{\pi i f_l(-\tau)} \tag{39}$$

Thus the half-shell t_l-matrix elements can be expressed directly in terms of the off-shell Jost functions. Equation (39) can lead to a series of successive approximations to the t_l-matrix (FW 73) via an iterative solution for $f_l(\tau, k)$.

2.4. The Reaction Matrix

Equation (12) shows that $T(z)$ [and $t_l(z)$] will be complex even when z is real. Especially for numerical calculations it is convenient to introduce the reaction matrix or K-matrix, which is real in this case,

$$K(\omega) = V - V \frac{P}{H_0 - \omega} K(\omega), \qquad \omega = \text{real} \tag{40}$$

P is the principal value operator. Knowing K, one can find T from the Heitler integral equation:

$$T(k^2 + i\varepsilon) = K(k^2) - \sqrt{s}\, K(s)\, \delta(s - k^2) T(s) \tag{41}$$

In a partial wave we have

$$t_l(p, q; k^2) = K_l(p, q; k^2) - \frac{ik\theta(k^2)K_l(p, k; k^2)K_l(q, k; k^2)}{1 + ikK_l(k, k; k^2)} \tag{42}$$

where $\theta(k^2)$ is a unit step function, zero for negative k^2. The on-shell version of Eq. (42) is

$$t_l(k, k; k^2) = \frac{K_l(k, k; k^2)}{1 + ikK_l(k, k; k^2)} \tag{43}$$

It is really only a matter of convenience to deal with the reaction matrix. From Eq. (42) it is seen that the half-shell t_l-matrix elements $t_l(p, k; k^2)$ have a common phase factor, $e^{i\delta_l(k)}$. Further, K_l is separable if t_l is so.

If we write the analog of Eq. (17) for the K_l-matrix elements we have

$$\psi^{(0)}_{k,k^2}(r) = j_l(kr) + \frac{2}{\pi} P \int_0^\infty \frac{j_l(k'r)K_l(k', k; k^2)}{k^2 - k'^2} k'^2\, dk' \tag{44}$$

The K_l-matrix element is given by

$$K_l(p, q; s) = \frac{1}{pq} \int_0^\infty j_l(pr)v_l(r, r')\psi^{(0)}_{q,s}(r')\, dr\, dr' \tag{45}$$

and satisfies the integral equation

$$K_l(p, q; s) = v_l(p, q) + \frac{2}{\pi} P \int_0^\infty \frac{v_l(p, k)K_l(k, q; s)}{s - k^2} k^2\, dk \tag{46}$$

It is real and Hermitian for fixed s. Its on-shell value is related to the phase shifts by

$$K_l(k, k; k^2) \equiv K_l(k) = -\{\tan \delta_l(k)\}/k \tag{47}$$

K_l has no cut from 0 to ∞ along the real axis in the z $(=k^2)$ plane and has poles in z at the actual position of resonances and not on the unphysical sheet as is the case with $t_l(k)$ (KF 61, KF 63).

At the origin or inside a hard core, $\psi^{(0)}_{k,s}(r)$ vanishes. At large r the behavior is determined by the fact that the principal value operator in Eq.

(44) is half the sum of incoming and outgoing waves. This implies

$$\lim_{r \to \infty} \psi_{k,s}^{(0)}(r) = \jmath_l(kr) - A_l(k, \tau) \nu_l(\tau r), \qquad s = \tau^2 > 0 \qquad (48)$$

where $\nu_l(x) = x n_l(x)$ is the Riccati Bessel function of the second kind and A_l is a constant determined by the equation [analogous to Eq. (18)]

$$\left\{ \frac{d^2}{dr^2} - \frac{l(l+1)}{r^2} + s \right\} \psi_{k,s}^{(0)}(r)$$
$$= \int_0^\infty v_l(r, r') \psi_{k,s}^{(0)}(r') \, dr' + (s - k^2) \jmath_l(kr) \qquad (49)$$

It turns out that $A_l(k, \tau) = kK_l(k, \tau; \tau^2)$.

It is convenient to introduce the wave defect

$$\chi_{k,s}(r) = \jmath_l(kr) - \psi_{k,s}^{(0)}(r) \qquad (50)$$

satisfying the equation (for a local potential)

$$\left\{ \frac{d^2}{dr^2} + s - \frac{l(l+1)}{r^2} \right\} \chi_{k,s}(r) = -v_l(r) \psi_{k,s}^{(0)}(r) \qquad (51)$$

or

$$\left\{ \frac{d^2}{dr^2} + s - v_l(r) - \frac{l(l+1)}{r^2} \right\} \chi_{k,s}(r) = -v_l(r) \jmath_l(kr) \qquad (52)$$

with the boundary conditions

$$\chi_{k,s}(r = c) = \jmath_l(kc) \qquad \text{at the hard core edge} \qquad (53a)$$

or

$$\chi_{k,s}(r = 0) = 0 \qquad \text{if there is no hard core} \qquad (53b)$$

and

$$\chi_{k,s}(r \to \infty) = A_l(k, \tau) \nu_l(\tau r) \qquad (54)$$

In the case $s < 0$, χ decays exponentially at large r.

From Eqs. (45) and (51) we can deduce

$$K_l(p, q; s) = \frac{p^2 - s}{pq} \int_0^\infty \jmath_l(pr) \chi_{q,s}(r) \, dr + \text{surface terms} \qquad (55)$$

which was much used in nuclear matter calculations, as $s < 0$ and the surface terms vanish. It is, however, not very useful for positive s.

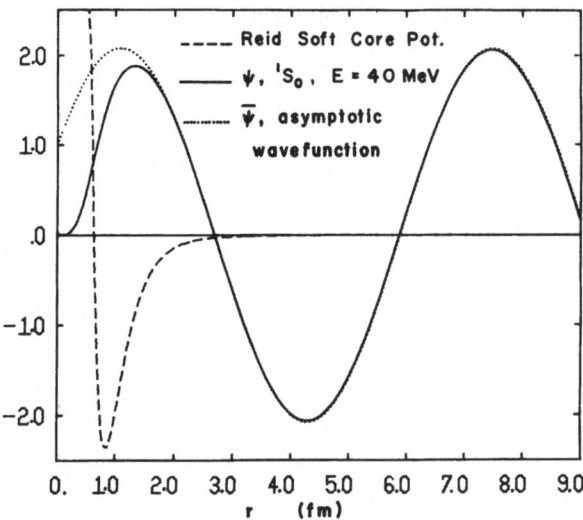

Fig. 1. Singlet S-state wave function ψ and its asymptotic form $\bar{\psi}$ at 40 MeV lab for the Reid soft core potential.

Equation (51), along with the equation satisfied by the free wave function and $\bar{\chi}$, the asymptotic limit of χ, gives a very useful expression for the off-shell K_l-matrix elements (SS 70)

$$K_l(p, q; k^2) = K_l(p, k; k^2) + \frac{p^2 - k^2}{pq} \int_0^\infty \mathscr{J}_l(pr)[\bar{\psi}^{(0)}_{q,k^2}(r) - \psi^{(0)}_{q,k^2}(r)] \, dr \tag{56}$$

Here $\bar{\psi}$ denotes for all r the function on the right-hand side of Eq. (48). It is the asymptotic limit of ψ, so their difference in the integral equation (56) vanishes at large r, as illustrated in Fig. 1. The contributions to the integral come only from within the range of the potential. A well-known special case of this result is an expression for the half-shell element (FS 59):

$$K_l(p, k; k^2) = \frac{-\tan \delta_l(k)}{k} + \frac{p^2 - k^2}{pk} \int_0^\infty \mathscr{J}_l(pr)\{\bar{\psi}^{(0)}_{k,k^2}(r) - \psi^{(0)}_{k,k^2}(r)\} \, dr \tag{57}$$

which can also be obtained by eliminating the potential from Eq. (45).

2.5. Separable Potentials

When the potential is separable things become considerably simpler. The Lippmann–Schwinger equation can be solved explicitly. Writing the potential in the rotationally invariant form

$$\langle \mathbf{k} \mid V \mid \mathbf{k}' \rangle = \frac{\hbar^2}{m} \frac{1}{2\pi^2} \sum_{l=0}^{\infty} (2l+1) \sum_{i=1}^{N_l} \sigma_{il} g_{il}(k) g_{il}(k') P_l(\cos \theta_{\mathbf{k}\mathbf{k}'}) \quad (58)$$

the Lippmann–Schwinger equation for the lth partial wave becomes

$$\psi_{k,k^2}^{(+)}(p) = \delta(p-k) + \frac{2}{\pi} \int_0^{\infty} pk' \sum_{i=1}^{N_l} \frac{\sigma_{il} g_{il}(p) g_{il}(k')}{k^2 - k'^2 + i\varepsilon} \psi_{k,k^2}^{(+)}(k') \, dk' \quad (59)$$

The sign factor σ_{il} is -1 for an attractive component and $+1$ for a repulsive one. The solution of (59) can be written as (GS 68)

$$\psi_{k,k^2}^{(+)}(p) = \delta(p-k) + [\det \mid \delta_{ij} + G_{ij,l}^{(+)}(k^2) \mid]^{-1}$$
$$\times \sum_{i,j=1}^{N_l} pk \frac{\sigma_{il}^{1/2} \sigma_{jl}^{1/2} g_{il}(p) g_{jl}(k)}{k^2 - p^2 + i\varepsilon} d_l(_i^j) \quad (60)$$

where

$$G_{ij,l}^{(+)}(k^2) = \frac{2}{\pi} \int_0^{\infty} \frac{\sigma_{il}^{1/2} \sigma_{jl}^{1/2} g_{il}(k') g_{jl}(k')}{k'^2 - k^2 - i\varepsilon} k'^2 \, dk' \quad (61)$$

and $d_l(_i^j)$ is cofactor of the element ij in the determinant $\mid \delta_{ij} + G_{ij,l}^{(+)}(k^2) \mid$. Noting that the scattered wave in momentum space is $t_l(p, k; k^2)/(k^2 - p^2 + i\varepsilon)$, we have

$$t_l(p, k; k^2) = \sum_{i,j=1}^{N_l} \frac{\sigma_{il}^{1/2} \sigma_{jl}^{1/2} g_{il}(p) g_{jl}(k)}{\det \mid \delta_{ij} + G_{ij,l}^{(+)}(k^2) \mid} d_l(_i^j) \quad (62)$$

The fully off-shell t_l-matrix element is given by

$$t_l(p, q; s) = \sum_{i,j=1}^{N_l} \frac{\sigma_{il}^{1/2} \sigma_{jl}^{1/2} g_{il}(p) g_{jl}(q)}{\det \mid \delta_{ij} + G_{ij,l}^{(+)}(s) \mid} d_l(_i^j) \quad (63)$$

and the Fredholm determinant by

$$D_l(-k) = \det \mid \delta_{ij} + G_{ij,l}^{(+)}(k^2) \mid \quad (64)$$

It should be noted that the t_l-matrix obtained from a separable potential is separable in the initial and final momenta. It is this feature which has made separable potentials so easy to use.

2.6. Restrictions on the Off-Shell T-Matrix Elements

The off-shell elements of T are not completely arbitrary. They are, for any Hermitian interaction, time-reversal invariant and satisfy off-shell two-body elastic unitarity (Mon 69d)

$$\text{Im } t_l(p, q; k^2) = -k t_l(p, k; k^2) t_l^*(q, k; k^2), \qquad k^2 > 0 \qquad (65)$$

The only singularities in the complex energy variable $(=k^2)$ are the bound state and resonance poles and the unitarity cut prescribed by Eq. (65).

Crucial constraints are imposed by the requirement of the completeness and orthonormality of the scattering states,

$$\int |\Psi_{\mathbf{k},k^2}^{(+)}\rangle \, d\mathbf{k} \, \langle \Psi_{\mathbf{k},k^2}^{(+)}| = 1 - P_B \qquad (66)$$

and

$$\langle \Psi_{\mathbf{k},k^2}^{(+)} | \Psi_{\mathbf{k}',k'^2}^{(+)} \rangle = \delta(\mathbf{k} - \mathbf{k}') \qquad (67)$$

where $P_B = \sum_b |\Psi_b\rangle\langle\Psi_b|$ is the projection onto the bound states of the system. What this implies is that the scattering, both on- and off-shell, is generated by a Hermitian Hamiltonian with scattering states satisfying the Bethe–Goldstone equation. Equations (66) and (67) along with Eq. (11) lead to the following relations (MST 71a):

(i) $\quad \langle \mathbf{p} | V | \mathbf{q} \rangle = \langle \mathbf{p} | T(q^2) | \mathbf{q} \rangle$

$$+ \int \langle \mathbf{p} | T(k^2) | \mathbf{k} \rangle \left(\frac{d\mathbf{k}}{k^2 - q^2 - i\varepsilon} \right) \langle \mathbf{k} | T^\dagger(k^2) | \mathbf{q} \rangle$$

$$+ \sum_b \frac{a_b(\mathbf{p}) a_b(\mathbf{q})}{w_b - q^2} \qquad (68)$$

where

$$H | \Psi_b \rangle = w_b | \Psi_b \rangle \qquad (69)$$

and the bound state form factor is

$$a_b(\mathbf{p}) = (w_b - p^2)\langle \mathbf{p} | \Psi_b \rangle = (w_b - p^2) g_b(\mathbf{p}) \qquad (70)$$

Equation (68) defines the interaction potential in terms of the bound state properties of the system and the half-shell elements of the T-matrix. Once V is determined, all elements of T are uniquely defined.

(ii) $\quad \langle \mathbf{p} | T(k^2) | \mathbf{q} \rangle = \langle \mathbf{p} | T(q^2) | \mathbf{q} \rangle + \int \langle \mathbf{p} | T(k'^2) | \mathbf{k}' \rangle \, d\mathbf{k}'$

$$\times \{(k^2 - k'^2 + i\varepsilon)^{-1} - (q^2 - k'^2 + i\varepsilon)^{-1}\}$$

$$\times \langle \mathbf{k}' | T^\dagger(k'^2) | \mathbf{q} \rangle$$

$$+ (q^2 - k^2) \sum_b \frac{a_b(\mathbf{p}) a_b(\mathbf{q})}{(w_b - q^2)(w_b - k^2)} \qquad (71)$$

This relation (a subtracted Low equation) expresses any fully off-shell matrix element in terms of the half-shell ones and the bound state form factors, which contain all the physics of the problem. Another relation of this type, showing explicitly the symmetry of the matrix element $\langle \mathbf{p} \mid T(k^2) \mid \mathbf{q} \rangle$ in p and q, has been obtained by Bishop (Bis 73):

$$\langle \mathbf{p} \mid T(k^2) \mid \mathbf{q} \rangle$$

$$= \frac{1}{q^2 - p^2 + i\varepsilon} [(k^2 - p^2)\langle \mathbf{p} \mid T(q^2) \mid \mathbf{q} \rangle - (k^2 - q^2)\langle \mathbf{p} \mid T^\dagger(p^2) \mid \mathbf{q} \rangle]$$

$$- (k^2 - p^2)(k^2 - q^2)$$

$$\times \int \frac{\langle \mathbf{p} \mid T(k'^2) \mid \mathbf{k}' \rangle \langle \mathbf{k}' \mid T^\dagger(k'^2) \mid \mathbf{q} \rangle}{(k'^2 - k^2 - i\varepsilon)(k'^2 - p^2 + i\varepsilon)(k'^2 - q^2 - i\varepsilon)} \, d\mathbf{k}' \qquad (72)$$

If the half-shell matrix element is separable, the fully off-shell is also separable.

(iii) $\langle \mathbf{k} \mid T(k'^2) - T^\dagger(k^2) \mid \mathbf{k}' \rangle$

$$= \langle \mathbf{k} \mid T^\dagger(k^2) \Big[\frac{P}{k'^2 - H_0} - \frac{P}{k^2 - H_0} - i\pi\delta(k^2 - H_0)$$

$$- i\pi\delta(k'^2 - H_0) \Big] T(k'^2) \mid \mathbf{k}' \rangle \qquad (73)$$

This is an off-shell extension of the optical theorem.

   Baranger *et al.* (Bar+ 69) have shown that the conditions (66) and (67) imply even more restrictive conditions on the off-shell matrix elements of T. They define, for any partial wave, a real half-shell matrix element

$$\phi(k, k') = e^{-i\delta(k)} t(k', k; k^2) \qquad (74)$$

and a real wave function

$$\langle k' \mid \psi_{k,k^2} \rangle = e^{-i\delta(k)} \langle k' \mid \psi_{k,k^2}^{(+)} \rangle$$

$$= \delta(k - k') \cos \delta(k) + \frac{2}{\pi} \frac{P}{k^2 - k'^2} kk' \phi(k, k') \qquad (75)$$

The subscript l has been omitted for clarity. By this choice the conditions (66) and (67) of completeness and orthogonality reduce to

$$WW^\dagger = 1 - P_B \qquad (76)$$

and

$$W^\dagger W = 1 \qquad (77)$$

where W is a real operator with matrix elements given by

$$\langle k' \mid W \mid k \rangle = \langle k' \mid \psi_{k,k^2} \rangle \tag{78}$$

and

$$\langle k' \mid P_B \mid k \rangle = \sum_b g_b(k')g_b(k) \tag{79}$$

Equation (75), when used in Eqs. (76) and (77), gives (Kow+ 71)

$$\phi(k, k') \cos \delta(k) - \phi(k', k) \cos \delta(k')$$
$$+ \frac{2}{\pi} P \int q^2 \, dq \, \phi(q, k')\phi(q, k)\left(\frac{1}{q^2 - k^2} - \frac{1}{q^2 - k'^2}\right)$$
$$= \frac{\pi}{2} \left(\frac{k'^2 - k^2}{kk'}\right) \sum_b g_b(k')g_b(k) \tag{80}$$

and

$$\phi(k, k') \cos \delta(k') - \phi(k', k) \cos \delta(k)$$
$$+ \frac{2}{\pi} P \int q^2 \, dq \, \phi(k', q)\phi(k, q)\left(\frac{1}{q^2 - k^2} - \frac{1}{q^2 - k'^2}\right) = 0 \tag{81}$$

These are necessary and sufficient conditions on the half-shell matrix elements ϕ for the validity of Eqs. (76) and (77).

The orthogonality of the bound states and the scattering states

$$\langle \psi_b \mid \psi_{k,k^2} \rangle = 0 \tag{82}$$

leads to another (though not independent) condition

$$\frac{g_b(k)}{k} \cos \delta(k) + \frac{2}{\pi} P \int \frac{\phi(k, q)g_b(q)}{k^2 - q^2} q \, dq = 0 \tag{83}$$

Fully off-shell t-matrix elements contain no new information since they can be determined from the half-shell ones using Eq. (71), which takes, in terms of ϕ, the following form:

$$t(p, q; s) = \phi(q, p) \cos \delta(p)$$
$$+ \frac{\pi}{2} \int_0^\infty k^2 \, dk \left(\frac{1}{s - k^2} - \frac{P}{q^2 - k^2}\right) \phi(k, p)\phi(k, q)$$
$$+ \frac{\pi}{2} \frac{q^2 - s}{pq} \sum_b \frac{a_b(p)a_b(q)}{(w_b - q^2)(w_b - s)} \tag{84}$$

It should be noted that the binding energy w_b for a fixed $|\psi_b\rangle$ is arbitrary (Kow+ 71); any change in w_b is merely reflected in a change in the potential.

By writing W and ϕ as the sum of a symmetric part and an antisymmetric part

$$W = W_S + W_A \tag{85}$$

$$\phi(k, k') = \sigma(k, k') + \alpha(k, k') \tag{86}$$

we find that Eq. (75) reduces to

$$\langle k' \mid W_S \mid k \rangle = \delta(k - k') \cos \delta(k) + \frac{2}{\pi} kk' \frac{\alpha(k, k')}{k^2 - k'^2} \tag{87}$$

$$\langle k' \mid W_A \mid k \rangle = \frac{2}{\pi} kk' \frac{P}{k^2 - k'^2} \sigma(k, k') \tag{88}$$

and Eqs. (76) and (77), in the absence of bound states ($P_B = 0$), to

$$W_S{}^2 - W_A{}^2 = 1 \tag{89}$$

$$W_A W_S - W_S W_A = 0 \tag{90}$$

Baranger *et al.* have shown that, in such a situation, the conditions (89) and (90) together with the relations (87) and (88) determine $\phi(k, k')$

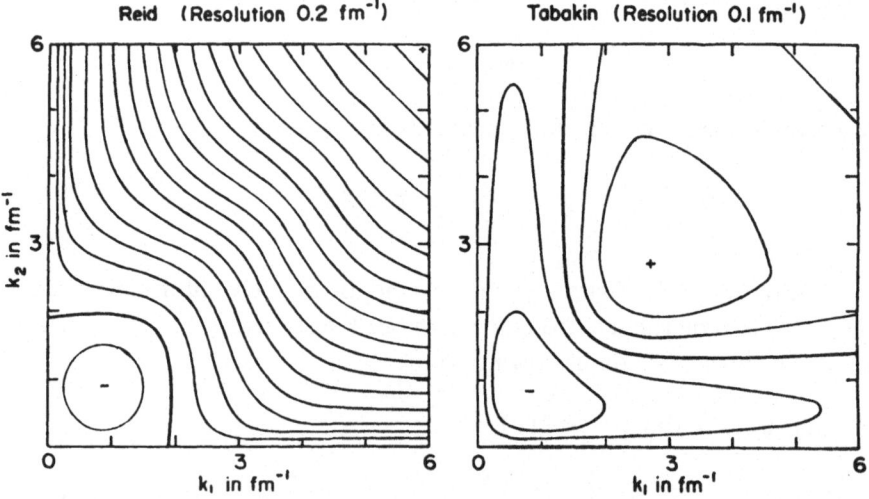

Fig. 2a. Equal-value contour plots of the function $(2/\pi)k_1 k_2 \sigma(k_1, k_2)$ **for the** $^1S_0$ **Reid and Tabakin potentials.** The minima ($-$) and maxima ($+$) are indicated. The maximum momentum 6 fm^{-1} corresponds to 2986 MeV in the lab system. [Taken from (Sau 73).]

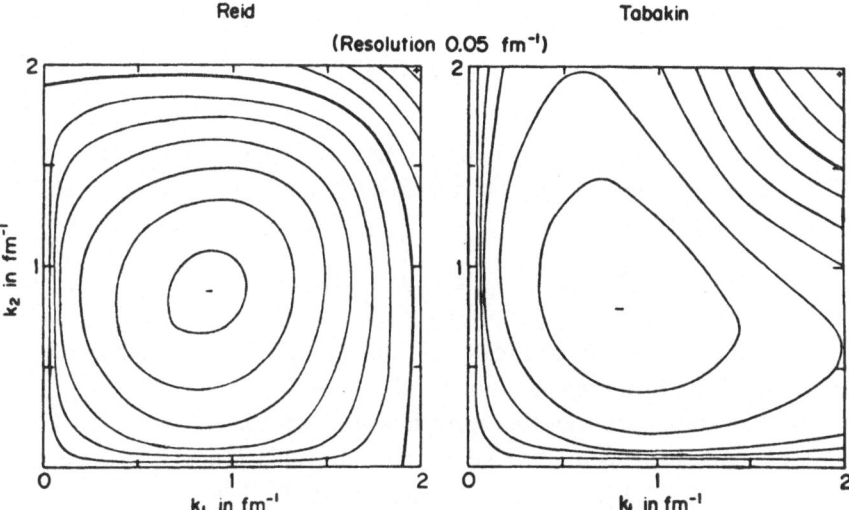

Fig. 2b. Same as Fig. 2a but on an expanded scale 0–2 fm^{-1}.

completely if $\sigma(k, k')$ is given for all k, k'. On-shell data give only $\sigma(k, k)$:

$$\sigma(k, k) = e^{-i\delta(k)}t(k, k; k^2) = -\{\sin \delta(k)\}/k \qquad (91)$$

Thus the arbitrariness in the continuation of t off the energy shell is isolated as an arbitrariness in the off-diagonal elements of the symmetric part of the t-matrix. This formalism has recently been extended to the impact parameter representation (Bha 73). Contour plots of the σ function for the Reid and Tabakin forces are illustrated in Fig. 2.

3. PHASE SHIFT EQUIVALENT POTENTIALS

In order to isolate effects due to locality versus nonlocality, it is desirable to construct several forces which are precisely equivalent on the energy shell. It will then be possible, at least in principle, to associate a particular feature in the off-shell behavior with the model assumed for the interaction. That infinitely many' such potentials exist was pointed out by Ekstein (Eks 60). Suppose that we have found one such potential V giving

$$H\Psi = (H_0 + V)\Psi = E\Psi \qquad (92)$$

Now if U is any unitary operator of finite range, i.e.,

$$U\Psi = \tilde{\Psi} \xrightarrow[r \to \infty]{} \Psi \qquad (93)$$

then with $\tilde{H} = UHU^\dagger$, we have

$$\tilde{H}\tilde{\Psi} = E\tilde{\Psi} \qquad (94)$$

Equations (93) and (94) show that $\tilde{\Psi}$ has the same phase shift as Ψ at all energies and the energy spectrum of $\tilde{H}$ is the same as that of H. The transformed potential

$$\tilde{V} = \tilde{H} - H_0 = U[H_0, U^\dagger] + UVU^\dagger \qquad (95)$$

is thus phase shift (or elastically or on-energy-shell) equivalent to the potential V. Its form depends on the choice of U.

The relations between the transition matrices corresponding to V and $\tilde{V}$ can be obtained from the two-potential formula (16), if we take V as V_1 and $\tilde{V} - V$ as V_2. These relations are (MST 71a)

$$\langle \mathbf{k}' | \tilde{T}(k^2) - T(k^2) | \mathbf{k} \rangle = (k^2 - k'^2)\langle \mathbf{k}' | (U - 1)\Omega(k^2 + i\varepsilon) | \mathbf{k} \rangle \qquad (96)$$

and

$$\langle \mathbf{k} | \tilde{T}(k^2) - T(k^2) | \mathbf{k}' \rangle = (k^2 - k'^2)\langle \mathbf{k} | \Omega^\dagger(k^2 - i\varepsilon)(U^\dagger - 1) | \mathbf{k}' \rangle \qquad (97)$$

for the half-shell T-matrix elements and

$$\tilde{\phi}(k, k') = \phi(k, k') + (k^2 - k'^2)\langle k' | (U - 1)W | k \rangle, \qquad U \text{ real} \qquad (98)$$

for the Baranger half-shell function. The difference between fully off-shell transition matrix elements is given by

$$\begin{aligned}
\langle \mathbf{p} | \tilde{T}(k^2) - T(k^2) | \mathbf{q} \rangle = {} & (k^2 - q^2)\langle \mathbf{p} | \Omega^\dagger(k^2 - i\varepsilon)(U^\dagger - 1) | \mathbf{q} \rangle \\
& + (k^2 - p^2)\langle \mathbf{p} | (U - 1)\Omega(k^2 + i\varepsilon) | \mathbf{q} \rangle \\
& + (k^2 - q^2)(k^2 - p^2) \\
& \times \langle \mathbf{p} | (U - 1)(k^2 - H + i\varepsilon)^{-1}(U^\dagger - 1) | \mathbf{q} \rangle
\end{aligned} \qquad (99)$$

The condition (93) ensures that

$$\lim_{k \to k'} (k^2 - k'^2)\langle \mathbf{k}' | (U - 1)\Omega(k^2 + i\varepsilon) | \mathbf{k} \rangle = 0 \qquad (100)$$

so that on-shell matrix elements remain unaltered.

A number of practical methods have been used to generate pairs or families of phase-shift-equivalent potentials. These are described below.

3.1. Isometric Point Transformation

Mittelstaedt and Ristig (MR 66), Ristig (Ris 67), Ristig and Kistler (RK 68), and Srivastava (Sri 70b) have used "isometric point transformations," which amount to a distortion of the radial scale $r \to y(r)$. This can be expressed in the above language of unitary transformations by a suitable choice of the operator U. The only requirements on the function $y(r)$ are that it be monotonic and approach r sufficiently fast at large distances. A special case of this type of transformation had been given earlier by Baker (Bak 62) to show that a potential with a hard core can be turned into a potential with no hard core but with an additional p^2- and l^2-dependent force that arises from the commutator in Eq. (95). To remove a hard core, one simply requires $y(c) = 0$. The possibilities are limitless. An example used by Kistler (Kis 69) and by Srivastava (Sri 70b, Sri 71) for a potential regular at the origin is

$$y(r) = r[1 + \varrho_0 e^{-\varrho_1 r}] \tag{101}$$

By choosing values of ϱ_1 large enough, say two to three times $m_\pi c/\hbar$, where m_π is the pion mass, one can ensure that the transformed potential remains short-ranged. The actual transformation is best written in terms of the inverse transformation $R(r) \to r$. If R is a smooth function such that

$$dR/dr = \mu^{-1/2}(r) > 0 \tag{102}$$

and $R - r$ vanishes at least as fast as $1/r$ for large r, the transformed wave function $\tilde{\psi}(r)$ is given by

$$\tilde{\psi}(r) = \mu^{-1/4}(r)\psi(R(r)) \tag{103}$$

For the partial wave Hamiltonian (Coe+ 70)

$$H_l = -\frac{1}{2}\left\{ w(R)\frac{d^2}{dR^2} + \frac{d^2}{dR^2}\,w(R) \right\} + \frac{l(l+1)}{R^2} + v_l(R) \tag{104}$$

the transformed Hamiltonian is

$$\tilde{H}_l = -\frac{1}{2}\left\{ \mu w\frac{d^2}{dr^2} + \frac{d^2}{dr^2}\,\mu w \right\} + \frac{1}{4}\,w\left\{ \frac{d^2\mu}{dr^2} + \frac{1}{4\mu}\left(\frac{d\mu}{dr}\right)^2 \right\}$$
$$+ \frac{l(l+1)}{R^2(r)} + v_l(R(r)) \tag{105}$$

3.2. Unitary Transformations

Coester *et al.* (Coe+ 70) have used for U short-range separable operators of the following general form:

$$\langle r' | (U - 1) | r \rangle = \sum_{i,j=1}^{N} g_i(r')(\lambda_{ij} - \delta_{ij})g_j(r) \qquad (106)$$

where the finite range functions $g_i(r)$ form an orthonormal set

$$\int_0^\infty g_i(r)g_j(r) \, dr = \delta_{ij} \qquad (107)$$

and the λ_{ij} are elements of a unitary matrix. For $N = 1$, Eq. (106) becomes

$$\langle r' | (U - 1) | r \rangle = -2g(r')g(r) \qquad (108)$$

For $N = 2$, it becomes the transformation of rank two described by Coester *et al.* (Coe+ 70), namely

$$\langle r' | (U - 1) | r \rangle = (g_1(r'), g_2(r'))\begin{pmatrix} \cos\theta - 1 & \sin\theta \\ -\sin\theta & \cos\theta - 1 \end{pmatrix}\begin{pmatrix} g_1(r) \\ g_2(r) \end{pmatrix} \qquad (109)$$

where $-\pi \leq \theta \leq \pi$. If, in addition, $g_1(r)$ and $g_2(r)$ are chosen such that the transforms

$$g_i(k) = (2/\pi)^{1/2} \int krj_l(kr)g_i(r) \, dr \qquad (110)$$

are easily obtained analytically, then the evaluation of the matrix elements

$$\langle k' | (U - 1)W | k \rangle = \int \langle k' | U - 1 | k'' \rangle \, dk'' \langle k'' | W | k \rangle \qquad (111)$$

required in Eq. (98) is simplified considerably.

By a suitable choice of U it is possible (i) to produce (MST 71b) any desired change in the bound state wave function ψ_b or to leave it unchanged, besides keeping the binding energy and the phase shifts unaltered, or (ii) to modify (MST 72) the energy eigenfunctions above some cutoff energy E_c together with (or without) any desired changes in the phase shifts. These features are sometimes desirable in analyzing the off-shell effects of the various changes in the interaction.

3.3. Fiedeldey's Procedure

Fiedeldey (Fie 69a), following the methods of Ghirardi and Rimini (GR 64) and Chadan (Cha 58, Cha 67), has shown how to construct a family of separable rank-two potentials by choosing rather arbitrarily one of the form factors and calculating the other so as to fit the on-shell data. These potentials have different bound state wave functions, although they are all asymptotically the same. The method can also be used to produce a potential with a prescribed bound state wave function.

3.4. Fuda's Method

Fuda (Fud 70) has suggested yet another method of constructing phase shift equivalent potentials including tensor forces. It is based on the procedure proposed earlier by Chadan (Cha 58, Cha 67) and on factorizing the Jost function. If the potential is taken as composed of parts

$$v_l = \sum_{i=1}^{n} v_l^{(i)} \tag{112}$$

and the Green's function $G_i(s) = \{s - H_0 - \sum_{j=1}^{i} v_l^{(j)}\}^{-1}$ is defined as the solution of

$$G_i(s) = G_{i-1}(s) + G_{i-1}(s)v_l^{(i)}G_i(s)$$
$$= G_{i-1}(s) + G_i(s)v_l^{(i)}G_{i-1}(s) \tag{113}$$

one can obtain

$$1 - G_0(s)v_l = \prod_{i=1}^{n} [1 - G_{i-1}(s)v_l^{(i)}] \tag{114}$$

leading to the following factorization of the Jost function:

$$f_l(-k) = \prod_{i=1}^{n} f_l^{(i)}(-k) \tag{115}$$

with

$$f_l^{(i)}(-k) = \det[1 - G_{i-1}(k^2)v_l^{(i)}] \tag{116}$$

Fuda takes $n = 2$. In this procedure one assumes, for any given on-shell data, a local potential* which may contain a hard core repulsion such

---

* This could also be a one-term separable potential. In that case the method becomes similar to that of Fiedeldey.

that the phase shifts $\delta_l^{(1)}$ produced by it satisfy the condition

$$\delta_l(k) - \delta_l^{(1)}(k) = \text{same sign for all } k$$

and then calculates a one-term separable potential which together with the assumed local potential produces the original phase shifts $\delta_l(k)$ and the prescribed bound state wave function (if any). The method is thus capable of producing phase shift equivalent mixtures of a local and a separable potential.

The advantage of the first two methods is that one can introduce short distance nonlocality into the potential while retaining such features as the long distance conformity to OPEP. On the other hand, the advantage of the Fiedeldey procedure is that all the potentials have the same type of nonlocality, and there is no obvious a priori reason to prefer one to another. All of these methods have the advantage that there are several adjustable parameters, and any number of forces can be generated, forming a family of phase shift equivalent potentials.

3.5. Method of Srivastava and Sprung

The authors (SS 70) have followed a different procedure. Starting with a rational S-matrix, we have analytically constructed pairs of phase shift equivalent potentials, one local and one separable. The potentials are constructed using the techniques of Gelfand and Levitan (GL 51), Marchenko (Mar 50), de Alfaro and Regge (AR 65), Omnes (Omn 61), Bolsterli and Mackenzie (BM 65), and Tabakin (Tab 69). The rational S-matrix gives a superposition of exponential terms in the potential of range determined by the singularities of the Jost function nearest the origin. Such potentials have been considered earlier by Bargmann (Bar 49), Jost and Kohn (JK 53), Newton and Jost (NJ 55), Newton (New 55), and Newton and Fulton (NF 57). This method has the disadvantage that only one pair of equivalent potentials results, but is of interest because the most realistic phenomenological forces have been local and one wants to know how different are the off-shell behaviors of equivalent local and separable forces.

By combining all these practical methods one can compare the off-shell behavior of a family of separable potentials, a family of mixtures of local and separable potentials, a local potential, and a family of p^2-dependent potentials, all being equivalent on the energy shell. Such comparisons, however, as pointed out earlier, are useful only in conjunction with many-body calculations so that one can analyze their manifestation in physical properties.

3.6. Coulomb–Nuclear Interference

Recently Sauer (Sau 74a, Sau 74b) has proposed using the principle of charge symmetry to constrain nuclear potential models in the $^1S_0$ state. In the past, it was assumed that Coulomb effects can be separated off in an essentially model-independent way. This belief was supported by plausible arguments as well as a large number of calculations [for example, (KS 69)]. What Sauer realized was that the unitary transform method allows one rather easily to generate a phase equivalent potential for which $\psi(r)$ at small $r \lesssim 0.5$ fm is not small, in contrast to almost all local potential models. For such transformed potentials, the Coulomb–nuclear interference can be very model dependent. From pp scattering data, the predicted nn scattering length would be very model dependent as well. Measurements of the nn scattering length, however, generally favor a value close to the old "model-independent" predictions. Sauer (Sau 74b) proposes to use this as a reason for rejecting the unitary transforms, which predict otherwise. As a working principle, one can require that the zero-energy $^1S_0$-state wave function is not altered by the unitary transform. This is quite analogous to demanding that the deuteron wave function be preserved. It may be possible, however, to settle for less; perhaps only the off-shell effective range need be conserved. Certainly this greatly restricts the freedom of choice of unitary transforms. In addition, Sauer has criticized the practice of making only the nuclear potential phase equivalent. In Eq. (95) one has usually used H_0 as the kinetic energy; properly it should also include the Coulomb potential. All of the past work on phase equivalent potentials is subject to revision on this account.

A rather nice illustration of nuclear–Coulomb interference has been pointed out by Kermode *et al.* (KNV 74); it is the Tabakin one-term separable potential (Tab 68) which changes the sign of the phase shift by introducing a pole of zero residue in the positive energy t-matrix. However, when the Coulomb force is added in, the balance is upset, the pole acquires a residue, and the phase shift is seen to increase by π over a finite width. Thus Tabakin's force could be useful for one but not both pp and nn scattering. This argument seems better than the principle of "polophobia" usually invoked against the Tabakin force (Bea 69, Bol 69).

4. AVOIDING POTENTIALS AS FAR AS POSSIBLE

In this section we discuss the extent to which one can avoid potentials altogether and still make meaningful calculations of nuclear processes. This question has been clarified by the work of Baranger et al. (Bar+ 69) and its further extension by Haftel (Haf 70), Amado (Ama 70), Van Dijk and Razavy (VR 70), Kowalski et al. (Kow+ 71), Sauer (Sau 71, Sau 73), and Kim and Tubis (KT 73a). It is still assumed that a potential exists, but one works directly with the on- and off-shell t-matrix elements, never actually calculating the potential. The question of whether the potential is local or nonlocal does not arise. Picker et al. (PRS 71, PRS 72, PRS 73), on the other hand, have exploited the lack of information about the short-range part of the interaction to obtain a simple procedure for off-shell extrapolations. A dispersion-theoretic approach has recently been proposed by Reiner (Rei 74) which allows the possibility of working directly from relativistic meson theory (PL 70, CDR 72, Cot+ 73) without the necessity of explicitly using a potential. This approach has the advantage that one is working from more fundamental principles and has no arbitrary parameters. In the following we discuss these methods.

4.1. Method of Baranger *et al.*

Baranger et al. (Bar+ 69), as pointed out in Section 2, confined the arbitrariness in the continuation of t off the energy shell (in the absence of any bound state) to an arbitrariness in the off-diagonal elements of the symmetric part of the t-matrix. Starting with an arbitrary symmetric function $\sigma(k, k')$ consistent with the phase shift data, one may follow this formalism to obtain the antisymmetric part $\alpha(k, k')$ and thereby $\phi(k, k')$ and a t-matrix [through Eq. (84)] which satisfies Eq. (12) for some (unspecified) potential. Since the diagonal elements of σ are directly related to the phase shifts, a correct fit to the experimental phase shifts is guaranteed by starting with σ instead of v to determine t. The variations in σ are generated by expressing it as

$$\sigma(k, k') = \sigma_M(k, k') + (k^2 - k'^2)\Xi(k, k') \tag{117}$$

and varying the antisymmetric function Ξ. Here σ_M is some initially chosen form of σ giving the correct phase shifts. It may well correspond to some preferred "model" or "realistic" potential. Sauer (Sau 73) has studied in detail the behavior of σ functions in the $^1S_0$ state for a wide class of changes in the interaction. He finds that (i) smooth off-shell extrapolations of σ keep the interaction short ranged, (ii) these extrapolations should pass over at low energies to a form determined almost uniquely by the effective range parameters of the on-shell data, and (iii) σ is almost independent of the shape of the interaction tail as long as the force is short ranged.

The off-shell variations could also be induced through orthogonal transformations U (Section 3) and use of Eq. (98). Monahan, Shakin, and Thaler (MST 71b) have given unitary transformations which leave bound state eigenfunctions either unchanged or change them in any preassigned manner for small distances of separation. In this way, by a proper choice of U the conditions initially imposed on the model potential v_m can also be satisfied. The implied potential need not be calculated in any application. The method of Sauer, though advantageous in the sense that changes are inserted in a quantity σ which is a convenient starting point for calculations, has the drawback that these influence T in a nonlinear way, and σ itself is a quantity for which we do not have, as yet, much physical intuition.

If a bound state is present in any partial wave, the real matrix W [Eq. (78)] is not unitary, i.e., Eqs. (89) and (90) are not valid and the above prescription breaks down. In this case one can choose a model potential v_m (Haf 70) which has a specified bound state energy and eigenfunction, and construct a model real matrix W_m which is unitary. The procedure of Baranger *et al.* can now be followed, the only difference being that the phase shift difference $\delta - \delta_m$ (where δ_m is the phase shift for the model potential) plays the role previously played by the phase shift δ. Sauer (Sau 71) has extended this method to include the case of a spin-dependent noncentral interaction and has proposed alternative methods for constructing and calculating W. Kim and Tubis (KT 73a) have generalized the procedure to incorporate the constraints due to singular core interactions. The problem of incorporating the Coulomb interaction is considered by Sauer and Walliser (SW 74).

In order to apply Haftel's method, one must begin by introducing the bound state wave function $| \psi_b \rangle$, for this is additional information not contained in the phase shifts. The problem is then to construct the basis of scattering states orthogonal to $| \psi_b \rangle$; the simplest way to generate these is by introducing the one-term separable potential whose form factor is $| \psi_b \rangle$, namely $v_m = \text{const} \times | \psi_b \rangle \langle \psi_b |$. Kowalski *et al.* (Kow+ 71) have given more complicated prescriptions for the choice of v_m, which among other things avoid introducing the bound state eigenvalue w_b until one wishes to calculate fully off-shell elements. .

4.2. Method of Picker *et al.*

Picker, Redish, and Stephenson (PRS 71, PRS 72, PRS 73) have given another simple and rather direct procedure for devising models of on-shell equivalent half-shell K-matrix elements. It is based on Eq. (57), which expresses a half-shell element in terms of the on-shell element plus a term depending on the Fourier–Bessel transform of the difference between the actual wave function and its asymptotic form. They exploit the lack of information about the short-range part of the interaction by varying the wave function in that region rather arbitrarily subject to the constraints of the boundary condition at the origin, smoothness, and the matching to the wave function generated by the known external part of the interaction and having the given asymptotic form. A drawback of the method is that the modified wave functions do not form a complete orthogonal set and one can not go completely off-shell. His matrix elements do not satisfy Eqs. (12), (13). This can be circumvented by using the method to generate models of the symmetric part of the half-shell t-matrix elements. Alternatively, the changes in the wave function for small r could be induced by short-range unitary transformations discussed earlier.

To start with, one could also use the Gelfand–Levitan (GL 51) or Marchenko (Mar 50) inverse scattering formalism to generate a complete orthonormal set of scattering wave functions from the phase shifts and directly calculate half-shell reaction matrix elements (Sob 68, Und 70, Kar 72). Other sets can then be generated by using unitary transformations. This merger of the unitary transformations with the inverse scattering theory thus eliminates the potential and provides a workable framework for analyzing the off-shell behavior of the interaction (PL 72).

4.3. Dispersion-Theoretic Approach

This formalism assumes that the half-shell t-matrix element $t(p, k; k^2)$, for fixed p, can be analytically continued throughout the complex energy plane (k^2 plane) and a dispersion relation can be written for it. Reiner (Rei 74) writes the following dispersion relation subtracted at the on-shell point:

$$t(p, k; k^2) = t(p) + \frac{k^2 - p^2}{\pi} \int_0^\infty dq^2 \frac{\text{Im } t(p, q; q^2)}{(q^2 - p^2)(q^2 - k^2)}$$

$$- (k^2 - p^2) \sum_b \frac{r(p, k_b)}{(k_b^2 - p^2)(k_b^2 - k^2)}$$

$$+ \frac{k^2 - p^2}{2\pi i} \int_C dq^2 \frac{\text{disc } t(p, q; q^2)}{(q^2 - p^2)(q^2 - k^2)} \tag{118}$$

The first term on the right-hand side is the on-shell t-matrix; the second term is the contribution from the unitarity cut, whose discontinuity is given by elastic off-shell unitarity; the third term is the contribution from possible poles due to bound states at $k_b{}^2$ with residue $r(p, k_b)$; and the fourth term is the contribution from cuts other than the unitarity cut, where C is the appropriate contour along these cuts. Using the off-shell unitarity, Eq. (65), this dispersion relation becomes an integral equation of the Muskhelishvili–Omnes type for the half-shell t-matrix element, which can be solved directly. The only inputs are the phase shifts (directly from experiment), bound state form factors, and the discontinuities across the dynamical cuts to be determined from the one- and two-boson exchange amplitudes (obtainable from the relativistic meson theory in terms of the coupling constants and the masses of the exchanged bosons). In actual practice one need consider only nearest poles and cuts. Fully off-shell matrix elements could be obtained from a further dispersion relation subtracted at a half-shell point. In general, additional singularities will occur, generated by the energy dependence of the underlying meson theory. These singularities must be taken into account.

Reiner has tested this method with Yukawa and Reid potentials in the $^1S_0$ state. He finds it very good up to fairly high values of p and k. Actual determination of the t-matrix directly from the relativistic meson theory by this method has not yet been done.

5. SEPARABLE APPROXIMATIONS TO THE T-MATRIX

A separable t-matrix can be written as the product of functions of the initial and final momenta

$$t^S(p, q; s) = g(p, s)g(q, s) \tag{119}$$

or more generally

$$t^S(p, q; s) = \sum_{\nu=1}^{N} g_\nu(p, s)g_\nu(q, s) \tag{120}$$

If the two-body potential is itself separable, then the solution of the Lippmann–Schwinger equation gives a separable t-matrix of the above form, with the additional feature that the function $g(p, s)$ can be written as the product of a function of p and a different function of s.

Even for a general t, one might approximate it by a sum of terms as on the right-hand side of Eq. (120). By taking N large enough, t can be very closely approximated,* although large N destroys the intent of simplifying calculations. In either case, it is wise to choose separable forms $g(p, s)$ which are easily integrated. This requirement is dictated by the awkward complexity of Faddeev-type calculations for the three-body problem.

Before proceeding further with this, let us be clear about what it is we are approximating and what should be called a good approximation. Since the "actual" off-shell behavior is not known, one might very well say that any off-shell behavior whatsoever is alright. Only a careful consideration of experiments involving three or more particles can help in discovering the true off-shell behavior. But such a study, as pointed out earlier, has not yet been performed. Separable approximations offer a way out of this impasse (LLS 69). Suppose that t-matrices obtained from a local potential and a separable potential agree well enough in some region of p, q, and s. It is then plausible to assume that all extrapolation procedures from on-shell to off-shell will agree well enough in that region, since locality and separability represent extreme assumptions. Then this separable t-matrix will be close enough to the "true" t-matrix in that region. This approach is limited in scope for two reasons. (i) This agreement could be achieved only in a restricted region (which, of course, could be varied) of p, q, and s space, and (ii) the local potentials among themselves give rise to a reasonably wide variation in the overall off-shell behavior. Nevertheless these approximations are useful because of the possibilities provided by them for exploration in nuclear physics calculations. We should also have, at least in principle, some criterion for judging the agreement between the separable t-matrices and those from a local potential. There is at present an appreciable experimental error in the determination of the t-matrix on the energy shell (SH 71). For example, the singlet effective range, $^3D_1$ percentage in the deuteron, and the $^1P_1$ part of the interaction are poorly known (Sig 69a, BB 74). An off-shell extrapolation error, small compared to the on-shell error, can be neglected.

In the following we discuss various separable approximations.

---

* It has been pointed out (Osb 73a) that no separable expansion of the t-matrix for a local or a partly local potential can converge, i.e., the mean square deviation between the actual t-matrix and its separable approximation, for any finite value of N, is infinite. But these difficulties have been shown to arise only from large values of momenta and are therefore not a cause for concern in any practical situation (SG 73, Lev 73).

5.1. Weinberg's Expansion

Weinberg (Wei 63) was the first to suggest expanding the t-matrix in terms of the eigenfunctions of the kernel of the Lippmann–Schwinger equation. These eigenfunctions and the related expansions have been studied by several workers (Mee 61, Rot 62, Tan 66a, Tan 66b, Tan 68). In particular, Ball and Wong (BW 68) have used such an expansion to calculate the low-energy properties of a system of three identical spinless particles interacting via a Yukawa potential. The expansion can be used even if the potential contains a hard core (Fud 68a).

Let $|\psi_n(s)\rangle$ be the eigenvector (a partial wave subscript l has been omitted) of the kernel $vG_0(s)$ [Eq. (4)] of the Lippmann–Schwinger equation belonging to the eigenvalue $\lambda_n^{-1}(s)$

$$vG_0(s) \mid \psi_n(s)\rangle = \lambda_n^{-1}(s) \mid \psi_n(s)\rangle \qquad (121)$$

For negative values of s, $\mid \psi_n(s)\rangle$ forms a complete orthonormal set, with normalization (BW 68)

$$\langle \psi_m(s) \mid G_0(s) \mid \psi_n(s)\rangle = -\delta_{mn} \qquad (122)$$

If we introduce a vector $\mid \xi_n(s)\rangle$ defined by

$$\mid \xi_n(s)\rangle = G_0(s) \mid \psi_n(s)\rangle \qquad (123)$$

or alternatively by

$$\mid \psi_n(s)\rangle = \lambda_n(s)v \mid \xi_n(s)\rangle \qquad (124)$$

then the homogeneous Lippmann–Schwinger equation (121) reduces to the Schrödinger-like form

$$\{H_0 + \lambda_n(s)v\} \mid \xi_n(s)\rangle = s \mid \xi_n(s)\rangle \qquad (125)$$

Equation (125) is not the eigenvalue equation for the Hamiltonian $H_0 + \lambda v$ in the usual sense but one in which λ takes discrete eigenvalues for a definite s.

Since $\{\mid \psi_n(s)\rangle\}$ constitute a complete set, the t-matrix can be expanded in the form

$$t(s) = \sum_n \zeta_n(s) \mid \psi_n(s)\rangle\langle\psi_n(s) \mid \qquad (126)$$

Substituting this in the Lippmann–Schwinger equation

$$t(s) = v + vG_0(s)t(s) \qquad (127)$$

and using Eqs. (121) and (122) gives the Weinberg expansion

$$t^{W}(s) = \sum_{n} \frac{1}{1 - \lambda_{n}(s)} \mid \psi_{n}(s) \rangle \langle \psi_{n}(s) \mid \tag{128}$$

This form really corresponds to the expansion

$$v = - \sum_{n} \frac{\mid \psi_{n}(s) \rangle \langle \psi_{n}(s) \mid}{\lambda_{n}(s)} \tag{129}$$

of the potential in terms of the energy-dependent orthogonal set $\{\mid \psi_{n}(s) \rangle\}$.

The convergence of the series depends on the nature of the potential. For example, a square well potential gives a rapid convergence compared to the Hulthen or the Yukawa shape (LLS 69). At positive energies, the Weinberg expansion has the disadvantage that it does not satisfy the off-shell unitarity relation when cut off at a finite N.

5.2. Unitary Pole Expansion

A separable expansion which does satisfy off-shell unitarity has been given by Harms (Har 70). Suppose in Eqs. (121)–(125) we fix s at some negative value $-B$ so that they become

$$vG_{0}(-B) \mid \psi_{n}(-B) \rangle = \lambda_{n}^{-1}(-B) \mid \psi_{n}(-B) \rangle \tag{130}$$

$$\langle \psi_{m}(-B) \mid G_{0}(-B) \mid \psi_{n}(-B) \rangle = -\delta_{mn} \tag{131}$$

$$\mid \xi_{n}(-B) \rangle = G_{0}(-B) \mid \psi_{n}(-B) \rangle \tag{132}$$

$$\mid \psi_{n}(-B) \rangle = \lambda_{n}(-B)v \mid \xi_{n}(-B) \rangle \tag{133}$$

$$[H_{0} + \lambda_{n}(-B)v] \mid \xi_{n}(-B) \rangle = -B \mid \xi_{n}(-B) \rangle \tag{134}$$

respectively. The separable expansion (129) now takes the form

$$v = - \sum_{n} \frac{\mid \psi_{n}(-B) \rangle \langle \psi_{n}(-B) \mid}{\lambda_{n}(-B)} \tag{135}$$

If we insert Eq. (135) (truncated at some finite N) into the Lippmann–Schwinger equation (127), the resulting t-matrix is given by

$$t^{UPE}(s) = \sum_{n,m}^{N} \mid \psi_{n}(-B) \rangle \Delta_{n,m}(s) \langle \psi_{m}(-B) \mid \tag{136}$$

where

$$-[\Delta(s)^{-1}]_{n,m} = \lambda_{n}\delta_{n,m} + \langle \psi_{n}(-B) \mid G_{0}(s) \mid \psi_{m}(-B) \rangle \tag{137}$$

This is the unitary pole expansion (UPE) of the t-matrix given by Harms.

The expansion (128) is simpler than (136), but is obtained at the expense of having to solve Eq. (121) for each desired s. Since in the three-body problem t enters in the form $t(s - \frac{3}{4}p^2)$ with $0 \le p^2 \le \infty$, use of the Weinberg series requires the solution of Eq. (121) many times, with the subsequent increased difficulty in computation. One of the greatest advantages of UPE is that this problem is removed. We solve for the form factors $| \psi_n(-B)\rangle$ once, and then the t-matrix is obtained by evaluating the integrals in Δ, a much easier task. If a hard core is present in the potential, one can solve Eq. (134) in the coordinate representation (MLM 72) for the eigenvalues $\lambda_n(-B)$ and eigenfunctions $\langle \mathbf{r} \mid \xi_n(-B)\rangle$ and then go over to the momentum representation to determine the form factors $\langle \mathbf{p} \mid \psi_n(-B)\rangle$.

Another advantage of this expression is that it is always unitary [no matter where the expansion (135) is truncated]. We discuss this below in relation to the unitary pole approximation (UPA) of Lovelace (Lov 64) and Fuda (Fud 68b).

5.3. Unitary Pole Approximation

Let us assume for the moment that there is a two-body bound state and take B to be the two-body binding energy. The equation (134) for $|\xi_n(-B)\rangle$ is then just the bound state Schrödinger equation with the potential $\lambda_n(-B)v$. Thus $| \xi_n(-B)\rangle$ may be interpreted as bound states of energy $-B$ of the potentials $\lambda_n(-B)v$. However, by our assumption, v has a bound state, $| B\rangle$, and hence we may choose $\lambda_1(-B)$ equal to one and $| \xi_1(-B)\rangle \propto | B\rangle$. If we retain only one term in Eq. (136), we find

$$t^{\mathrm{UPA}}(s) = | \psi_1\rangle\Delta(s)\langle\psi_1 | \tag{138}$$

$$\Delta(s) = -[1 + \langle\psi_1 | G_0(s) | \psi_1\rangle]^{-1} \tag{139}$$

This is the unitary pole approximation. From Eqs. (131), (138), and (139) we see that the UPA t-matrix has a pole at the two-body bound state energy, $s = -B$, corresponding to the bound state pole in the actual t-matrix. The residues of the exact and UPA t-matrices at $s = -B$ are also the same and hence, in the neighborhood of the pole, the exact and UPA t-matrices agree.

We see from Eq. (131) that retaining N terms in the expansion (135) does not change the bound state energy or wave function for the separable potential. As a result, the N-term UPE t-matrix has the same pole position as the UPA t-matrix and the same residue as the actual t-matrix. However, since retaining additional terms in Eq. (135) gives a better approximation to the potential, the N-term t-matrix gives a better approximation to the actual t-matrix. For any number of separable terms, Eq. (135) gives a separable potential which is real and symmetric. The UPE t-matrix will therefore always be unitary. It thus gives a systematic method of improving upon the UPA, while still retaining the proper pole behavior and satisfying unitarity.

When there is no two-body bound state, we may choose B to have any convenient value. If there is an antibound state near $s = 0$ (as, for example, in the singlet-S two-nucleon system), B may be taken as zero. For partial waves without poles near the physical region, the expansion method is still applicable. However, the pole dominance aspect of the expansion is then absent, and more terms would probably be required.

5.4. Approximation of Bhatia and Walker

Bhatia and Walker (BW 72) have extended the one-pole approximation to take into account the effect of the presence of additional bound state and resonance poles. Instead of introducing additional separable terms, they construct an energy-dependent state $|Z(E)\rangle$ by superposing the eigenstates $|\psi_i\rangle$ corresponding to the various bound states and resonances of the Hamiltonian H

$$| Z(E)\rangle = \sum_{i=1}^{n} w_i \alpha_i(E) | \psi_i\rangle \tag{140}$$

where n is the number of poles whose effect one would like to include in the expansion, w_i are free parameters to be determined by comparison with the exact t-matrix, and the energy-dependent coefficients α_i are such that $| Z(E_i)\rangle$ at any bound state or resonance pole E_i is identical to $| \psi_i\rangle$. The one-term separable potential is then expressed in the form

$$v^{\mathrm{BW}} = \lambda v \, | Z(E)\rangle\langle \bar{Z}(E) | \, v \tag{141}$$

with

$$\lambda^{-1} = \langle \bar{Z}(E) | \, v \, | Z(E)\rangle \tag{142}$$

giving the separable t-matrix

$$t^{\mathrm{BW}}(E) = \frac{v \, | Z(E)\rangle\langle \bar{Z}(E) | \, v}{\lambda^{-1} - \langle \bar{Z}(E) | \, v G_0(E) v \, | Z(E)\rangle} \tag{143}$$

Here $G_0(E)$ is the free particle Green's function and the state $\langle \bar{Z}(E) |$ is a linear combination of the solutions to H with Hermitian conjugate boundary conditions. This t-matrix satisfies unitarity and guarantees by construction the correct position of and residues at the various poles.

5.5. Separable Approximation of Ernst *et al.*

Ernst, Shakin, and Thaler (EST 73) have recently proposed a rank-N separable approximation which has the property that the resulting t-matrix is exact on-shell and half-shell at N chosen bound state and/or continuum energies. They propose, in any partial wave, a separable potential of the form

$$v^{\text{EST}} = \sum_{i,j}^{N} v \mid \psi_i \rangle \langle \psi_i \mid M \mid \psi_j \rangle \langle \psi_j \mid v \tag{144}$$

where v is the given local potential, $| \psi_i \rangle$ stands for its eigenfunctions (scattering states $| \psi^{(+)}_{k_i, k_i^2} \rangle$ and/or bound states $| \psi_{b_i} \rangle$) at N predetermined energies, and the $N \times N$ matrix M is defined by the relation

$$\delta_{im} = \sum_{j}^{N} \langle \psi_i \mid M \mid \psi_j \rangle \langle \psi_j \mid v \mid \psi_m \rangle = \sum_{j}^{N} \langle \psi_i \mid v \mid \psi_j \rangle \langle \psi_j \mid M \mid \psi_m \rangle \tag{145}$$

At the energy $E_n = k_n^2$ ($1 \leq n \leq N$), $| \psi_n \rangle$ is an eigenstate of both $H = H_0 + v$ and $H = H_0 + v^{\text{EST}}$, since

$$v^{\text{EST}} \mid \psi_n \rangle = v \mid \psi_n \rangle \tag{146}$$

and

$$\langle \psi_n \mid v^{\text{EST}} = \langle \psi_n \mid v \tag{147}$$

The two half-shell t-matrices $t(p, k_n; k_n^2)$ and $t^{\text{EST}}(p, k_n; k_n^2)$ are thus equal. Diagonalizing the matrix M by means of a unitary transformation U, the interaction [Eq. (144)] can be written as

$$v^{\text{EST}} = \sum_{i}^{N} v \mid \tilde{\psi}_i \rangle \langle \tilde{\psi}_i \mid M \mid \tilde{\psi}_i \rangle \langle \tilde{\psi}_i \mid v = \sum_{i}^{N} \mid \tilde{v}_i \rangle \lambda_i \langle \tilde{v}_i \mid \tag{148}$$

where

$$\mid \tilde{\psi}_i \rangle = \sum_{j}^{N} U_{ij} \mid \psi_j \rangle \tag{149}$$

$$\lambda_i = \langle \tilde{\psi}_i \mid M \mid \tilde{\psi}_i \rangle \tag{150}$$

and

$$\tilde{v}_i = v \mid \tilde{\psi}_i \rangle \tag{151}$$

The scattering state wave function and the half-shell t-matrix for the separable interaction [Eq. (148)] can be easily obtained and are given by

$$| \psi_{k,k^2}^{\mathrm{EST}(+)} \rangle = | k \rangle + G_0^{(+)}(k^2) \sum_{i,j}^{N} | \tilde{v}_i \rangle \Gamma_{ij}(k^2) \langle \tilde{v}_j | k \rangle \qquad (152)$$

and

$$t^{\mathrm{EST}}(p, k; k^2) = \sum_{i,j}^{N} \langle p | \tilde{v}_i \rangle \Gamma_{ij}(k^2) \langle \tilde{v}_j | k \rangle \qquad (153)$$

where $\Gamma_{ij}(k^2)$ is defined by

$$\sum_{j}^{N} \Gamma_{ij}(k^2) \langle \psi_j | v - v G_0^{(+)}(k^2) v | \psi_m \rangle = \delta_{im} \qquad (154)$$

If k^2 is equal to E_n ($1 \leq n \leq N$), ψ^{EST} and the half-shell t-matrix elements above are respectively equal to ψ and t for the potential v. In the case where v^{EST} is taken to be of rank one, Eqs. (152)–(154) reduce to

$$| \psi_{k,k^2}^{\mathrm{EST}(+)} \rangle = | k \rangle + \frac{G_0^{(+)}(k^2) v | \psi_{k_1,k_1^2}^{(+)} \rangle \langle \psi_{k_1,k_1^2}^{(+)} | v | k \rangle}{\langle \psi_{k_1,k_1^2}^{(+)} | v - v G_0^{(+)}(k^2) v | \psi_{k_1,k_1^2}^{(+)} \rangle} \qquad (155)$$

and

$$t(p, k; k^2) = \frac{\langle p | v | \psi_{k_1,k_1^2}^{(+)} \rangle \langle \psi_{k_1,k_1^2}^{(+)} | v | k \rangle}{\langle \psi_{k_1,k_1^2}^{(+)} | v - v G_0^{(+)}(k^2) v | \psi_{k_1,k_1^2}^{(+)} \rangle} \qquad (156)$$

If the potential supports only one bound state and E_1 ($= k_1^2$) is taken as $-B$, Eq. (156) reduces to the UPA.

This separable representation is found to correctly reproduce the fully off-shell t-matrix in the vicinity of the N predetermined energies (EST 73). If the set of the function ψ_i includes the off-shell scattering state functions $| \psi_{k_i,s_i}^{(+)} \rangle$, the resulting separable representation correctly reproduces the off-shell t-matrix in the neighborhood of these specified on-shell/off-shell points k_i and s_i (EST 74). Its accuracy has been studied by Ernst et $al.$ (Ern+ 73) for S-wave scattering from a square well potential. They constructed a rank-two separable potential taking the energies E_1 and E_2 (at which the separable potential and the square well have identical wave functions) as -2.225 MeV (deuteron bound state energy) and 250 MeV. They found that this approximation fits the phase shifts quite well below 300 MeV. It also fits the off-shell t-matrix over the entire range where the t-matrix itself is not small. Pieper (Pie 74) has used this method in the $^3S_1$–$^3D_1$ channel and found it to be better than any other available separable potential or the UPE (Har 70, AR 73) of the same rank.

5.6. Kowalski–Noyes Approximation

Another separable approximation which satisfies the off-shell unitarity relation and contains the entire resonance content of the exact t has been proposed by Kowalski (Kow 65) and Noyes (Noy 65). They show that the t-matrix element $t(p, q; k^2)$ can be written as

$$t(p, q; k^2) = \frac{t(p, k; k^2)t(q, k; k^2)}{t(k, k; k^2)} + \frac{\pi}{2} \frac{(k^2 - q^2)}{q^2} R(p, q; k^2) \qquad (157)$$

where $R(p, q; k^2)$ satisfies the integral equation

$$R(p, q; k^2) = \Lambda(p, q; k^2) + \int_0^\infty R(p, p'; k^2)\Lambda(p', q, k^2)\, dp'$$

$$= \Lambda(p, q; k^2) + \int_0^\infty \Lambda(p, p'; k^2)R(p', q; k^2)\, dp' \qquad (158)$$

with

$$\Lambda(p, q; k^2) = \frac{2}{\pi} \frac{q^2}{k^2 - q^2} \left[v(p, q) - \frac{v(p, k)v(k, q)}{v(k, k)} \right] \qquad (159)$$

and

$$\frac{q^2 - k^2}{q^2} R(p, q; k^2) = \frac{p^2 - k^2}{p^2} R(q, p; k^2) \qquad (160)$$

The relation (157) is valid for all p, q, and k^2 and the first term is separable in p and q. The nonseparable residue is completely determined by the kernel $\Lambda(p, q; k^2)$. The Kowalski–Noyes (KN) approximation consists of neglecting the nonseparable term,

$$t^{\text{KN}}(p, q; k^2) = \frac{t(p, k; k^2)t(q, k; k^2)}{t(k, k; k^2)} \qquad (161)$$

$$= g^{\text{KN}}(p, k)g^{\text{KN}}(q, k) \qquad (162)$$

where

$$g^{\text{KN}}(p, k) = t(p, k; k^2)/[t(k, k; k^2)]^{1/2} \qquad (163)$$

From the definition of $\Lambda(p, q; k^2)$ it follows that the half-shell $\Lambda(p, k; k^2)$ is identically zero and so, consequently, is $R(p, k; k^2)$. The approximation is thus exact on-shell and half-shell but has a serious drawback. It has an unphysical singularity when used in the neighborhood of a zero of the on-shell t-matrix. Osborn (Osb 69) has suggested a procedure to remedy this problem. He expresses the nonseparable residual term in Eq. (157) in a Weinberg-type expansion involving the eigenfunctions of the kernel $\Lambda(k^2)$. These additional separable terms then cancel the unphysical poles of the first term in Eq. (157). The number of additional separable terms needed is just the number of distinct zeros of the on-shell t-matrix. However, handling this cancellation of singularities in a practical numerical calculation could be awkward. An alternative way to avoid this difficulty has been suggested by Fuda (Fud 69b) for the cases when the vanishing of the on-shell t-matrix is brought about by the presence of a hard core in the potential. He suggests using one expansion for the part of the total t-matrix arising from a pure hard core potential and another expansion for the rest of the total t-matrix. Yet another method to avoid this difficulty of unphysical singularities has recently been given by Kowalski (Kow 72). By generalizing the Sasakawa theory of scattering (Sas 63), he has obtained a third-rank separable approximation to the fully off-shell t-matrix. This approximation has all the good attributes of the Kowalski–Noyes approximation and is free of singularities no matter how many values of k exist for which $t(k, k; k^2) = 0$. This is in contrast to the procedures suggested above by Osborn (Osb 69) and Fuda (Fud 69b), which require an additional separable term for each zero of $t(k, k; k^2)$. See, however, Picker and Stephenson (PS 75), who criticize the method for distorting the pole contributions to the fully off-shell T-matrix.

5.7. Other Approximations

Wong and Zambotti (WZ 67) write the t-matrix element in the form

$$t(p, q; s) = N(p, q; s)/D(s) \tag{164}$$

where $D(s)$ contains all the zeros corresponding to bound state poles and has a branch cut for positive s and $N(p, q; s)$ has no singularity in the region p^2 and q^2 positive and s negative. For negative p^2 and q^2, N will in general have singularities. They exploit this simple analytic structure and write N as a sum of separable terms

$$N(p, q; s) = \sum_{n,m=0} C_{nm}(s)\left(\frac{p^2}{p^2 + \mu^2}\right)^n\left(\frac{q^2}{q^2 + \mu^2}\right)^m \{(p^2 + \mu^2)(q^2 + \mu^2)\}^{-1} \tag{165}$$

For a given s and a fixed number of terms in Eq. (165), the parameters C_{nm} and μ^2 can be chosen to optimize the fit of $N(p, q; s)$ for the relevant values of p and q. For low energies a two-parameter formula obtained by taking only one term in the above is found to be quite good.

Other types of separable approximation can be found in the papers of Guennéguès (Gue 66), Purkayastha, Banerjee, and Sil (PBS 71), and Oryu, Ishihara, and Shioyama (OIS 74).

5.8. Comparison of Various Approximations

Reiner (Rei 67) has compared the Guennéguès, Lovelace, and KN approximations for a central attractive local square well potential. He finds the KN approximation an order of magnitude better than others in the region $0 \leq p$, $q \leq 1.5$ fm^{-1} and $0 \geq s \geq -30$ MeV. Another calculation (HHO 68) using the KN approximation shows 10% error in ground state binding energy in the n–p triplet interaction. The authors (SS 70) find it quite accurate in the case of $^1S_0$-state local potentials phase-shift equivalent to separable ones of Mongan (Mon 68, Mon 69a) and Tabakin (Tab 64) over a considerable range. It is found to be excellent when $s \leq 0.25$ fm^{-2} (20 MeV lab). A possible reason for this is the existence of the $^1S_0$-anti-bound state, which, along with the repulsion setting in at 240 MeV, dominates the structure of the matrix elements at low energy.

The unitary pole approximation has been investigated for the square well potential by Levinger, Lu, and Stagat (LLS 69), for the Hulthen potential by Kok, Erens, and Van Wageningen (KEW 68), and by Harms (Har 70), Harms and Levinger (HL 69), Harms and Laroze (HL 71), Harms and Newton (HN 70), and Brady et al. (Bra+ 69) for a number of examples. Levinger, Lu, and Stagat have compared the UPA and other separable approximations with the exact t-matrix for a square well potential deep enough to have a single, weakly bound state. They find that the UPA gives a good approximation both for phase shifts at modest energies ($\sim$40 MeV lab) and for the off-shell t-matrix in the range of momenta and negative energies relevant for the calculation of the trinucleon binding energy. The one-term Weinberg and the KN approximations are of similar or slightly inferior accuracy. In the coupled $^3S_1$–$^3D_1$ channel, the UPA is found (SLH 72) to be less successful for t_{02}, t_{20}, and t_{22}. At positive energies Levinger and O'Donoghue (LO 73) have investigated it for the Reid soft core singlet potential and the Tabakin two-term separable potential. They find it very accurate for momenta up to about 1.5 fm^{-1} and energies up to about 3 fm^{-2}. The quality of the fit is found (SL 74) not to depend strongly on features such as locality or presence of spin–orbit forces in the potential.

Srivastava and Sirohi (SS 73b) have recently investigated the effect of the strength of the repulsion in the interaction on the accuracy of UPA. Square well potentials are taken and the height and the range of the repulsive part are varied keeping the two-body binding energy fixed. It is found, in agreement with the findings of Kok, Erens, and Van Wageningen (KEW 68), Harms and Levinger (HL 69), and Harms (Har 70), that the UPA is best in the case of potentials which contain some repulsion whose strength depends on the overall nature of the interaction. The accuracy can, of course, be improved by more terms of the UPE. Harms has reported excellent agreement between the three-term UPE and the exact t-matrices.

The UPE has also been used to calculate the properties of the three-nucleon system. The three-term UPE gives reasonable results for the trinucleon binding energy and the n–d doublet scattering length (Har 70, BLH 72, AR 73, AB 74, Har 74). The quality of the trinucleon wave function has also been looked into by calculating the form factor for electron–trinucleon elastic scattering and the two-body correlation function in the trinucleon (HHN 72). It is found to be in good agreement with the form factor calculations of Tjon, Gibson, and O'Connell (TGC 70).

The energy-dependent separable approximation of Bhatia and Walker has been tested for the case of an attractive square well potential in the S-wave. It is found (BW 72) that the one-pole separable potential (UPA) is a good approximation to the local potential when the bound state or antibound state is near the physical region. The closer it is, the better is the approximation. Fortunately, this is the situation in nuclear physics. When this pole is rather far, the one-pole approximation is poor even at low energies. The inclusion of other poles is found to improve the agreement substantially.

The success of the unitary pole approximation even for a limited range of p, q, and s emphasizes the importance of the bound state (antibound state) pole. This suggests (Fud 70) that a two-term separable potential model which reproduces not only the phase shifts and the two-body binding energy but also the bound state wave function of some realistic potential should be a very good choice. Such a potential could be constructed following the methods of Fiedeldey (Section 3.3). If a further improvement is desired, the method of Ernst, Shakin, and Thaler (EST 73, EST 74) could be used to construct a rank-three or rank-four separable potential.

6. APPROXIMATE METHODS OF AVOIDING POTENTIALS

There are many approximations which attempt to use only the on-shell variation, ignoring or at any rate evading the off-shell variation of matrix elements. Elliott, Mavromatis, and Sanderson (EMS 67), Elliott *et al.* (Ell+ 68), and Koltun (Kol 67), for example, have proposed to calculate the reduced matrix elements needed for the harmonic oscillator shell model. Similar methods have been considered by Galonska, Faessler, and Appel (GFA 70), Bollini and Giambiagi (BG 69), and Koo *et al.* (Koo+69). Srivastava, Jopko, and Sprung (SJS 69) extended the approximation of Elliott, Mavromatis, and Sanderson to all partial waves, studied its numerical effectiveness, and made some applications. Elliott and Jackson (EJ 68) used these matrix elements to calculate properties of the deuteron and Dey *et al.* (Dey+ 69) to calculate the first-order binding energies and spectra of several nuclei. These methods were further improved by Ripa and Maqueda (RM 71). Mavromatis and Singh (MS 69a) have calculated matrix elements in a plane wave basis by expressing the spherical Bessel function in terms of a finite number of oscillator wave functions and using the technique of Elliott *et al.* The idea that only on-shell information might be adequate to describe a dilute, many-body system is an old one. It was among the motivations of the Brueckner theory. Beg (Beg 61) clarified the problem somewhat by considering a model where a third particle scatters from a pair of fixed particles a distance $2a$ apart. Supposing that the nuclear force has a finite range $r_n < a$, whenever the wave function for the third particle overlaps one of the scattering centers, it has already taken on its asymptotic form relative to the other. This is described by the phase shift. So this simple problem is soluble in terms of on-shell information when $r_n < a$. Off-shell information becomes important because (i) the nuclear force has a tail and (ii) the particles in a many-body system move around and sometimes are so close together that on any reasonable basis $r_n > a$. In some cases, such as for nuclear valence states, it may be a good first approximation to suppose $r_n < a$. This would depend also on the physical property considered. For average binding energy, the OPEP, with its $(\sigma_1 \cdot \sigma_2)(\tau_1 \cdot \tau_2)$ exhange character, often averages to zero, leaving only shorter range forces as the important ones.

Koltun's approximation is of this general type. He considers the relative wave function ψ for a pair of interacting nucleons. For very close separations, the nuclear force will be strong and dominate the behavior of ψ, but the nuclear one-body field, being something like an oscillator potential, will be nearly constant and have little effect. At large separations just the opposite will be true; the nuclear potential will have gone to zero. One can therefore write down ψ in these two extreme limits, and match the logarithmic derivatives at an intermediate distance where both the nuclear force and the one-body field are relatively constant. The basic assumption of the method is that such an intermediate distance, of order 1–2 fm, exists. At this point the inside wave function is determined completely by the phase shifts. Koltun used oscillator wave functions for the large separation zone. He showed that in general the matching can be made only for a particular energy of the interacting pair, and this energy implies a value for the two-body matrix element. The method is further simplified by going to the limit of a zero-range nuclear force, so the matching is done at $r = 0$.

The original approximation of Elliott *et al.* was phrased in the language of Born approximation, but it can be regarded as an exact relation provided the free particle K_l-matrix is local in coordinate space:

$$K_l(k, k; k^2) = (1/k^2) \int_0^\infty \not{j}_l^2(kr) K_l(r)\, dr \qquad (166)$$

The latter is certainly not true, yet it might be a useful approximation. For the nuclear matter G-matrix, Negele (Neg 70), Siemens (Sie 70), and Banerjee (Ban 69) have shown how to construct a local-in-r approximation that reproduces the diagonal G-matrix elements. In this view, the Elliott method produces matrix elements of the free reaction matrix K between oscillator wave functions, writing each one as a weighted integral of the plane wave diagonal $K_l(k, k; k^2)$ matrix elements,

$$\langle nl \mid K \mid n'l \rangle \approx \int_0^\infty e^{-E} K_l(E) F_{nn'l}(E)\, dE \qquad (167)$$

Srivastava (Sri 70a) proposed to calculate the half-shell K_l-matrix elements from the on-shell ones under the same approximation as above, namely that $K_l(r)$ is local in r. Writing

$$K_l(p, k; k^2) = (1/pk) \int_0^\infty \not{j}_l(pr) K_l(r) \not{j}_l(kr)\, dr \qquad (168)$$

he eliminated the "potential" $K_l(r)$ between the $p = k$ and $p \neq k$ cases to obtain

$$K_l^{(A)}(p, k; k^2) = \int_0^\infty W_l(p, k; q) \frac{d}{dq} \{q^2 K_l(q, q; q^2)\}\, dq \qquad (169)$$

where the kernel

$$W_l(p, k; q) = - \frac{8}{\pi} \frac{1}{pk} \int_0^\infty \frac{dr}{r} \, j_l(pr) j_l(kr) j_l(qr) v_l(qr) \quad (170)$$

has the remarkable property of being zero except for $|(p - k)/2| < q < (p + k)/2$. Thus to go some distance off-shell one needs to know the on-shell information only over a finite range of energies. In the $l = 0$ case, Eq. (169) reduces to

$$K_0^{(A)}(p, k; k^2) = \frac{1}{pk} \left. \{q^2 K_0(q, q; q^2)\} \right|_{|p-k|/2}^{(p+k)/2} \quad (171)$$

The tensor coupling case is not much more complicated (Sri 70a). The relation is clearly exact when $p = k$. The defect that it makes $K_l^{(A)}(p, k; k^2)$ a symmetric function of p and k could perhaps be improved by considering it to produce the symmetric function $\sigma(k, p)$ of Baranger et al. (Bar+ 69) and using their equations to generate the corresponding antisymmetric part $\alpha(k, p)$. However, for $l = 2$ and 3 it is found to give quite satisfactory estimates ($\lesssim 10\%$ error) of half-shell matrix elements for a hard core plus square well potential. Using this method, Sprung et al. (Spr+ 70) were able to estimate the higher partial wave ($j = 3, 4,$ and 5) G-matrix elements in nuclear matter and thus make a model-independent estimate of their contribution to nuclear matter binding.

It is unlikely that methods of this type will be useful for studying the strong S-state interactions. For P- and D-waves the approach from phase shifts may be adequate, since in nuclear matter calculations there is little difference between the results of different potentials fitting the same data. One of the aims of these approximate methods is to clarify what energy range of on-shell information is important for a given problem, and how significant are the errors in the input data.

7. POWER SERIES EXPANSIONS OF THE K-MATRIX

In many cases [for example, $(p, 2p)$ reactions, proton–proton bremsstrahlung, etc.] K_l-matrix elements $K_l(p, k; k^2)$ for which $|p - k| \leq 2$ fm^{-1} are of primary interest. The on-shell momenta may vary from essentially zero (in some bremsstrahlung experiments) to ~ 5 fm^{-1} [in some $(p, 2p)$ experiments]. With this in view, expansions of the half-shell K_l-matrix elements in a power series about the on-shell point have been proposed. The coefficients of the series serve as parameters for a description of the half-shell behavior.

7.1. Fuda's Expansion

If we add and subtract $(p/k)^l$ times the on-shell K_l-matrix element from Eq. (45) for the half-shell element, we obtain (Fud 70)

$$
\begin{aligned}
K_l(p, k; k^2) &= -\left(\frac{p}{k}\right)^l \frac{\tan \delta_l(k)}{k} + \frac{1}{k} \\
&\quad \times \int_0^\infty \left\{\frac{1}{p}\, \jmath_l(pr) - \frac{1}{k}\left(\frac{p}{k}\right)^l \jmath_l(kr)\right\} v_l(r)\psi_l^{(0)}(k, r)\, dr \\
&= -\left(\frac{p}{k}\right)^l \frac{\tan \delta_l(k)}{k} + \left(\frac{p}{k}\right)^l \sum_{n=1}^\infty \frac{(-\tfrac{1}{2})^n(p^{2n} - k^{2n})}{n!\,(2l + 1 + 2n)!!} \\
&\quad \times \int_0^\infty r^{2n+l+1} v_l(r)\psi_l^{(0)}(k, r)\, dr
\end{aligned}
\tag{172}
$$

To lowest order in the energies k^2 and p^2, it leads to the following expansion for the Kowalski–Noyes half-shell function (Section 1.2)

$$
f_l(p, k) = (p/k)^l\{1 + \tfrac{1}{2}\lambda_l^{(1)}(k^2 - p^2) + \cdots\}
\tag{173}
$$

where $\lambda_l^{(1)}$, the off-shell scattering length, is given by

$$
\lambda_l^{(1)} = \frac{\int_0^\infty r^{l+3} v_l(r)\psi_l^{(0)}(0, r)\, dr}{(2l + 3) \int_0^\infty r^{l+1} v_l(r)\psi_l^{(0)}(0, r)\, dr}
\tag{174}
$$

It is obvious that $\lambda_l^{(n)}$ for n greater than some N are determined by the long-range (and known) part of the force. For $n < N$, the λ_l can be used as parameters to describe the off-shell behavior. For higher partial waves one can estimate the off-shell lengths by approximating the wave function $\psi_l^{(0)}(0, r)$ by its asymptotic form or simpler still by the free wave function

$$
\lambda_l^{(1)} \approx \frac{\int_0^\infty r^{2l+4} v_l(r)\, dr}{(2l + 3) \int_0^\infty r^{2l+2} v_l(r)\, dr}
\tag{175}
$$

7.2. Expansion of Redish *et al.*

Redish, Stephenson, and Picker (RSP 72) have obtained another expansion by using Eq. (57). For the $l = 0$ case it gives

$$
K_0(p, k; k^2) = K_0(k, k; k^2) + \frac{k^2 - p^2}{pk} \int_0^\infty (\sin pr)\Delta_0(k, r)\, dr
\tag{176}
$$

where

$$\Delta_0(k, r) = \psi_0^{(0)}(k, r) - \{\sin kr + \tan \delta_0(k) \cos kr\} \tag{177}$$

Introducing the variable $q = p - k$ (in order to expand in p about k) and expanding all functions of q in a power series about the on-shell point $q = 0$, they find

$$K_0(k + q, k; k^2) = K_0(k, k; k^2) + \sum_{m=1}^{\infty} \bar{A}_m(k)q^m \tag{178}$$

where

$$\bar{A}_m(k) = (-1)^m k^{-m} \sum_{n=0}^{[m/2]-1} (-1)^n(\bar{D}^{(2n)} - \bar{D}^{(2n+1)})$$

$$+ (1 + \delta_{(m+1)/2,[(m+1)/2]})(-1)^{[m/2]}\bar{D}^{(m-1)} \tag{179}$$

The notation $[n]$ means the largest integer contained in n and the functions $\bar{D}^{(n)}(k)$ are given by the moments of the difference function $\Delta_0(k, r)$ weighted with a trigonometric function,

$$\bar{D}^{(n)}(k) = \frac{k^n}{n!} \int_0^{\infty} dr \, r^n \Delta_0(k, r) \begin{Bmatrix} \sin kr \\ \cos kr \end{Bmatrix} \quad \text{for} \quad \begin{Bmatrix} n \text{ even} \\ n \text{ odd} \end{Bmatrix} \tag{180}$$

Here again, advantage can be taken of the known long-range part of the interaction by writing Eq. (178) as

$$K_0(p, k; k^2) = K_0(k, k; k^2) + K_0^{\text{ext}}(p, k; k^2) + \sum_{m=1}^{\infty} A_m(k)q^m \tag{181}$$

where

$$K_0^{\text{ext}}(p, k; k^2) = \frac{k^2 - p^2}{pk} \int_R^{\infty} (\sin pr)\Delta_0(k, r) \, dr \tag{182}$$

The coefficients $A_m(k)$ are given by Eq. (179) with the moments $\bar{D}^{(n)}(k)$ replaced by the moments $D^{(n)}(k)$, defined by

$$D^{(n)}(k) = \frac{k^n}{n!} \int_0^{R} dr \, r^n \Delta_0(k, r) \begin{Bmatrix} \sin kr \\ \cos kr \end{Bmatrix} \quad \text{for} \quad \begin{Bmatrix} n \text{ even} \\ n \text{ odd} \end{Bmatrix} \tag{183}$$

where R is some distance of the order of one pion Compton wavelength. The variation of the coefficients $A_m(k)$ with energy is found to be simple and quite smooth.

These expansions thus replace the half-shell K-matrix, a function of two variables, by a discrete set of functions of one variable, namely the energy-dependent coefficients of the power series. The coefficients λ or A thus offer a simple way of parameterizing the off-shell behavior of the half-shell K-matrix.

8. GENERAL FEATURES OF OFF-SHELL K-MATRIX ELEMENTS

Some general features of the off-shell matrix elements can be understood, especially the behavior at small energy or momentum and for $p \sim 4$ fm^{-1}. We first consider the former.

8.1. Low-Energy Behavior

The behavior for very low energies can be understood in terms of an off-shell effective range formula obtained by the authors (SS 70). It is based on the exact relation $(l = 0)$

$$K_0(p, q; k^2) = K_0(p, k; k^2) + \frac{p^2 - k^2}{pq} \int_0^\infty (\sin pr)\{\bar{\psi}_{q,k^2}^{(0)}(r) - \psi_{q,k^2}^{(0)}(r)\} \, dr$$

obtained in Section 2. When p, q, and k^2 are all close to zero, to the accuracy of quadratic terms, the integral can be replaced by its value at $p = q = k^2 = 0$,

$$I_0 = \frac{1}{pq} \int_0^\infty (\sin pr)\{\bar{\psi}_{q,k^2}^{(0)}(r) - \psi_{q,k^2}^{(0)}(r)\} \, dr$$

$$\to \int_0^\infty r \lim_{q,k^2 \to 0} \frac{\psi_{q,k^2}^{(0)}(r) - \psi_{q,k^2}^{(0)}(r)}{q} \, dr \tag{184}$$

At zero energy $\bar{\psi}_{q,k^2}^{(0)}(r)/q = r - a_0$, where a_0 is the scattering length. To this degree of accuracy, we find

$$K_0(p, q; k^2) \approx K_0(k, k; k^2) + I_0(p^2 + q^2 - 2k^2)$$

$$\approx \frac{-1}{-(1/a_0) + \frac{1}{2}r_0k^2} + I_0(p^2 + q^2 - 2k^2) \tag{185}$$

or

$$[K_0(p, q; k^2)]^{-1} \approx \frac{1}{a_0} - \frac{1}{2} r_0 k^2 - \frac{I_0}{a_0^2} (p^2 + q^2 - 2k^2) \tag{186}$$

where we have used the effective range form of the on-shell K_0-matrix element; r_0 is the effective range.

The on-shell effective range formula is also sometimes used for negative k^2 to relate the deuteron binding energy to the $^3S_1$-state a_0 and r_0. The present off-shell formula can also be extended to negative k^2. However, we should keep in mind that $[K_0(p, q; k^2)]^{-1}$ is not analytic at $k^2 = 0$. The step function in Eq. (42) indicates just this. If $\tilde{K}_0^{-1}$ defines the analytic continuation of K_0^{-1} from $k^2 > 0$ to $k^2 < 0$, Eqs. (42) and (43) give

$$K_0^{-1} = t_0^{-1} = \tilde{K}_0^{-1} - | k | \tag{187}$$

for $k^2 < 0$. The expression (186) gives $\tilde{K}_0$ instead of K_0. The generalization (Sri 70c)

$$[K_0(p, q; k^2)]^{-1} \approx \frac{1}{a_0} - \frac{1}{2} r_0 k^2 - \frac{I_0}{a_0^2} (p^2 + q^2 - 2k^2) - | k | \theta(-k^2) \tag{188}$$

obtained by using Eq. (187) in Eq. (186) is applicable both for positive and negative k^2.

Thus one additional constant describes the off-shell behavior of the K_0- or t_0-matrix elements near zero energy–momentum. The accuracy of the approximation is comparable to that of the usual effective range formula. We expect the integral in Eq. (184) to be over a distance of the order of the healing distance $d \approx 1$ fm, so we have an estimate of $I_0 \approx | a_0 d^2/2 |$. It can be called a zero-energy "wound integral" (DOB 69). Another thing to be noted is that, for the $^1S_0$ state, the behavior of the off-shell matrix elements at zero momentum p or zero energy k^2 is dominated by the large, negative scattering length (Spr 70, Sau 73). The close agreement observed in off-shell matrix elements near $p = 0$ for different potentials indicates that they have similar values of I_0.

A similar expansion has been obtained by Fuda (Fud 70, BF 72). He has obtained yet another expansion by exploiting the dominant influence (at low energy) of the bound (antibound) state t_0-matrix pole in the $^3S_1$ ($^1S_0$) state. He has suggested fitting the low-energy elastic scattering data by a single-term separable potential. This gives a simple relation for the Kowalski–Noyes half-shell function in terms of the Jost functions

$$f(p, k) \approx \left(\frac{k \operatorname{Im} f(p)}{p \operatorname{Im} f(k)} \right)^{1/2} \tag{189}$$

The expansion (convergent for small values of k even for potentials with OPE tail) of the Jost function in a power series about the origin and the use of the property

$$f(-k) = f^*(k) \tag{190}$$

leads to an expansion of the following form:

$$\mathscr{F}(p, k) = 1 - \tfrac{1}{2}\Lambda_1{}^2(p^2 - k^2) - \tfrac{1}{4}\Lambda_1{}^4 k^2(p^2 - k^2)$$
$$+ (\tfrac{1}{2}\Lambda_2{}^4 - \tfrac{1}{8}\Lambda_1{}^4)(p^4 - k^4) + \cdots \qquad (191)$$

where

$$\Lambda_n^{2n} = -\frac{(i)^{2n+1}}{(2n+1)!} \left\{ \frac{d^{2n+1}}{dk^{2n+1}} f(k) \right\}_{k=0} \qquad (192)$$

In this approximation the expansion coefficients are given by the on-shell data. Phase shift equivalent potentials will lead to identical off-shell behavior for low energies.

8.2. Behavior for $p \sim 4$ fm^{-1}

Consider Eq. (55). Since most nuclear forces contain some short distance repulsion, let us consider a pure hard core potential. Because $v(r)$ is infinite for $r \leq c$, the hard core radius, $\psi^{(0)}$ vanishes and the wave defect χ is $j_l(kr)$. Outside $r = c$, v is zero and so the solution is

$$\chi_{k,s}(r) = \frac{j_l(kc)}{n_l(\tau c)}\, n_l(\tau r), \qquad r \geq c, \quad \tau^2 = s \qquad (193)$$

Providing $\tau c \ll (l + 1)$, $|\, n_l(\tau c)\,| \approx \text{const} \times (\tau c)^{-l} \gg 1$, so $\chi_{k,s}(r)$ will decay rapidly outside the core radius and the general picture will be a triangular function peaked at $r = c$ with long-range oscillations of small amplitude. According to Eq. (55), the K_l-matrix element will be determined by the Fourier transform of this function, and will be large when $j_l(pr)$ has its maximum coincide with the maximum of χ. In the case $l = 0$, this requires $\sin pc \approx 1$ and so at $p \approx \pi/2c$ we expect a maximum in the K_0-matrix element. Since $c \approx 0.4$ fm, this should occur for $p \sim 3\text{–}4$ fm^{-1}. This argument was originally given by Bethe, Brandow, and Petschek (BBP 63) for the negative energy case. For a more realistic force we still expect to see this repulsive maximum because at such a high energy ($p = 4$ fm^{-1} corresponds to 1200 MeV lab) the short-range repulsion should dominate over the weak, long-range attraction. Local potentials do show such a repulsive maximum, corresponding to a localization of their repulsion at $r \approx 0.5$ fm. The separable forces may or may not show positive matrix elements for large p, but in any case they are much flatter, for most of the proposed models, corresponding to a diffuse repulsion. This "flat" behavior, though, is not an inherent characteristic of the separable potentials. Sirohi and Srivastava (SS 72, SS 73a) have shown that an off-shell

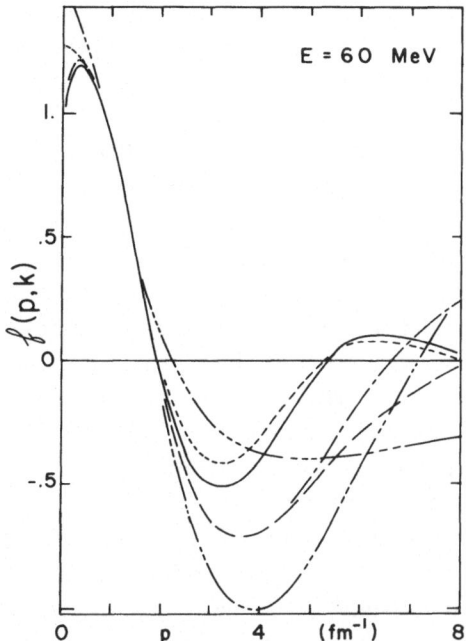

Fig. 3. The Kowalski–Noyes half-shell function plotted against the final state momentum p. The initial state momentum corresponds to 60 MeV lab. Several potentials are shown as follows: (a) SSC-PP-1 (——), (b) SSC-NP-2 (- - -), (c) Reid soft core (— —), (d) Bressel, Kerman, and Rouben (— · —), (e) Hamada–Johnston (— · · —), (f) the separable potential of Mongan (— · · · · —). [Taken from (SS 69).]

behavior typical of the local potentials can also be generated for separable ones by a suitable choice of the form factors. The hard core Hamada–Johnston force shows the greatest repulsive hump, next are the Reid and Bressel–Kerman "soft core" forces, and least are the "super soft core" (SSC) forces (SS 70). This is illustrated in Fig. 3, showing the KN half-shell function for a number of these potentials.

8.3. General Behavior and Comments

At the on-shell point, $K_0(k, k; k^2)$ is equal to $-[\tan \delta_0(k)]/k$ and so is negative at energies $k^2 < 240$ MeV lab. Hence the half-shell elements have to change sign between $p = k$ and $p \approx 4$ fm^{-1}. In almost all the cases this sign change occurs in the region $p \sim 1.5$–2 fm^{-1}.

For $k^2 \lesssim 1.5$ fm^{-2} (120 MeV lab) and over a range in p of about 1.5 fm^{-1}, all the potentials give very similar half-shell matrix elements unless they have rather different values of I_0 [Eq. (184)].

The general conclusion is that in the $^1S_0$ state the off-shell matrix elements are dominated by the large negative scattering length, the on-shell point, the zero near $p = 2$ fm^{-1}, and by a repulsive hump near $p = 4$ fm^{-1} of height governed by the strength of the repulsion. Within these constraints quite wide variations could be produced in the off-shell behavior. For example, the height of the repulsive hump in the half-shell matrix elements could be increased or decreased and its position shifted by suitably introducing momentum dependence in the potential [on-shell matrix elements remaining unchanged (Sri 70b, Sri 71)]. Similarly, for separable potentials, large variations in matrix elements could be produced by suitably choosing one of the form factors in the Fiedeldey procedure (Fie 69a).

In the $^3S_1$ case, the off-shell behavior is dominated by the deuteron pole and form factor, the on-shell point, and (like the $^1S_0$ case) a repulsive hump near $p = 4$ fm^{-1}. It is this pole dominance which is behind all the success of the UPA at low energies. Large changes in the off-shell behavior are not possible if the deuteron form factor is kept fixed.

Higher partial waves $(l \geq 1)$ have not been studied much (VR 73, Sri 74). In general the magnitude of the change in the off-shell matrix elements is small here compared to what one can get in the case of S-waves.

8.4. Interplay between On-Shell and Off-Shell Matrix Elements

The effect of varying the on-shell matrix elements on the half-shell reaction matrix elements has been investigated by Srivastava (Sri 73a). Using Fiedeldey's procedure for generating second-rank separable potentials, he has studied which off-shell matrix elements are mainly affected by the various changes in the on-shell behavior. He finds that $K(p, k; k^2)$ is primarily determined by the on-shell data up to p or k, whichever is larger. The data beyond this have in general very little effect, especially if the deuteron wave function is kept fixed or nearly so (Fie 74). Thus in calculations such as proton–proton bremsstrahlung or $(p, 2p)$ reactions requiring half-shell reaction matrix elements one need not bother very much about phase shifts, etc., beyond this limit.

9. METHODS FOR CALCULATING THE T- OR K-MATRIX

T- or K-matrix elements for any partial wave are given respectively by the Lippmann–Schwinger integral equation (21) or (46). Solving this equation is difficult because the kernel is singular. The choice of the method depends on the nature of the interaction. With a separable potential the calculations are quite simple. They are comparatively more involved and tedious with local, momentum-dependent, and nonlocal nonseparable potentials, particularly in the presence of a hard core repulsion and for positive energies. In the following we describe various methods, pointing out their applicability and special features.

9.1. Method for Separable Potentials

For a general separable potential of the form (58), the t_l-matrix element is given by Eq. (62). It essentially involves the evaluation of an integral of the form

$$G^+(k^2) = \frac{2}{\pi} \int_0^\infty \frac{g_i(k')g_j(k')}{k'^2 - k^2 - i\varepsilon} k'^2 \, dk' \tag{194}$$

Using the identity

$$\frac{1}{k'^2 - k^2 - i\varepsilon} = P \frac{1}{k'^2 - k^2} + i\pi\delta(k'^2 - k^2) \tag{195}$$

it can be reduced to the principal value integral

$$G(k^2) = \frac{2}{\pi} P \int_0^\infty \frac{g_i(k')g_j(k')}{k'^2 - k^2} k'^2 \, dk' \tag{196}$$

which occurs in the calculation of the reaction matrix elements.

If the potential form factors are given as analytic functions, this integral can be evaluated analytically, provided (GS 68) (i) the form factors $g(k)$ are nonsingular and real for real k, (ii) there exists a unique analytic continuation of $g(k)$ into the complex k-plane, such continuation being an even meromorphic function of k, and (iii)

$$\lim_{|\bar{k}| \to \infty} \int_C g_i(k)g_j(k) \, dk = 0 \tag{197}$$

along a semicircle C of radius $\bar{k}$ in the upper half k-plane. All the separable potentials in use satisfy this criterion. In case the form factors are the result of some numerical calculation and are given as numbers at mesh points, Eq. (196) is evaluated numerically. It can be regularized by subtracting a vanishing principal value integral

$$P \int_0^\infty \frac{dk'}{k'^2 - k^2} = 0 \tag{198}$$

multiplied by an appropriate constant. Thus

$$G(k^2) = \frac{2}{\pi} \int_0^\infty \frac{g_i(k')g_j(k')k'^2 - g_i(k)g_j(k)k^2}{k'^2 - k^2}\, dk' \qquad (199)$$

A similar device to improve the numerical accuracy has been suggested by Chiang and Lee (CL 72). If k^2 is negative, this operation is not required. The integral (199) is of the type

$$\int_0^\infty F(t)\, dt \qquad (200)$$

One can now use Gaussian quadrature by transforming it to the form

$$\int_{-1}^1 \mathscr{F}(z)\, dz \qquad (201)$$

For example:

1. Put (Fie 69a)

$$t = \frac{z}{1-z} + \frac{1}{2}, \qquad \frac{dt}{dz} = \frac{1}{(1-z)^2} \qquad (202)$$

to get

$$\int_0^\infty F(t)\, dt = \int_{-1}^1 F(t(z))\, \frac{dt}{dz}\, dz = \int_{-1}^1 F\left(\frac{z}{1-z} + \frac{1}{2}\right) \frac{dz}{(1-z)^2} \qquad (203)$$

2. Put (BJK 69)

$$t = C \tan\left\{\frac{\pi}{4}(1+z)\right\}, \qquad \frac{dt}{dz} = C\,\frac{\pi}{4} \sec^2\left\{\frac{\pi}{4}(1+z)\right\} \qquad (204)$$

to get

$$\int_0^\infty F(t)\, dt = C\,\frac{\pi}{4} \int_{-1}^1 F\left(C \tan\left\{\frac{\pi}{4}(1+z)\right\}\right) \sec^2\left\{\frac{\pi}{4}(1+z)\right\} dz \qquad (205)$$

The value of the integral is independent of the constant C, which can be profitably used to adjust the positions of the Gauss ordinates (half of the points from zero to C and the remaining half from C to infinity).

Mongan (Mon 69c) has suggested yet another procedure. Instead of writing $G(k^2)$ in the form of Eq. (199), he suggests breaking the principal value integral as follows:

$$P \int_0^\infty \frac{F(t)}{t-k}\, dt = \int_0^{k-\Delta} \frac{F(t)}{t-k}\, dt + \int_{k+\Delta}^\infty \frac{F(t)}{t-k}\, dt + \int_{k-\Delta}^{k+\Delta} \frac{F(t)}{t-k}\, dt \qquad (206)$$

Now making the transformation

$$t = \tfrac{1}{2}(k - \varDelta)(1 + \alpha), \qquad dt/d\alpha = (k - \varDelta)/2 \tag{207}$$

in the first integral on the right and

$$t = 2(k + \varDelta)/(1 - \beta), \qquad dt/d\beta = 2(k + \varDelta)/(1 - \beta)^2 \tag{208}$$

in the second, both the integrals reduce to the form (201) and can be integrated. Since the Gauss quadrature rule does not use the end points, the limit $\varDelta \to 0$ can be taken immediately with the third integral contributing nothing. Alternatively (AC 67), one can choose a small value for $\varDelta$, integrate the first and the second regular integrals by Gauss quadrature or otherwise, and approximate the third by

$$\int_{k-\varDelta}^{k-\varDelta} \frac{F(t)}{t - k}\, dt \approx 2\varDelta \left[\frac{\partial F(t)}{\partial t} \right]_{t=k} \tag{209}$$

In both these methods the choice of the magnitude of $\varDelta$ is a delicate thing. If we let $\varDelta$ approach zero, the error in the third term of Eq. (206) approaches zero but in the first two terms we have to calculate the difference of two large numbers. As result, the rounding-off errors become large. If the magnitude of $\varDelta$ is increased, the error in the third term would become large. Another disadvantage of both these methods is that the choice of the Gauss ordinates depends on the input k.

9.2. Method for Hard Core Potentials

If the potential contains a hard core repulsion, Eq. (21) or (46) cannot be solved directly in momentum space, since the potential matrix elements are infinite. Coordinate space equation (18) or (49) is more useful in this case to calculate first $\psi^{(+)}$ or $\psi^{(0)}$ and then to obtain matrix elements using Eq. (20) or (45). Let us concentrate on reaction matrix elements and assume that the potential is local.

Inside the hard core $v_l(r)\psi_{k,s}(r)$ (a superscript 0 and the partial wave subscript l on ψ have been suppressed) has the indeterminate form $\infty \times 0$ but this must include a term of the form $\psi'_{k,s}(c)\,\delta(r - c)$, which comes from the discontinuous change in the slope of $\psi_{k,s}(r)$ at the core boundary $r = c$. It is given by (BBP 63, LS 68)

$$v_l(r)\psi_{k,s}(r) = \psi'_{k,s}(c)\delta(r - c) + (k^2 - s)\mathcal{J}_l(kr) \tag{210}$$

for $r \leq c$. Boundary conditions on $\psi_{k,s}(r)$ appropriate to Eq. (49) are

$$\psi_{k,s}(r) \begin{cases} = 0, & r \leq c \\ \xrightarrow[r \to \infty]{} j_l(kr) + A_l(k, \tau) n_l(\tau r), & s = \tau^2 > 0 \\ j_l(kr) + A_l(k, \gamma) \mathcal{H}_l(\gamma r), & s = -\gamma^2 < 0 \end{cases} \tag{211}$$

where $\mathcal{H}_l(\gamma r)$ is the solution of the free equation

$$\left\{ \frac{d^2}{dr^2} - \frac{l(l+1)}{r^2} - \gamma^2 \right\} \mathcal{H}_l = 0 \tag{212}$$

with the boundary condition

$$\mathcal{H}_l(\gamma r) = i^{l+2} \gamma r h_l^{(+)}(i\gamma r) \xrightarrow[\text{large } r]{} e^{-\gamma r} \tag{213}$$

$h_l^{(+)}$ is the spherical Hankel function for outgoing waves and A_l is a constant, which, along with $\psi'_{k,s}(c)$, is determined by the equations

$$\left\{ \frac{d^2}{dr^2} - \frac{l(l+1)}{r^2} + s \right\} \psi_{k,s}(r)$$

$$= \psi'_{k,s}(c)\, \delta(r - c), \qquad r \leq c$$
$$= v_l(r)\psi_{k,s}(r) + (s - k^2) j_l(kr), \qquad r > c \tag{214}$$

Once $\psi_{k,s}(r)$ is known, the K_l-matrix element can be readily constructed using Eq. (45):

$$K_l(p, k; s) = \frac{1}{pk} \left[j_l(pc)\psi'_{k,s}(c) + (k^2 - s) \int_0^c j_l(pr) j_l(kr)\, dr \right.$$

$$\left. + \int_c^\infty j_l(pr)v_l(r)\psi_{k,s}(r)\, dr \right] \tag{215}$$

The first term on the right is the contribution from the core surface, the second from the core interior, and the third from the outer potential. $\psi'_{k,s}(c)$ is given by (BBP 63)

$$\psi'_{k,s}(c) = j_l'(kc) - \mathcal{H}_l'(c) - [j_l(kc)]^{-1} \int_c^\infty \mathcal{H}_l(r)v_l(r)\psi_{k,s}(r)\, dr \tag{216}$$

where

$$\mathcal{H}_l(r) = \begin{cases} j_l(kc) n_l(\tau r)/n_l(\tau c), & s = \tau^2 > 0 \\ j_l(kc) \mathcal{H}_l(\gamma r)/\mathcal{H}_l(\gamma c), & s = -\gamma^2 < 0 \end{cases} \tag{217}$$

$\psi_{k,s}$ can be obtained by solving Eq. (214) or, better still, by solving the simpler equation (52) for the wave defect $\chi_{k,s}(r) = \mathscr{j}_l(kr) - \psi_{k,s}(r)$ with appropriate boundary conditions:

$$\chi_{k,s}(c) = \mathscr{j}_l(kc), \qquad \chi'_{k,s}(d) + \Gamma_l\chi_{k,s}(d) = 0 \qquad (218)$$

where d is some distance outside the range of interaction and

$$\Gamma_l = -(d/dr)\{\ln \mathscr{H}_l(r)\} \qquad (219)$$

Equation (52) represents a two-point boundary value problem and can be solved in several ways (Har 58, Tak 72, Tak 73). For $s < 0$, Bhargava and Sprung (BS 67) have shown that the Ridley method (Rid 57) is very convenient and efficient. The same applies for positive s, but with some additional complications.

In the ordinary Ridley method, one factorizes the left-hand side of Eq. (52), reducing the problem to the solution of three first-order equations

$$(dy/dr) - y^2 = f \qquad (220)$$

$$(dw/dr) - yw = g \qquad (221)$$

$$(d\chi/dr) + y\chi = w \qquad (222)$$

where

$$f(r) = s - v_l(r) - \{l(l+1)/r^2\}, \qquad g(r) = -v_l(r)\mathscr{j}_l(kr) \qquad (223)$$

The boundary conditions (218) are satisfied if $y(d) = \Gamma_l$ and $w(d) = 0$. Using these initial values, Eqs. (220) and (221) can be integrated inward to $r = c$ and then, using the boundary condition at $r = c$, Eq. (222) can be solved back to $r = d$ generating the desired solution. There is no danger of building any spurious solution of χ on the outward integration because a first-order equation is being solved.

The complication when $s > 0$ arises because over most of the interval $c \leq r \leq d$ we have

$$f(r) = s - v_l(r) - [l(l+1)/r^2] > 0 \qquad (224)$$

which causes $y(r)$ to have poles. This problem is overcome by introducing the Ridley-alternate method. Three new functions are defined (SS 70)

$$Y = 1/y, \qquad W = w/y, \qquad Z = w - y\chi$$

or

$$y = 1/Y, \qquad w = W/Y, \qquad \chi = W - YZ \qquad (225)$$

which satisfy

$$dY/dr = -1 - YfY \tag{226}$$

$$dW/dr = Y(g - fW) \tag{227}$$

$$dZ/dr = g - f(W - YZ) \tag{228}$$

Whenever the solution $y(r)$ begins to grow, one can switch over to the alternate method. Since the operation (225) is reflexive, this is very convenient on the computer; one needs only to keep a record of the step at which a flip-flop was performed, so that on the reverse integration for χ (or Z) the corresponding operation is effected. Equations (225)–(228) apply equally well to the case of coupled partial waves, where f, g, y, Y, w, W, χ, and Z become 2×2 matrices.

Equation (215) for the matrix element involves the spherical Hankel transform of $v_l \psi_{k,s}$. This can be evaluated by an extension of Filon's method (Fil 28), which is a generalisation of Simpson's rule, in which the wiggles of $\sin qr$, etc., are exactly accounted for, so the mesh required is determined by the smoothness of $v_l \psi_{k,s}$.

For soft core potentials it is probably easier to evaluate the matrix elements in momentum space, but for hard core forces the coordinate space solution is necessary. It is also adaptable to momentum-dependent potentials (Sri 70b).

The Orsay group (BHB 74, L'Hu 74) have used essentially the above method to build a practical computer code for the calculation of the off-shell T-matrix. Since they calculate T rather than K, complex arithmetic is required, but the poles inherent in working with $\tan \delta$ are avoided. In their paper they give the proper equations including spin and coupled partial waves. They also show contour maps of T-matrix elements in various partial waves for a variety of hard and soft local potentials.

9.3. Boundary Condition Model of Nuclear Forces

Feshbach, Lomon, and collaborators have developed an alternative representation of the two-nucleon interaction in which the shortest ranged parts of the force are represented not by a potential but by a (in practice, energy-independent) boundary condition on the two-body wave function at a distance c called the boundary radius:

$$c\psi_l'(c) = \psi_l(c)f_l \tag{229}$$

Outside c, usually taken as half a pion Compton wavelength or 0.7 fm, the interaction is represented as usual by a potential model. In this region a good deal of theoretical basis can be supplied for the potential. In certain circumstances, it may be advantageous to apply such a model omitting the exterior potential, in which case it is called the pure BCM (PBCM).

On-shell matrix elements, which are expressed in terms of phase shifts, clearly do not require any information about the interaction inside the boundary radius. Fuda (Fud 71) has shown that the half-shell matrix elements are also well defined. For off-shell matrix elements this is no longer the case, so there is an apparent ambiguity in off-shell matrix elements of the BCM. Various proposals have been made to replace the boundary condition by a pseudopotential which, in a limiting case, forces the wave function to satisfy the boundary condition. The appropriate off-shell matrix elements are then taken to be those obtained from the pseudopotential. One of the simplest was proposed by Razavy and Sprung (RS 64). Their pseudopotential consists of an infinite repulsive core of radius $b = c - \varepsilon$, followed by a deep square well of width ε and depth

$$v_l = -(\pi/2\varepsilon)^2 + (2f_l/c\varepsilon) \tag{230}$$

In the limit $\varepsilon \to 0$, the wave function is zero inside the boundary radius c, but has a finite discontinuity at c and satisfies the boundary condition. In the exterior region $r > c$, one still solves Eq. (52) but with the boundary conditions (218) altered so that at $r = c$, ψ satisfies Eq. (229). The K-matrix element is given by Eq. (215), except that the first or "core boundary term" is replaced by

$$j_l(pc)\psi'_{k,s}(c) \to (1/c)\{j_l(pc)f_l - kc\,j_l'(pc)\}\psi_{k,s}(c) \tag{231}$$

Of course, if $|f_l| \to \infty$, one recovers the case of the hard core potential.

In the paper by Razavy and Sprung, this method is extended to coupled partial waves. This or equivalent methods have been rediscovered by Kim and Tubis (KT 70a, KT 70b), Fuda (Fud 72), and Brayshaw (Bra 71). Hoenig and Lomon (HL 66) have used a more general pseudopotential which does not include an infinite repulsion inside the boundary radius, but which introduces an "interior boundary condition" b_l applied to the wave function at $r = c - \varepsilon$. As $|b_l| \to \infty$, their result goes over to the preceding one. It was shown by Brayshaw that the Hoenig–Lomon pseudo-potential leads to unphysical poles in the fully off-shell t-matrix elements and this favors the use of the off-shell matrix elements described here.

Hoenig (Hoe 73) has attempted to resurrect the interior boundary parameter, which would in principle be a nice device for investigating the sensitivity of many-body calculations to the off-shell arbitrariness.

9.4. Matrix Inversion Method

This method (HTT 66, Har 67, BJK 69, HT 70, Wal 71) can be applied to any nonsingular potential either local or nonlocal, central or noncentral. Let us consider, for simplicity, Eq. (46) for the uncoupled case and regularize the principal value integral by subtracting a zero term to get

$$K_l(p, q; k^2) = v_l(p, q) + \frac{2}{\pi} \int_0^\infty \{v_l(p, k')K_l(k', q; k^2)k'^2$$
$$- v_l(p, k)K_l(k, q; k^2)k^2\} \frac{dk'}{k^2 - k'^2} \tag{232}$$

or by using the Kowalski–Noyes trick to transform it to a Fredholm form (Kow 65, Noy 65, TGG 71). If $k^2 < 0$, this need not be done. Now the integrand in Eq. (232) has a finite limit even for $k' = k$; however, it will be better to avoid these points. If the integrand is a sufficiently smooth function of k', the integral can be replaced by a sum over a small set of points k_i with weights W_i (Gaussian or Laguerre ordinates and weights). Writing Eq. (232) with p and q equal to each of the points k_i as well as $k_{N+1} = k$, we obtain the following $N + 1$ linear equations:

$$K_l(k_i, k_j; k^2) = v_l(k_i, k_j) + \frac{2}{\pi} \sum_{m=1}^N W_m \frac{v_l(k_i, k_m)K_l(k_m, k_j; k^2)k_m^2}{k^2 - k_m^2}$$
$$- \frac{2}{\pi} \left(\sum_{m=1}^N \frac{W_m}{k^2 - k_m^2} \right) v_l(k_i, k_{N+1})K_l(k_{N+1}, k_j; k^2)k_{N+1}^2 \tag{233}$$

or

$$v_l(k_i, k_j) = \sum_{m=1}^{N+1} P_l(k_i, k_m; k)K_l(k_m, k_j; k^2) \tag{234}$$

where

$$P_l(k_i, k_m; k) = \delta_{im} - v_l(k_i, k_m)w_m(k) \tag{235}$$

with

$$w_m(k) = \begin{cases} \dfrac{2}{\pi} \dfrac{k_m^2 W_m}{k^2 - k_m^2}, & m < N \\[3mm] -\dfrac{2}{\pi} \sum_{n=1}^N \dfrac{W_n}{k^2 - k_n^2} k^2, & m = N + 1 \end{cases} \tag{236}$$

The nonsingular matrix P_l can be inverted to yield the reaction matrix elements both on and off the energy shell

$$K_l(k_i, k_j; k^2) = \sum_{m=1}^{N+1} P_l^{-1}(k_i, k_m; k) v_l(k_m, k_j) \tag{237}$$

A grid of about 25 points is found to be sufficient for most potentials. Greater N may be required for some potentials with a lot of structure in momentum space (Fie 69a). The justification of the numerical procedures applied to Eq. (232) has been given by Heller and Reinhardt (HR 73).

An advantage of the method is that it is indifferent as to whether the potential is local or nonlocal. For the local case the matrix element $\langle \mathbf{k}' \mid V \mid \mathbf{k} \rangle$ depends only on the momentum transfer $\mathbf{k}' - \mathbf{k}$. For a non-local force this is not true but the partial wave matrix elements $v_l(k', k)$ are, in general, no more complicated. The principal problem in applying the method is to compute $v_l(k', k)$ in an analytic form. Such a form is almost essential since the $v_l(k', k)$ are needed for very large k' and k, where numerical methods become impractical.

9.4.1. Method of Osborn

Osborn (Osb 71) has given a novel method of reducing the Lippmann–Schwinger equation to a nonsingular Fredholm equation which has only a finite range of integration in momentum space, and whose kernel is square integrable for bounded and smooth potentials (Fad 65).

Let us consider Eq. (46),

$$K_l(p, p'; k^2) = v_l(p, p') - \frac{2}{\pi} P \int_0^\infty \frac{v_l(p, p'') K_l(p'', p'; k^2)}{p''^2 - k^2} p''^2 \, dp'' \tag{238}$$

By changing the momentum variables according to

$$p(x) = k(1 + x)/(1 - x), \qquad x \in [-1, 1] \tag{239}$$

this equation becomes

$$
\begin{aligned}
K_l(x, x'; k^2) \\
= v_l(x, x') - \frac{1}{\pi k} P \int_{-1}^{+1} \frac{v_l(x, x'') K_l(x'', x'; k^2)}{x''} k^2 \left(\frac{1 + x''}{1 - x''} \right)^2 dx''
\end{aligned}
\tag{240}
$$

Osborn now defines even and odd projection operators

$$
\begin{aligned}
\mathscr{E}(x) f(x) &= [f(x) + f(-x)]/2 \\
\mathscr{O}(x) f(x) &= [f(x) - f(-x)]/2
\end{aligned}
\tag{241}
$$

and exploits the identities

$$P \int_{-1}^{+1} \frac{dx''}{x''} \, \mathscr{E}(x'')f(x'') = 0 \tag{242}$$

$$P \int_{-1}^{+1} \frac{dx''}{x''} \, \mathscr{O}(x'')f(x'') = 2 \int_{-1}^{0} \frac{dx''}{x''} \, \mathscr{O}(x'')f(x'') \tag{243}$$

to obtain

$$K_l(x, x'; k^2) = v_l(x, x') - \frac{2}{\pi k} \int_{-1}^{0} \frac{dx''}{x''} \, \mathscr{O}(x'') \Big[k^2 \Big(\frac{1 + x''}{1 - x''} \Big)^2 v_l(x, x'')$$

$$\times K_l(x'', x'; k^2) \Big] \tag{244}$$

Mapping this equation back to momentum variables gives

$$K_l(p, p'; k^2) = v_l(p, p') - \frac{4}{\pi} \int_{0}^{k} \frac{dp''}{p''^2 - k^2}$$

$$\times \mathscr{O}(p'')[p''^2 v_l(p, p'')K_l(p'', p'; k^2)] \tag{245}$$

$$= v_l(p, p') - \frac{4}{\pi} \int_{0}^{k} \{ [\mathscr{O}(p'')p''^2 v_l(p, p'')]K_l^e(p'', p'; k^2)$$

$$+ [\mathscr{E}(p'')p''^2 v_l(p, p'')]K_l^o(p'', p'; k^2) \} \frac{dp''}{p''^2 - k^2} \tag{246}$$

where $\mathscr{E}(p)$ and $\mathscr{O}(p)$ are the transformations of the even and odd projection operators in momentum space

$$\mathscr{E}(p)f(p) = [f(p) + f(k^2/p)]/2$$
$$\mathscr{O}(p)f(p) = [f(p) - f(k^2/p)]/2 \tag{247}$$

and

$$K_l(p, p'; k^2) = [\mathscr{E}(p) + \mathscr{O}(p)]K(p, p'; k^2)$$

$$= K_l^e(p, p'; k^2) + K_l^o(p, p'; k^2) \tag{248}$$

Multiplying Eq. (246) by $\mathscr{E}(p)$ and $\mathscr{O}(p)$ will now give us two coupled nonsingular equations for K_l^e and K_l^o. The boundedness and smoothness conditions on the potential ensure the square integrability of the kernels.

9.4.2. Use of Padé Approximants

Padé approximants can be used to continue analytically a negative energy solution of the partial wave Lippmann–Schwinger equation (21) to positive energies (Ste 72). The method thus attempts to avoid the problems associated with singular integrals at positive energies. Since the t-matrix element $t_l(p, q; k^2)$ is an analytic function (apart from possible poles on the negative real axis) in the k^2 plane cut along the positive real axis and has the asymptotic behavior

$$t_l(p, q; k^2) \to v_l(p, q) \qquad (249)$$

at high energies, it can be represented at both positive and negative energies by an $[N, N]$ Padé approximant

$$t_l(p, q; k^2) \simeq \frac{P_N(\gamma)}{Q_N(\gamma)} = \frac{\sum\limits_{n=0}^{N} a_n(p, q)\gamma^n}{\sum\limits_{n=0}^{N} b_n(p, q)\gamma^n}, \qquad \gamma = ik \qquad (250)$$

with

$$\frac{a_N}{b_N} \simeq v_l(p, q) \qquad (251)$$

The zeros of the denominator correspond to bound state or resonance poles, if any. If the t_l-matrix is known at $2N + 1$ negative energy values for given p and q, the coefficients a_n and b_n can be calculated. The approximant can then be employed to continue analytically the matrix element from negative energies to positive energies through the upper half k^2 plane. These approximants preserve the analytic properties of the functions.

9.4.3. Moment Methods

Moment methods (Vor 65) allow a reduction in the dimension of the matrices required to solve a singular integral equation such as the Lippmann–Schwinger equation by the matrix inversion method (Osb 73b). Kim used this method in a three-body bound state calculation (Kim 69) and Harms for n–d scattering (Har 72). The advantage of the method is that the problem of representing the singular integral is separated from that of representing the kernel of the integral equation. The result is a gain in efficiency and accuracy. A special feature of the method is that the final solution comes out in an analytic form.

Consider an arbitrary one-dimensional equation

$$f(x) = g(x) + \int K(x, y)f(y)\, dy \qquad (252)$$

The usual method of solving this equation is to turn it into a matrix equation by some quadrature rule

$$f(x_i) = g(x_i) + \sum_{j=1}^{N} K(x_i, x_j)f(x_j)W_j, \qquad i = 1, \ldots, N \qquad (253)$$

where W_j and x_j are the weights and the ordinates of the quadrature rule. The size of the matrix N is determined by the number of integration points required to evaluate the integrals

$$\int K(x, y)f(y)\, dy \qquad (254)$$

accurately. In the moment method one selects a set of linearly independent functions $\{g_i, \ i = 1, \ldots, N\}$ such that a linear combination of g_i can provide a good approximation to f:

$$f(x) \approx \sum_{i=1}^{N} C_i g_i(x) \qquad (255)$$

One evaluates the moments

$$I_i(x) = \int K(x, y)g_i(y)\, dy \qquad (256)$$

and sets up a matrix equation to determine the expansion coefficients C_i. All the singular aspects of the original equation have been absorbed in the calculation of the moments I_i (which, in some cases, can be done analytically). The size of the resulting matrix depends on the number of functions g_i necessary to reproduce the functional structure of f. If the functions g_i have a functional behavior similar to the solution of the integral equation, the size N of the matrix can be reasonably curtailed. Osborn (Osb 73b) has tested this method by calculating the Kowalski–Noyes half-shell functions $\mathscr{f}(p, k)$ for a Yukawa potential. He obtains an accuracy of better than 99.9% with just 7×7 matrices for $k = 0.1$ fm^{-1} and $p \leq 5$ fm^{-1}.

9.4.4. Method of Optimized Polynomial Approximation

Chao and Jackson (CJ 73a, CJ 73b) have recently applied the method of optimized polynomial approximation (CD 68, Ciu 69) to reduce the dimensions of the matrices required to solve the Lippmann–Schwinger equation. The dimension really depends on the rate of convergence of the approximation scheme. Consider Eq. (252). In this method the singularities of $f(x)$ are first studied through the perturbation series

$$f = g + Kg + KgKg + \cdots \qquad (257)$$

After the singularities are located, a conformal mapping $Z = Z(x)$ is constructed so as to obtain the largest possible domain of convergence and hence the optimum variable for a polynomial expansion. Equation (252) is given in terms of this new variable by

$$\tilde{f}(Z) = \tilde{g}(Z) + \int \tilde{K}(Z, Z') \tilde{f}(Z') \frac{dy}{dZ'} \, dZ' \qquad (258)$$

If $\tilde{f}(Z)$ is now expanded in terms of an orthonormal set of polynomials in Z

$$\tilde{f}(Z) \simeq \sum_{i=1}^{N} C_i P_i(Z) \qquad (259)$$

and Eq. (259) is inserted into Eq. (258), one has linear equations of the following form for the coefficients C_i:

$$C_n = g_n + \sum_{m=1}^{N} U_{nm} C_m \qquad (260)$$

The dimensions of the matrix necessary for some given accuracy is minimized because of the optimal convergence of the expansion (259). Chao and Jackson claim that for comparable accuracy this method reduces the dimensions by a factor of approximately three.

This procedure can be looked upon as an improvement over the moment method in the sense that by optimized conformal mapping an attempt is made to optimize the expansion (255).

9.5. Evaluation

The matrix inversion method is certainly the most powerful except when the potential contains a hard core. Even this can be handled, providing the infinite core is replaced by a large finite repulsion. However, rigorous calculations must be carried out in coordinate space where the Ridley method has been found to be convenient (BS 67). The two problems which one must face in the matrix method are those connected with the singularity of the Lippmann–Schwinger kernel for positive energies, and the size of the matrix equation for a desired accuracy.

The simplest and the most commonly used method of dealing with the first problem is to subtract a vanishing principal value integral as in Eq. (232). Osborn's method has not been applied but is expected to be particularly useful when the potential matrix elements $v(p, q)$ are difficult to obtain accurately for large p and q, e.g., when $v(p, q)$ is not given in an analytic form. Padé approximants seem to be useful only when one is studying the energy dependence of the t-matrix elements. An approximant must be constructed for every value of p and q.

Considerable reduction in the size of the matrix equation is possible using the moment method. With this method, one can make use of any available information regarding the nature of the solution. The optimized polynomial method is a further refinement, but works only when the analytic structure of the solution is known, which it is not in most cases. A side advantage of these methods is that the singular integrals involved can sometimes be done analytically. These methods are likely to be used in future.

10. MANY-BODY CALCULATIONS

In the foregoing sections we have collected a great deal of technical information regarding the off-shell behavior of the nucleon–nucleon interaction. It is only in conjunction with an analysis of many-body calculations that this information becomes physically meaningful. Out of such an analysis one hopes to deduce essential features of the nuclear interaction. The suitability of any particular many-body problem for this purpose depends on the sensitivity of the calculated/measured observable quantities to the off-shell variations. Ideally, such calculations should be free from the complications caused by extraneous features such as the effects of meson exchanges (in deuteron magnetic properties), three-body forces (in three-nucleon problem), etc., and the doubts created by the various approximations made in arriving at the final result.

Though present-day calculations do not satisfy all these requirements, they do provide valuable information. We shall now present these calculations, classified in three groups:

1. Effects of variations of the phase shifts in the experimentally inaccessible high-energy region.
2. Effects of off-shell variations.
3. Effects of having a definite deuteron wave function. This is discussed in conjunction with effects 1 and 2.

10.1. Variations in the High-Energy Phase Shifts

Some attempts (Fie 70, Sri 70d, Fie 71, Cho+ 72, Fie 72, Fie 74, Sau 73) have been made to investigate the sensitivity of many-nucleon calculations to the ambiguities in the high-energy behavior of the two-nucleon interaction. These attempts are of interest because, as pointed out earlier, without recourse to a fundamental theory, the high-energy part of the interaction cannot be fixed and even a careful analysis of future nucleon–nucleon scattering data will not resolve this difficulty. The variations in the nuclear matter and triton binding energies have been studied using second-rank separable potentials. The low-energy phase shifts are kept fixed, while in the high-energy region large, arbitrary changes are made. The potentials are generated by the method of Fiedeldey (Fie 69a). It is found (Cho+ 72) that both the nuclear matter binding energy and the triton binding energy have a strong dependence on the high-energy phase shifts, at least for second-rank separable potentials, when the form factors are chosen to have a long range in momentum space. When this range is chosen to be reasonable, about 1 fm^{-1}, very little sensitivity is found. The variations in binding energy are of the same size as those associated with changes in the off-shell behavior of the interaction. For critical potentials (in the sense of Fiedeldey) the results appear to be rather independent of the phase shifts, but do depend on the choice of the attractive form factor of the potential. The off-shell sensitivity found for one set of on-shell information may not be the same as for another set, i.e., there is a strong coupling between the two effects (Cho+ 72, Fie 72). The changes in the phase shifts at high energies significantly affect the triton and nuclear matter binding energies only when the deuteron wave function is also modified by them (Fie 72, Fie 74). For critical potentials, the deuteron wave function remains unchanged. Another result is that a more repulsive phase shift may not necessarily be associated with reduced binding energy in nuclear matter, as pointed out earlier by Law (Law 68). Sauer (Sau 73) has used the method of Baranger et al. (Bar+ 69) to investigate the changes in the nuclear matter and ^{16}O binding energies for arbitrary changes in the high-energy $^1S_0$ phase shifts. He finds that, in agreement with the above results, the changes are insignificant. Only the matrix elements corresponding to the highly excited states (in ^{16}O) show some effect. These calculations are model dependent and limited in scope in the sense that the changes in the high-energy phase shifts are reflected in the potentials over a wide range in k or r. Ideally, the potential for each arbitrary variation in the high-energy phase shifts should be so designed that the modifications (local or nonlocal) occur only at small r, the long-range part (preferably OPEP) remaining unaltered.

10.2. Off-Shell Variation

10.2.1. Nucleon–Nucleon Bremsstrahlung

Nucleon–nucleon bremsstrahlung ($p + p \rightarrow p + p + \gamma$ and $n + p$ $\rightarrow n + p + \gamma$) has been quite widely studied during the last several years to extract off-shell information. This process has a number of advantageous features.

(i) The electromagnetic interaction is well understood and is relatively weak. This weakness of the electromagnetic coupling makes the multiple scattering series converge rapidly. In fact only first- and second-order terms in the two-body amplitude appear (CS 66), and if the second-order term is neglected, the cross section turns out to be a simple algebraic function of four half-shell amplitudes.

(ii) One need not consider the problem of three-body nuclear interactions.

(iii) Multiple scattering terms are much simpler than those involved in, for example, quasifree p–d scattering. If one includes all powers of k_γ (photon momentum) in the "external amplitude" (which is first order in the T-matrix) and neglects orders higher than k_γ^0 in the "internal amplitude", the bremsstrahlung cross section can be used directly to determine the off-shell T-matrix (LS 72).

(iv) One may consider events which are well off the energy shell.

(v) The cross section depends on half-shell matrix elements, which are somewhat easier to compute than fully off-shell scattering amplitudes.

But in practice the debit side is also pretty strong (Hal 72, Hel 72, Nym 74). The calculations necessarily involve several approximations regarding the contributions from internal scattering, exchange currents, Coulomb effects, noncoplanar scattering, magnetic interactions, gauge terms, modifications of the static electromagnetic form factors, relativistic effects, especially in the magnetic moment terms, and the influence of the negative energy states and covariant corrections. Some of these corrections have been estimated and analyzed (DM 68a, DMW 69, MS 69b, HR 71, McG 70b). For example, in the case of $pp\gamma$: the rescattering effects tend to increase the integrated cross section by less than 5% and the asymmetry by less than 20% (Bro 69); the Coulomb and relativistic effects account for up to about 30% of the cross section and these effects increase as the angle of scattering between the two protons decreases (SM '68a, Cel+ 73); the contribution from the gauge terms is usually very small (MS 69b).

A number of proton–proton bremsstrahlung ($pp\gamma$) calculations have been done using Hamada–Johnston (HJ 62), Bryan–Scott (BS 64), Tabakin (Tab 64, Tab 68), and Reid (Rei 68) potentials (Bro 67, Bro 69, PGD 67, DM 68a, DM 68b, MS 69b, LC 71, Bro 72). McGuire and collaborators (MCS 69, MC 69, MP 71a) have used experimental elastic phase shifts and generated the off-shell behavior from the OPE contribution. Felsner (Fel 67), Nyman (Nym 67, Nym 68), Pearce, Gale, and Duck (PGD 67), Signell and Marker (SM 68a, SM 68b), and Signell (Sig 69a, Sig 69b) have done approximate calculations in which off-shell t-matrix elements were replaced by on-shell ones. Baier, Kuhnelt, and Urban (BKU 69) have used a relativistic OBE model rather than a potential.

The neutron–proton bremsstrahlung ($np\gamma$) calculation is considerably more difficult than $pp\gamma$ (Bro 70). The identity of the protons in $pp\gamma$ requires totally antisymmetric nuclear states, which restrict orbital angular momenta to even (odd) values in singlet (triplet) states and only isotopic triplet states enter. In the case of $np\gamma$, however, isotopic singlet as well as isotopic triplet states must be included. As a result, there are fewer $np\gamma$ calculations. Baier, Kuhnelt, and Urban (BKU 69) have used a relativistic OBE model; McGuire (McG 70a) and McGuire and Pearce (MP 71b) used the experimental phase shifts with off-shell extrapolation through a modified OPE contribution; Pearce, Gale, and Duck (PGD 67) used the Tabakin potential; and Brown (Bro 70, BF 73b) used the Hamada–Johnston and Bryan–Scott potentials.

The main features which have emerged from all these calculations are the following:

(i) The internal scattering amplitude is strongly model dependent (McG 70a). The contribution from the pole term (external emissions) goes inversely as the photon momentum k_γ while that from the internal scattering is independent of k_γ. The influence of the internal scattering therefore decreases as the photon momentum goes to zero. Its contribution to $np\gamma$ is considerably more than in the case of $pp\gamma$ (Bro 70).

(ii) The magnetic terms dominate the $pp\gamma$ cross section while the electric terms dominate the $np\gamma$ cross section (McG 70a).

(iii) For a nuclear potential that does not include tensor and spin–orbit forces, magnetic internal off-shell terms cancel the pole off-shell terms to

$O(k_\gamma)$ (MCS 69). For the same potential, off-shell-electric effects cancel to $O(k_\gamma^0)$.

(iv) Above about 50 MeV and for $\theta \lesssim 30°$, the $pp\gamma$ cross section is dominated by the magnetic interaction, which in turn is strongly influenced by the tensor and spin–orbit forces (MCS 69). Precise experiments here may be able to distinguish off-shell effects. The noncentral forces are not as influential in the $np\gamma$ case (McG 70a).

(v) The gauge terms arising from the explicit dependence of the potential upon momentum and/or angular momentum operators are important in the study of the off-shell effects (Lio 70).

(vi) The $np\gamma$ cross section is much more sensitive to the S-wave phase shifts than the $pp\gamma$ cross section, which is quite sensitive to the higher partial waves (McG 70a).

(vii) The $np\gamma$ cross section increases as one goes out of the plane (noncoplanarity), in contrast with the $pp\gamma$ situation (BKU 69).

(viii) The photon angular distributions in the case of $np\gamma$ appear to provide a considerably more sensitive basis for testing the off-shell behavior of np interaction models than the cross sections integrated over θ_γ (BY 73). This is illustrated in Fig. 4.

(ix) The shapes of the cross sections and asymmetries are less sensitive (MP 71a, MP 71b) to off-shell influence than the corresponding quantities integrated over θ_1 or θ_2 or both, as well as θ_γ. Such integrated cross sections or asymmetries allow better experimental statistics and may also possibly

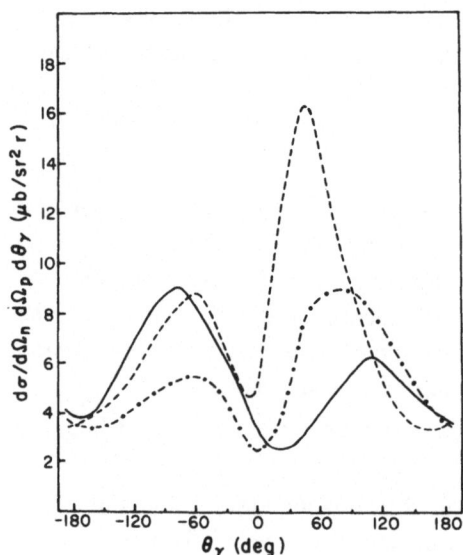

Fig. 4. Calculated $np\gamma$ photon angular distribution for $\theta_p = \theta_n = 30°$ and $E_{\text{Lab}} = 200$ MeV. Solid curve: Hamada–Johnston potential from BF 73b; dashed curve: OBE model of BKU 69; dash-dot curve: off-shell quasiphases from McG 70a. [Taken from (BY 73).]

help in relaxing the requirements on target design. At 150 MeV and above, the $np\gamma$ cross sections and asymmetries are about four times as large as the $pp\gamma$ cross sections and asymmetries (MP 71b). The reason is that the pp system, unlike the np case, does not have an electric dipole moment, so the lowest order radiation is of $M1/E2$ type.

(x) The most interesting difference between the $np\gamma$ and $pp\gamma$ cases is that the former shows greater off-shell variations (MP 71b). Furthermore, the cross-section itself tends to be relatively quite large just where the off-shell effects themselves are large, namely for forward scattering in the Gottschalk geometry and for k (momentum of the final state nucleon) near $k_{\max}$ in the Thorndike geometry. In contrast to this, the $pp\gamma$ cross section decreases monotonically.

The principal uncertainties in the calculations are connected with the proper handling of (i) the relativistic corrections to the electomagnetic vertex, (ii) the interaction currents, and (iii) the internal structure of nucleons. There are serious difficulties on the experimental side, too. First, the error bars in the existing experimental data are relatively large. Basically this is because the bremsstrahlung yield is about 10^6 times weaker than the elastic counting rate. The most precise data to date are those taken by Sannes, Trischuk, and Stairs (STS 68) at 99 MEV, Jovanovich *et al.* (Jov+ 71) at 42 MeV, and Willis *et al.* (Wil+ 72) at 156 MeV. The result is that the cross sections obtained with different potential models, though different, are all in general agreement with the experimental data. The $np\gamma$ experiments are even more difficult: There are presently only three measurements (BYB 68, BY 70, Fur+ 70, Edg+ 74). The $np\gamma$ asymmetry has not been measured at all. Second, the experiments have not been sufficiently far off the energy shell, with the result that the possible off-shell effects are not very prominent.

It is thus clear that bremsstrahlung will be able to provide useful information about the nuclear interaction only when more precise experiments are performed. Some typical recent data are shown in Fig. 5, illustrating the need for greater precision. What is needed are experiments in the 200–300 MeV range and sufficiently far off shell, i.e., at small proton exit angles so that most of the energy goes into the photon. An $np\gamma$ experiment near 300 MeV, preferably with a polarized neutron beam at small angles, would be still better (McG 70c). Besides the total cross section, one should also look more closely at the details of the differential cross section as a function of the various kinematical variables such as the noncoplanarity angle and the asymmetry in proton polar angles. These details are more

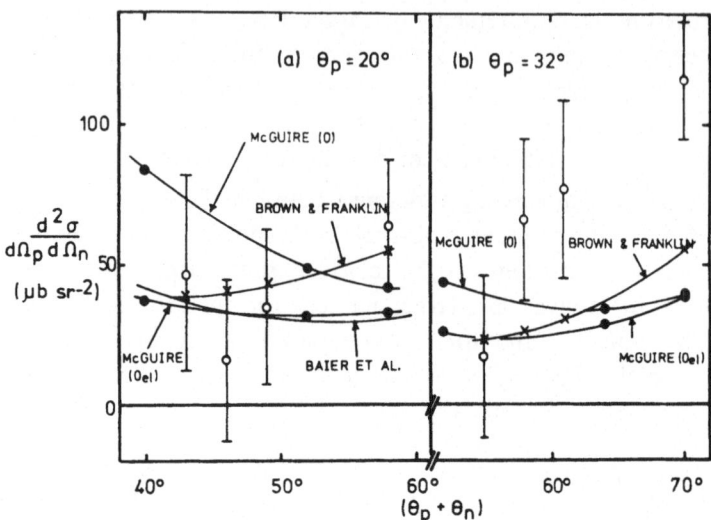

Fig. 5. *npγ* **cross-section data (open circles) at 131 MeV** (from Edg+ 74). Other lines
are the theoretical calculations indicated in Fig. 4. (a) $\theta_p = 20°$; (b) $\theta_p = 32°$.

likely to be the determining factors in distinguishing between one model and
another (LS 72). All this would require considerable improvement in experimental techniques. Together with this, the accuracy of the calculations
requires substantial improvement. It appears that further work in this
field will lead to a unique set of quasiphases and thereby provide information
about many aspects of the nucleon–nucleon interaction.

10.2.2. Nuclear Matter Calculations

The off-shell effects in nuclear matter have been quite widely studied
in recent years. They are especially important in view of the disagreement
between the binding energy obtained from "realistic" local potentials
(10–11 MeV/A) and the semiempirical mass formula prediction of 16
MeV/A. Some of this work was discussed in a previous article (Spr 72).

Srivastava, Banerjee, and Sprung (SBS 70) have used, in the $^1S_0$ state,
local potentials phase-shift equivalent to the separable potentials of Mongan
and Tabakin (both one-term and two-term), and found variations of up
to 5 MeV in the binding energy per particle. This calculation is rather
incomplete but is interesting because, given the separable interaction, the
equivalent local one is unique and because the separability represents some
sort of extreme nonlocality. Several workers (Mil+ 69, Sri 70b, Coe+ 70,
HT 71) have employed unitary transformations (Section 3.2) and isometric

point transformations (Section 3.1) to generate phase shift equivalent potentials in the S state. These potentials have p^2-dependent terms of varying strength and range, and differ in their smoothness. Large variations in binding energy have been reported, using these potentials. As shown in Fig. 6, taken from (Coe+ 70), one is dealing with a family of saturation curves and is moving the saturation point along a locus, where higher saturation density is associated with greater binding energy. Most of this gain in binding simply reflects the increase in size of the average matrix element with density. It has not been found easily possible to shift the locus of saturation points and gain energy at a fixed density.

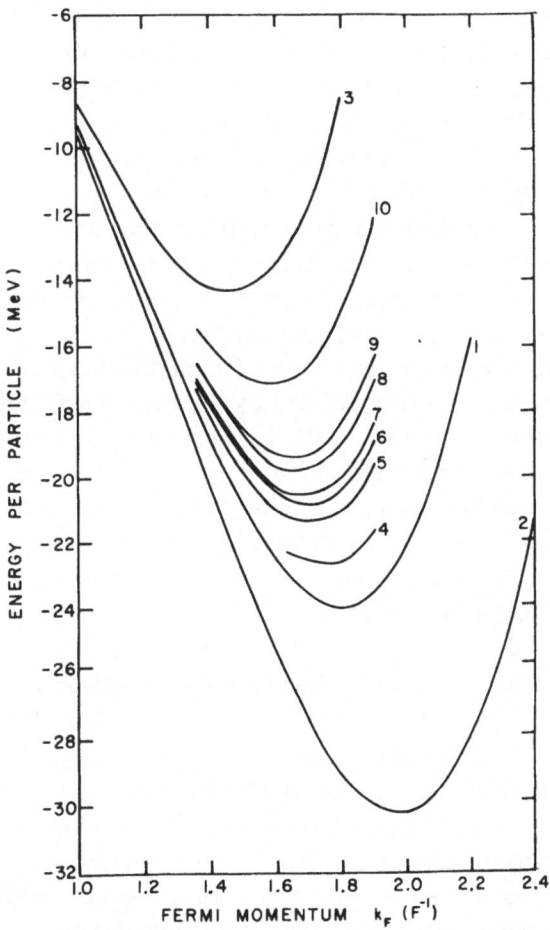

Fig. 6. A family of nuclear matter saturation curves obtained by a series of transformations on a Yukawa core potential. These curves can be related to each other by a scaling law involving the wound integral $\varkappa$. [Taken from (Coe+ 70).]

A detailed investigation of these effects has been made by Haftel and Tabakin (HT 71). They have used phase shift equivalent potentials generated from the Reid soft core potential by short-range unitary transformations. All these potentials have the required OPE tail and roughly the same deuteron electric form factor as the initial Reid potential. These potentials produce variations of up to 9.5 MeV in the binding energy and 0.33 fm^{-1} in the saturation density. This variation in binding energy is greater than the current 5–6 MeV disagreement between the binding energies of "realistic" local potentials and the binding energy predicted by the semiempirical mass formula and is also probably greater than the contribution attributable to the three- and higher-body clusters in the Goldstone expansion for realistic potentials. Similar results have been obtained by Wong and Sawada (WS 72). They used phase shift equivalent potentials generated from the Hamada–Johnston potential keeping the deuteron binding energy, D-state probability, and the potential beyond ~ 1 fm unchanged. They find a variation of about 4.5 MeV in the binding energy. Larger variations are possible only if the above empirical constraints on the deuteron observables and meson-theoretic constraints on the long-range part of the potential are lifted. The interesting features which emerge from these calculations are (i) the variations in energy at a given density are greater at higher densities, (ii) the principal quantity that governs the energy variations in nuclear matter is the wound integral $\varkappa$ of the Brueckner theory, (iii) the $^3S_1$ contribution to $\varkappa$ is controlled by the deuteron D-state probability, and (iv) the inclusion of the contributions of three- and four-hole diagrams does not affect much the magnitude of the variations in the binding energy [in agreement with the results of Coester, Day ,and Goodman (CDG 72)]. For a fixed starting energy and density, the binding energy is found to be nearly linear in $\varkappa$, with an increase in repulsion being associated with larger $\varkappa$, for potentials with approximately the same electric form factor. This is illustrated in Fig. 7. The electric form factor, however, does not uniquely determine the binding energy result. Furthermore, the result is also not found to be overly sensitive to the details of the wound. The wound integral (through the reference spectrum wound integral) can be expressed in terms of the half-shell K-matrix elements. It is found that the binding energy is sensitive to a large region of the off-shell K-matrix with far-off-shell elements playing a significant role. Since $\varkappa$ is taken to be the expansion parameter for the Goldstone series, one may reasonably expect significant off-shell variations in the many-body terms.

Recently Singh, Warke, and Bhaduri (SWB 72) have reported a nuclear matter calculation with phase shift equivalent mixtures of local and one-term

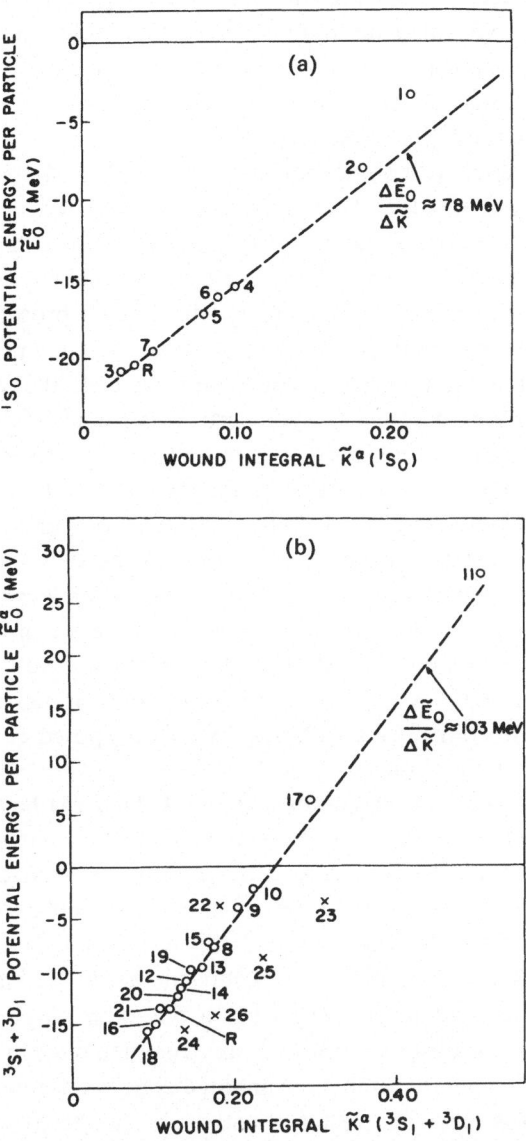

Figs. 7. Linear dependence of the potential energy on the wound integral $\varkappa$ for (a) the $\alpha = {}^1S_0$ and (b) the $\alpha = {}^3S_1 - {}^3D_1$ states. The numbers indicate the phase shift equivalent potentials considered in (HT 71). Potentials marked with $\times$ do not give the same electric form factors to within the experimental error as the Reid potential. [Taken from (HT 71).]

separable potentials. They take an attractive two-parameter potential for the local part and calculate the repulsive separable part by following Fuda's procedure (Section 3.4). They have shown that the wound integral $\varkappa$ decreases with increase in saturation density for all the potentials and as the limiting situation (in the sense of Fiedeldey) is approached, both the attractive and repulsive components become weaker, thereby making the wound integral $\varkappa$ smaller. They have also considered another class of potentials in which the attractive, two-parameter Bargmann potential is replaced by a phase shift equivalent, one-term, attractive, separable potential while the repulsive one is calculated by Fuda's procedure as before. These potentials are thus just two-term separable potentials. For these potentials, as the limiting condition is approached, the wound integral increases and the binding energy per particle is found to drop sharply. A similar sharp fall in binding energy per particle was observed by Srivastava, Singh, and Bhaduri (SSB 70) with limiting, second-rank, separable, phase shift equivalent potentials. This feature is thus peculiar to separable potentials and is associated with the behavior of the two-body bound-state wave function of separable potentials. It is absent for local or partly non-local potentials. In the light of Sauer's work discussed in Section 3.6, all of the work on phase equivalent $^1S_0$-state potentials is subject to revision. Some of the extreme variations found using transformed potentials may be associated with transforms violating Sauer's proposed constraint on the zero-energy wave function.

Recently Green and Haapakoski (GH 74, Haa 74) have found a new mechanism of great importance for the off-shell behavior seen in nuclear matter. Starting from the premise that the intermediate-range NN attraction is due to virtual intermediate states of a nucleon and a $\Delta(1236)$ isobar, they treat the NN and $N\Delta$ channels explicitly as a coupled channel problem. Since the Δ has $T = 3/2$, this only occurs in the $T = 1$ states, especially the $^1S_0$ state. The coupling potential has a one-pion range and a strength determined from the observed $\pi N\Delta$ vertex. One then finds in nuclear matter a strong saturating effect, completely analogous to that known in the tensor-coupled $^3S_1$–$^3D_1$ states. The result would be a significant lowering of the saturation density of nuclear matter, as compared to previous calculations where the $N\Delta$ channel was eliminated, along with other mesonic degrees of freedom, to give a normal potential model. There is no doubt that this process warrants further study. The long-range coupling potential is needed to give saturation near normal nuclear matter density.

10.2.3. Finite Nucleus Calculations

Recently some attempts have been made to study the off-shell effects in finite nuclei with phase shift equivalent potentials. The aim is to find out whether these effects, which are so important in nuclear matter, persist in finite nuclei. It is interesting to find out whether improved results for binding energies, energy spectra, and nuclear radii could be obtained by modifying the potential. Earlier attempts in this direction simply used different potentials or the phase shift approximation. The off-shell variations were not properly scanned or were ignored altogether.

Haftel, Lambert, and Sauer (HLS 72) have studied off-shell effects in ^{16}O. The occupied single-particle wave functions are taken to be oscillator wave functions of fixed oscillator energy. An "angle-averaged" Pauli operator is used, the particle–particle potential energy for the unoccupied states is set equal to zero, and the G-matrix elements are calculated according to the method of Sauer (Sau 70). The effects of the three-body clusters are ignored. These approximations are expected to be reasonably accurate for the purpose of comparing different potentials. The potentials used are those obtained by Haftel and Tabakin (HT 71) by a unitary transformation of the Reid soft core potential. It is found, as shown in Fig. 8, that the regular dependence of the binding energy predictions for phase shift equiv-

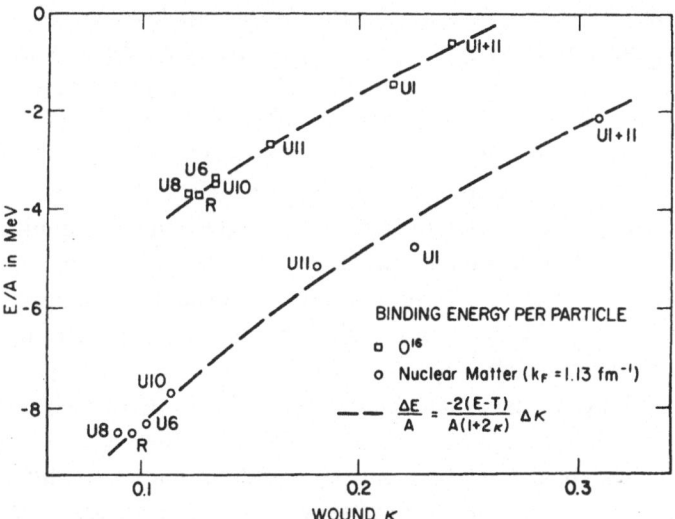

Fig. 8. Binding energy versus wound integral for a number of unitarily transformed potentials in nuclear matter and ^{16}O. [Taken from (HLS 72).] The dashed lines follow a simple theoretical estimate.

alent potentials on the wound integral $\varkappa$, observed in nuclear matter, persists in ^{16}O. This regular dependence occurs even if one considers potentials that produce extremely unusual short-range correlations. The magnitude of the changes is comparable to those in nuclear matter if the effects of the nuclear surface are taken into account. These are the lowering of the average density and the suppression (due to weight factors) of the D-state interactions. It is found that the defect wave functions in ^{16}O and nuclear matter show very good agreement state by state in spite of the differences in the Pauli operator and the hole spectrum. This similarity of correlations is of interest in deriving the density-dependent effective nucleon–nucleon interaction in finite nuclear systems from nuclear matter calculations.

Another finite nucleus calculation to investigate the off-shell effects has been done by Pradhan, Sauer, and Vary (PSV 72). It is a shell model calculation for ^{18}O and ^{18}F with the same set of phase shift equivalent potentials and basically the same approximations as in ^{16}O above. Core polarization corrections to the effective interaction have been considered by including all excitations up to the $1h$ oscillator orbital. The variations in matrix elements are found to be strongly state dependent and are much greater than those indicated by the calculations of Lynch and Kuo (LK 67) with the Yale, Reid, and Hamada–Johnston potentials and of Elliott *et al.* (Ell+ 68) in phase shift approximation. As in the nuclear matter and ^{16}O calculations, an approximately linear relationship between G-matrix elements and the wound integral $\varkappa$ at a fixed starting energy is also observed here. Regarding the spectra, Pradhan *et al.* (PSV 72) find strong sensitivity of the low-lying levels with respect to the off-shell variations, but these effects do not show up simply as a displacement of a whole group of levels. The reasons are that (i) the changes in the matrix elements are strongly state dependent and (ii) the different shell model states tend to exploit the components of the effective interaction with varying weights. A force producing small changes in ^{18}O or ^{18}F may produce substantial changes in other cases. Sauer (Sau 73) has come to similar conclusions generating off-shell variation by the method of Section 4.1.

These finite nucleus calculations have been performed at a fixed oscillator length and hence at a fixed rms radius. However, the phase shift equivalent potentials used here saturate nuclear matter at different densities. These potentials will, in fact, yield different rms radii and different charge distributions in a proper Brueckner–Hartree–Fock calculation (Tab 72).

10.2.4. Three-Body Bound State Calculations

The three-nucleon bound state calculation is another problem providing an excellent opportunity for investigating the consequences of the off-shell uncertainty (and thereby learning something about the interaction). It also broaches the question of the adequacy of two-body forces for the description of more complex nuclear phenomena (Mit 69). The three-body problem has been a spawning ground for separable potential models and a testing ground for the various separable approximations. Unlike most nuclear many-body problems, an exact theory is available, namely the Faddeev formalism (Fad 61), and the calculations are manageable for potentials that act in a small number of partial waves (HHO 68, MT 69a, Kim 69, Bra 69, Tjo 70, MT 70, HKT 70, HKT 72a, HKT 72b, KT 72).

Earlier studies of the off-shell effects in the triton indicated that the binding energy E_T is moderately sensitive to off-shell properties of the interaction. Recently two extensive calculations have been reported, one by Fiedeldey (Fie 69b, FM 72), using phase shift equivalent second-rank separable potentials, and the other by Haftel (Haf 73), employing unitary transformations to generate equivalent potentials. In the following we outline these calculations and discuss the main results.

In the Fiedeldey scheme the potentials are generated, as pointed out earlier, by choosing one of the form factors rather arbitrarily and calculating the other. For this study Fiedeldey (Fie 69b, FM 72) used the average S-state potential of Tabakin (Tab 65) as the starting potential, and took the attractive form factor of the Yamaguchi type

$$g(k) = \gamma/(k^2 + \beta^2) \tag{261}$$

He generated equivalent potentials while varying the parameters γ and β ove quite a wide range. It is found that in the cases where E_B (the two-body binding energy with the attractive form factor alone) is significantly different from E_D (the binding energy with the complete potential), the triton binding energy E_T is nearly proportional to the zero-energy wound integral I_0 [Eq. (184)] with small deviations depending on the far-off-shell behavior. This feature is similar to that observed by Haftel and Tabakin in nuclear matter (HT 71). In the limit $E_B \gg E_D$ it is found that $E_T(\beta, \gamma)$ is only weakly dependent on γ [Eq. (261)]. The UPA is a good approximation in these cases and the two-body bound-state wave function does not differ very much from the bound state wave function of the starting potential. In the cases where $E_B \approx E_D$, large off-shell effects are observed (E_T varying from 8.2 to 16.2 MeV). Here the UPA fails completely and the near linear relation between I_0 and E_T does not hold, though they are found to change in the same direction. The weak repulsive term in the potential in such cases is more strongly felt in nuclear matter ($s \approx -3.0$ fm^{-2}) than in the triton, due to its complete suppression at the nearby deuteron pole ($s = -mE_D/\hbar^2$). The result is that the nuclear matter and the triton binding energies show large but opposite variation (SSB 70, Fie 69b) as the potentials approach criticality ($E_B = E_D$). This is a special effect which can only occur for separable potentials.

McGurk *et al.* (McG+ 74) have recently extended this calculation to include the effect of the tensor force in the $^3S_1$–$^3D_1$ channel. They find, in agreement with the results of Malfliet and Tjon (MT 69b), that E_T decreases monotonically with decreasing D-state probability. The nuclear matter binding energy also shows a similar trend (CSA 69).

Haftel (Haf 73) has taken a two-term Yukawa potential in the relative S state as his starting potential and has analyzed the role played by the deuteron wave function. He finds that the principal factors dominating the off-shell variations in the triton binding energy are changes in the pole behavior of the T-matrix. Most of the variation ($\sim 80\%$) is found to be attributable to the changes in the deuteron wave function, in accordance with the findings of Fiedeldey and McGurk (FM 72), Afnan and Serduke (AS 73), Fiedeldey (Fie 74), and Lavine and Stephenson (LS 74). For potentials whose deuteron wave functions differ mainly for separations of less than about 1.5 fm, the change in triton binding energy is found to be associated with the changes in the Fourier components for $1 \lesssim k \lesssim 2$ fm^{-1}. Potentials that yield more compressed deuteron wave functions for $k \lesssim 2$ fm^{-1}, i.e., fall off more rapidly in k, give less binding in the triton. Figure 9 shows some results from Haftel's paper. In cases where potentials yield virtually the same deuteron wave function, changes in the low- and intermediate-momentum components ($k \lesssim 2$ fm^{-1}) of the low-energy half-shell T-matrix seem to govern the off-shell changes in triton results. The low-energy half-shell T-matrix, shown in Fig. 10 as a function of k, thus seems to play a role analogous to the deuteron wave function. Lavine *et al.* (LMS 73) have also studied the sensitivity of the triton binding energy and of scattering below breakup to the T-matrix in different regions of its momentum arguments by directly modifying the T-matrix through multiplication by a function of energy and momentum that maintains off-shell unitarity. They find, in agreement with the above results of Haftel, a strong sensitivity when both momenta are less than 2 fm^{-1}, a weak potential-dependent sensitivity when either argument is between about 2 and 5 fm^{-1}, and essentially no dependence if either argument exceeds 5 fm^{-1}. These

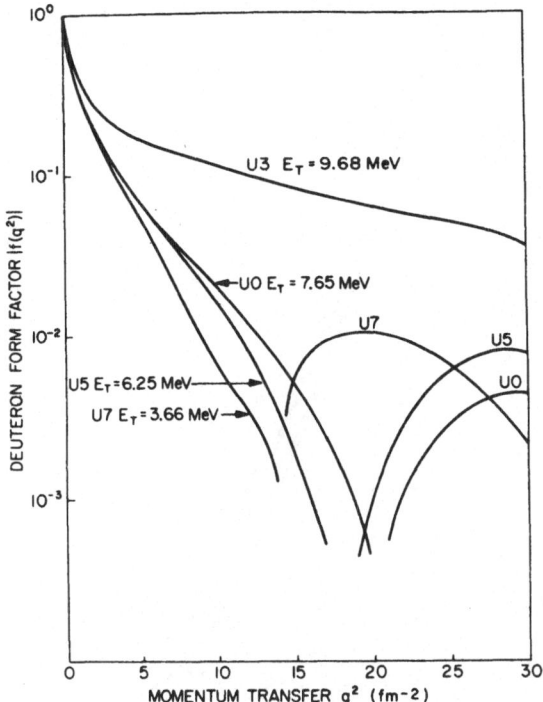

Fig. 9. Deuteron form factors for some phase shift equivalent potentials along with the corresponding triton binding energy. [Taken from (Haf 73).]

results are also supported by the recent work of Kharchenko *et al.* (KSS 71), who studied separable approximations to square well potentials with repulsive cores.

A similar pattern is observed between the triton binding energy and the deuteron form factor $|f(q^2)|$ as a function of the momentum transfer. It is found that the observed changes in E_T are attributable to the changes in $|f(q^2)|$ over a range of momentum transfers q centered at about $q^2 \approx 12$ fm^{-2}. This result is interesting since the form factor is more closely related to the experiment (i.e., e–d elastic scattering experiments) than is the wave function.

In r space the triton bound state properties are found to be rather sensitive to those parameters of the interaction that dominate the behavior at intermediate distances (about 1.5–2.0 fm). They are insensitive to its behavior near the origin (WE 71, Wag+ 71, Zoh 73, BF 73a, Sri 73b).

For potentials that give virtually the same deuteron wave functions the variations in E_T are in the same direction and are comparable to the

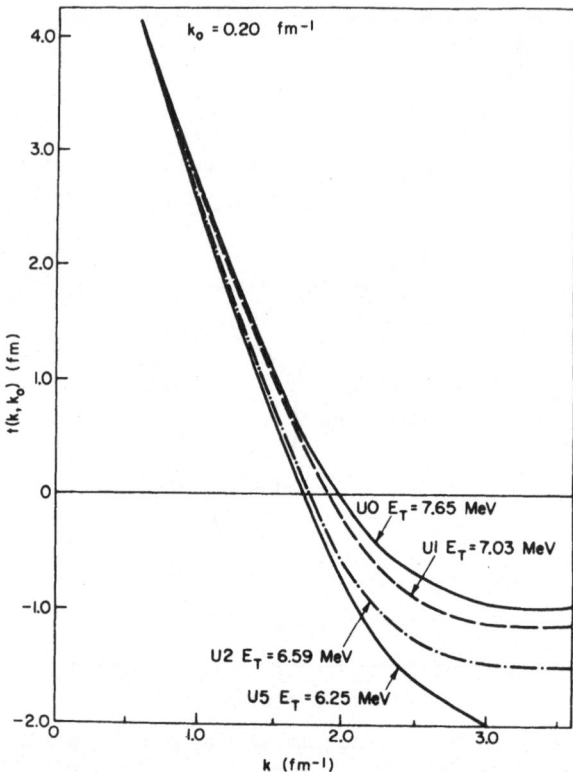

Fig. 10. Low-energy half-shell t-matrix elements and the triton binding energy for some phase shift equivalent potentials giving roughly the same deuteron wave function. [Taken from (Haf 73).]

variations in nuclear matter binding energy calculated at densities $k_F \approx 0.6$ fm^{-1}. In the cases where the deuteron wave functions are different, the variations are much larger but are not always in the same direction as those in nuclear matter binding energy.

Haftel (Haf 73) has also looked at the effects of off-shell variations in the $^1S_0$ state on the triton. The $^1S_0$-state interaction, while almost as strong as the $^3S_1$-state interaction, does not support a bound state. Instead, the virtual bound state resonance dominates the $^1S_0$ t-matrix at low energies. Direct experimental measurements of the "singlet deuteron" form factor (or low-energy half-shell $^1S_0$ t-matrix elements) do not exist. Applying the method of Picker et al. (PRS 71) to the two-nucleon scattering wave function at distances of less than 1.4 fm, he has shown that $^1S_0$ t-matrix variations could account for about 4 MeV uncertainty in E_T. For a given realistic deuteron wave function, most of the uncertainty in E_T thus comes from the $^1S_0$ interaction. Further experimental constraint on it is necessary. Alternatively, E_T itself could provide some constraint on the low-energy $^1S_0$ interaction.

Another important result of the above calculation, in agreement with findings of Harper, Kim, and Tubis (HKT 72c), Laverne and Gignoux (LG 73), and Sauer and Tjon (ST 73a), is that it appears impossible to find a potential that simultaneously predicts the correct phase shifts, triton binding energy, and the ^{3}He form factor near and past the diffraction minimum. Phase shift equivalent potentials that predict reasonable binding in the triton predict the diffraction minimum at too large values of momentum transfer. Inclusion of the tensor force, higher partial waves, and relativistic corrections does not change the situation much (BKT 74). Improving the binding energy appears to worsen the form factor behavior of realistic potential models [e.g., the Reid potential: $E_T = 6.9$ MeV, diffraction minimum at $q^2 \approx 17.7$ fm^{-2}, secondary diffraction maximum of about 0.7×10^{-3} at $q^2 \approx 22$ fm^{-2} (BKT 74) vs. experiment (Col+ 65, McC+ 70, Ber+ 72): $E_T = 8.5$ MeV, diffraction minimum (if indeed one exists) at $q^2 \approx 11.0 \pm 1$ fm^{-2}, secondary diffraction maximum of about 0.6×10^{-2} at $q^2 \approx 18$ fm^{-2}]. The variational calculations (YJ 71, HJ 72, HD 72), on the other hand, of the ^{3}He form factor yielded better agreement with experiment than the Faddeev calculation. These calculations, however, have doubtful reliability. The Jackson–Lande–Sauer method (JLS 70) was pursued at Orsay in an attempt to make it converge. Even with 4654 states included, the binding energy, charge radius, and form factor up to the diffraction minimum are only reliable to within 10% (SS 74). Nevertheless, the systematic nature of this calculation may permit reliable extrapolation to the true result.

The above incompatibility between the binding energy and the form factor results with phase shift equivalent potentials has prompted further study in looking for additional physical ingredients such as three-body forces, charge-dependent forces, meson-exchange currents, etc. Kloet and Tjon (KT 74a, KT 74b) have recently claimed (see Fig. 11) that the discrepancy between theory and experiment is considerably reduced by including the effect of meson exchange currents on the form factor.

Brayshaw (Bra 73) has investigated the possibility of three-body forces starting from the other end. He constructs a model of the triton wave function from the existing data for the charge form factors of ^{3}He and ^{3}H and solves the Schrödinger equation for the effective local potential. The discrepancy shown in Fig. 12 between this potential and those predicted by typical pair-interaction models suggests the presence of a strong at-

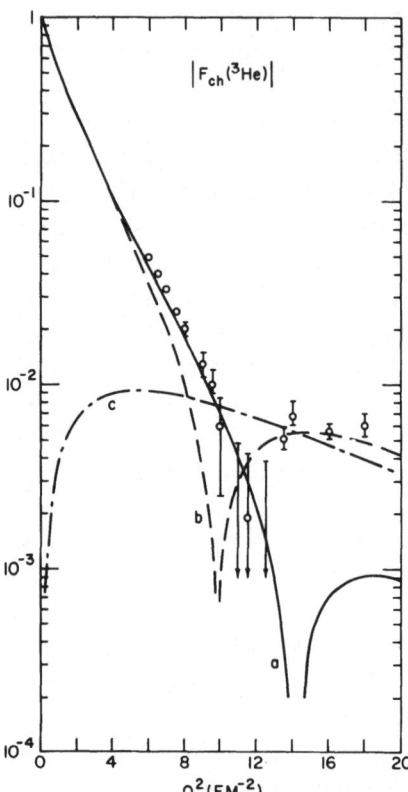

Fig. 11. The ³He charge form factor for the Reid soft core potential. (a) Without and (b) including meson exchange corrections (curve c) as calculated by Kloet and Tjon (KT 74b). The data points are from (McC+ 70).

Fig. 12. Comparison of the empirical and theoretical effective local potentials in the triton. Curve (a) corresponds to the wave function analysis of Brayshaw, while (b) and (c) are generated by the two-nucleon potentials considered by him. The discrepancy suggests the presence of an appreciable three-body interaction. [Taken from (Bra 73).]

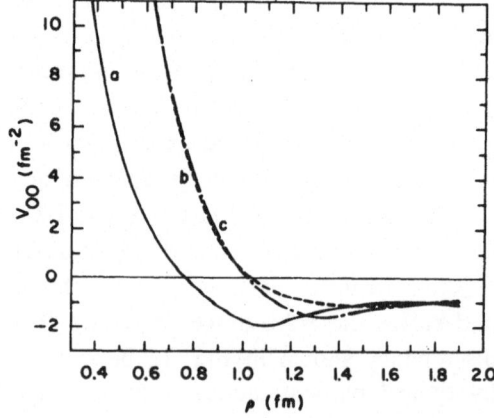

tractive three-body force. Loiseau and Nogami (LN 67) have extimated an extra binding of 1.5 MeV due to three-body forces.

The normalization constant of the N–d tail of the trinucleon wave function is another parameter whose variation should be studied along with those of the binding energy and the form factor (Lim 73).

10.2.5. Deuteron Breakup Calculations

Other simple systems where off-shell effects are important offer the possibility of extracting some information. Such is deuteron breakup by: (a) photodisintegration, (b) the $d(p, 2p)n$ reaction, and (c) muon capture.

Some of the earlier photodisintegration calculations (LR 68, RS 71) done with the Hamada–Johnston, Yale, Nestor (Nes+ 68) (velocity dependent), and Reid potentials indicated that experiment is unable to distinguish among the various potentials. Newton (New 57), on the other hand, employing phase shift equivalent Bargmann potentials (Bar 49), showed that the experimental disintegration cross section singles out the n–p central triplet potential with the shortest exponential tail. Recently a more extensive calculation to study the off-shell dependence of the disintegration cross section has been done by Van Dijk and Razavy (VR 73) using phase shift equivalent potentials generated through unitary transformations. They find that the cross section is sensitive to the off-shell t-matrix. The integrated cross section can be varied by as much as 30% or more [in agreement with an earlier result of Kistler (Kis 69)], and the matrix element for the $E1$ transition by a factor of two. However, this sensitivity is not strong enough to determine a unique off-shell extrapolation of the t-matrix. Van Dijk and Razavy have obtained a class of phase shift equivalent interactions with different off-shell elements such that all of them yield the same $E1$, $E2$, and $M1$ transition probabilities. Thus the result can certainly be used to limit the arbitrariness in the extrapolation.

The $d(p, 2p)n$ reaction has been quite extensively studied (Mor+ 69, Val+ 71, MT 71, Wal 72). It is found that the calculated cross sections (taking into account the first- and the second-order terms in a multiple scattering expansion: L'Hu 74) are in good agreement with the experimental results if the kinematics are such that the recoil neutron energy is small. In this case the interaction is nearly on-shell. Recently, Morlet *et al.* (Mor+ 72) have measured the cross sections for a 155-MeV proton beam and for high momentum transfer to the neutron (50 MeV). The interaction (pp or pn) in this case is far off-shell. The calculations carried out (L'Hu 74, BHB 74) with the Hamada–Johnston and Sprung–de Tourreil (TS 73) potentials showed that the difference between the results can be as much as 50%. The experiment is thus sensitive to the off-shell behavior of the interaction. Similar results have been obtained by Oryu (Ory 71). Analyzing the experimental results (GK 64, Sla+ 66, Bra+ 70) for incident energies in the range 38–50 MeV, he finds that the incorporation of off-shell amplitudes improves the energy distribution of the emitted proton. Another experiment has recently been done off the energy shell at 58.5 MeV (DBP 74). However, further study is necessary to pick suitable kinematical variables (JRS 73), choose sensitive regions in phase space (KT 73b), and investigate the importance of the deuteron D state and the higher order terms in the multiple scattering expansion. The possibility of three-body forces is another complicating factor.

The reaction $\bar{\mu} + d \rightarrow 2n + \nu_\mu$ has recently been studied by Sotona and Truhlík (ST 73b) using (i) several simple local and nonlocal potentials fitted to the low-energy scattering data and (ii) the Reid soft core potential and its phase shift equivalents obtained by using the rank-one unitary transformation of Haftel and Tabakin. It is found that any value of the $^1S_0$ contribution to the doublet transition rate in the range 100–250 sec^{-1} can be obtained by a suitable choice of the parameters in the unitary transformation. The main contribution comes from the region $p \leq 2$ fm^{-1}. In nuclear matter, on the other hand, far-off-shell matrix elements play a significant role. Thus this reaction is found to be quite sensitive to the off-shell behavior and can provide useful information about the nn interaction.

10.2.6. Knockout Reactions

Knockout reactions, which proceed via quasi-elastic scattering, can also be used, at least in principle, to provide some information about the off-shell behavior of the interaction (Pug+ 69, Wat+ 71). A detailed investigation in this direction has been initiated by Redish *et al.* (RSL 70) and Stephenson *et al.* (Ste+ 72). They have considered the $(p, 2p)$ reaction, treating it as a three-body process and using the plane wave approximation. The target is taken as a two-body bound state with a given wave function. The complexities due to the target structure, distortion effects, multiple scattering corrections, and inelastic effects have been ignored. With these simplifications the amplitude factorizes into a product of the off-shell two-body t-matrix and the target bound-state wave function in momentum space.

They find that the reactions above 300 MeV are not sensitive to the off-shell behavior. All the on-shell prescriptions studied by them give equivalent results. On the other hand, for incident proton energies below 200 MeV, the use of off-shell amplitudes is found to change the cross section significantly as compared with the on-shell approximations. The effect increases with the binding energy of the struck proton and with the recoil momentum. The reactions below 200 MeV are thus sensitive to the off-shell behavior of the interaction.

The extraction of off-shell information is, however, difficult because of distortion effects. The coupling between the distortion and the off-shell effects needs further sorting out. Some progress in this direction has recently been reported by Redish (Red 73). He has constructed an approximate factorized distorted-wave impulse approximation (DWIA) together with off-shell effects.

11. SUMMARY AND COMMENTS

The many-body studies described above have clarified the picture considerably. The relationship between the off-shell behavior and the characteristics of the interaction is pretty well understood. Effective separable approximations and powerful methods of calculating the T-matrix have been developed. We now summarize some of the main features which have emerged from these studies and which give an inkling of the directions for future work.

(i) Fixing the deuteron wave function (or the form factor) is found to restrict the off-shell variations considerably. Many-body calculations in general [for example: triton binding energy (MT 69b, McG+ 74), nuclear matter binding energy (CSA 69, WS 72), n–d scattering (Dol 74), (d, p) reactions (Knu+ 73), pion production (TA 71, PS 73)] are found to be quite sensitive to the D-state probability in the deuteron. Therefore the choice of the phase shift equivalent potentials in the $^3S_1$–$^3D_1$ channel should be restricted to those conforming to the known deuteron properties. The low-energy, energy-dependent phase shift analysis should also manifestly contain deuteron constraints (AR 70).

(ii) There is need for a similar restriction on the $^1S_0$-state interaction, which is almost as strong as the $^3S_1$-state interaction. The low-energy t-matrix is dominated here by the virtual bound state, but the direct experimental measurements of low-energy half-shell matrix elements do not exist. The $pp\gamma$ may be very useful in providing some information on the low-energy $^1S_0$-state interaction. Sauer has given arguments (Section 3.6) for constraining the zero-energy $^1S_0$ wave function. This should, as in the deuteron case, restrict the range of off-shell variations significantly.

(iii) The high-energy on-shell data are not very effective in most of the problems.

(iv) It appears that the short- and intermediate-range alterations (generated by the unitary transformations, etc.) of the potential will not by themselves produce simultaneously the correct nuclear binding and rms radius. For the triton an incompatibility between the binding energy and form factor seems well established.

(v) A major result of the studies with phase shift equivalent potentials has been to promote attempts to look for additional physical ingredients in the problem. For example: (a) falling back on the mesonic origin of the potential to decide on the proper charge and current density operators for $NN\gamma$, (b) the recent consideration of meson exchange effects in resolving the discrepancy in the triton binding energy and form factor results, and (c) the role and properties of the three-body force. In this connection a recent method (Rei 74) of obtaining off-shell T-matrix elements directly from the known on-shell data and the relativistic meson theory is worth recalling (Section 4.3).

(vi) On the experimental side, the $NN\gamma$ and the $d(p, 2p)n$ reactions appear to be the most promising. They provide a lot of maneuverability in fixing the kinematics (so that the interaction is quite off-shell) and exploring the phase space, and are simpler to understand. Of course, considerable improvement in experimental techniques will be required to reduce the error bars.

Throughout this chapter, we have proceeded on the premise that a nonrelativistic potential model is an adequate framework in which to discuss off-shell behavior, at least in a first approximation. The mesonic origin of the force would provide corrections to this picture. The work of Green and Haapakoski (GH 74, Haa 74) shows that when the virtual isobars are treated explicitly, drastic effects on the off-shell behavior are produced. During the next two years the meson factories will be producing new data, which may also show up inadequacies of the nonrelativistic potential model. Probably we have much to learn about the off-shell behavior of the nucleon–nucleon interaction.

ACKNOWLEDGMENTS

We would like to thank our colleagues who made suggestions for improvements in this chapter, and Mrs. Helen Kennelly for typing and retyping successive versions of it. Continued research support from the National Research Council of Canada under operating grant A-3198 is gratefully acknowledged. The Department of Atomic Energy, Government of India is also thanked for research support (MKS).

REFERENCES

AB 74	E. O. Alt and B. L. G. Bakker, Preprint (1974).
AC 67	A. Ahmadzadeh and V. Chung, *Phys. Rev.* **161**:1602 (1967).
ACS 71	I. R. Afnan, D. M. Clement, and F. J. D. Serduke, *Nucl. Phys.* **A170**:625 (1971).
Ama 70	R. D. Amado, *Phys. Rev. C* **2**:2439 (1970).
AR 65	V. de Alfaro and T. Regge, *Potential Scattering*, North-Holland, Amsterdam (1965), p. 150.
AR 70	R. A. Arndt and L. D. Roper, *Phys. Rev. D* **1**:129 (1970).
AR 73	I. R. Afnan and J. M. Read, *Phys. Rev. C* **8**:1294 (1973).
AS 66	M. Abramowitz and I. A. Stegun, *Handbook of Mathematical Functions*, National Bureau of Standards, Washington, D.C. (1966).
AS 73	I. R. Afnan and F. J. D. Serduke, *Phys. Lett.* **44B**:143 (1973).
AT 68	I. R. Afnan and Y. C. Tang, *Phys. Rev.* **175**:1337 (1968).
Bak 62	G. A. Baker, Jr., *Phys. Rev.* **128**:1485 (1962).
Ban 69	P. K. Banerjee, Thesis, McMaster University (1969), unpublished.
Bar 49	V. Bargmann, *Rev. Mod. Phys.* **21**:488 (1949).
Bar+ 69	M. Baranger, B. Giraud, S. K. Mukhopadhyay, and P. U. Sauer, *Nucl. Phys.* **A138**:1 (1969).
BB 74	J. Binstock and R. Bryan, *Phys. Rev. D* **9**:2528 (1974).
BBP 63	H. A. Bethe, B. H. Brandow, and A. G. Petschek, *Phys. Rev.* **129**:225 (1963).
Bea 69	J. E. Beam, *Phys. Lett.* **30B**:67 (1969).
Bec 67	R. L. Becker, *Nuclear Physics: An International Conference*, Academic Press, New York (1967), p. 673.
Beg 61	M. A. Baqi Beg, *Ann. Phys.* (*N.Y.*) **13**:110 (1961).
Ber+ 72	M. Bernheim, D. Blum, W. McGill, R. Rishalla, C. Trail, T. Stoval, and D. Vinciguerra, *Lett. Nuovo Cimento* **5**:431 (1972).
BF 72	O. P. Bahethi and M. G. Fuda, *Phys. Rev. C* **6**:1956 (1972).
BF 73a	O. P. Bahethi and M. G. Fuda, *Phys. Rev. C* **7**:1845 (1973).
BF 73b	V. R. Brown and J. Franklin, *Phys. Rev. C* **8**:1706 (1973).
BG 57	H. A. Bethe and J. Goldstone, *Proc. Roy. Soc.* **A238**:531 (1957).
BG 69	C. G. Bollini and J. J. Giambiagi, *Phys. Lett.* **30B**:441 (1969).
BGG 68	G. E. Brown, A. M. Green, and W. J. Gerace, *Nucl. Phys.* **A115**:435 (1968).
Bha 73	S. C. Bhatt, *Nucl. Phys.* **A217**:491 (1973).
BHB 74	J. L. Ballot, M. L'Huillier, and P. Benoist-Gueutal, Orsay preprint IPNO/TH 74-23.
Bis 73	R. F. Bishop, *Phys. Rev. C* **7**:479 (1973).
BJK 69	G. E. Brown, A. D. Jackson, and T. T. S. Kuo, *Nucl. Phys.* **A133**:481 (1969).
BKR 69	C. N. Bressel, A. K. Kerman, and B. Rouben, *Nucl. Phys.* **A124**:624 (1969).

BKT 74   R. A. Brandenburg, Y. E. Kim, and A. Tubis, *Phys. Lett.* **49B**:205 (1974).

BKU 69   R. Baier, H. Kühnelt, and P. Urban, *Nucl. Phys.* **B11**:675 (1969).

BLH 72   S. C. Bhatt, J. S. Levinger, and E. Harms, *Phys. Lett.* **40B**:23 (1972).

BM 65   M. Bolsterli and J. Mackenzie, *Physics (U.S.)* **2**:141 (1965).

Bol 69   M. Bolsterli, *Phys. Rev.* **182**:1095 (1969).

Bra 69   D. D. Brayshaw, *Phys. Rev.* **182**:1658 (1969).

Bra 71   D. D. Brayshaw, *Phys. Rev. C* **3**:35 (1971).

Bra 73   D. D. Brayshaw, *Phys. Rev. C* **7**:1731 (1973).

Bra+ 69   T. Brady, M. G. Fuda, E. Harms, J. S. Levinger, and R. W. Stagat, *Phys. Rev.* **186**:1069 (1969).

Bra+ 70   W. J. Braithwaite, J. M. Cameron, D. W. Storm, D. J. Margaziotis, G. Paic, J. G. Rogers, J. W. Verba, and J. C. Young, *Three-Body Problem in Nuclear and Particle Physics*, North-Holland, Amsterdam (1970).

Bro 67   V. R. Brown, *Phys. Lett.* **25B**:506 (1967).

Bro 69   V. R. Brown, *Phys. Rev.* **177**:1498 (1969).

Bro 70   V. R. Brown, *Phys. Lett.* **32B**:259 (1970).

Bro 72   V. R. Brown, *Phys. Rev. C* **6**:1110 (1972).

BS 64   R. A. Bryan and B. L. Scott, *Phys. Rev.* **135**:434 (1964).

BS 67   P. C. Bhargava and D. W. L. Sprung, *Ann. Phys. (N.Y.)* **42**:222 (1967).

BW 68   J. S. Ball and D. Y. Wong, *Phys. Rev.* **169**:1362 (1968).

BW 72   R. P. Bhatia and J. F. Walker, *Nucl. Phys.* **A192**:658 (1972).

BY 70   F. P. Brady and J. C. Young, *Phys. Rev. C* **2**:1579 (1970).

BY 73   F. P. Brady and J. C. Young, *Phys. Rev. C* **7**:1707 (1973).

BYB 68   F. P. Brady, J. C. Young, and C. Badrinathan, *Phys. Rev. Lett.* **20**:750 (1968).

CAM 70   M. Coz, L. B. Arnold, and A. D. MacKellar, *Ann. of Phys.* **59**:219 (1970).

CD 68   R. E. Cutkosky and B. B. Deo, *Phys. Rev.* **174**:1859 (1968).

CDG 72   F. Coester, B. Day, and A. Goodman, *Phys. Rev. C* **5**:1135 (1972).

CDR 72   M. Chemtob, J. W. Durso, and D. O. Riska, *Nucl. Phys.* **B38**:141 (1972).

Cel+ 73   L. S. Celenza, M. K. Liou, M. I. Sobel, and B. F. Gibson, *Phys. Rev. C* **8**:838 (1973).

Cha 58   K. Chadan, *Nuovo Cimento* **10**:892 (1958).

Cha 67   K. Chadan, *Nuovo Cimento* **47A**:510 (1967).

Cho+ 72   K. F. Chong, Y. Singh, D. W. L. Sprung, and M. K. Srivastava, *Phys. Lett.* **38B**:132 (1972).

Ciu 69   S. Ciulli, *Nuovo Cimento* **62A**:301 (1969).

CJ 73a   Y. A. Chao and A. D. Jackson, *Phys. Lett.* **43B**:449 (1973).

CJ 73b   Y. A. Chao and A. D. Jackson, *Nucl. Phys.* **A215**:157 (1973).

CL 72   H. C. Chiang and T. Y. Lee, *Chinese J. Phys.* **10**:35 (1972).

Coe+ 70   F. Coester, S. Cohen, B. Day, and C. M. Vincent, *Phys. Rev. C* **1**:769 (1970).

Col+ 65   H. Collard, R. Hofstadter, E. B. Hughes, A. Johansson, M. R. Yearian, R. B. Day, and R. T. Wagner, *Phys. Rev.* **138**:B57 (1965).

Cot+ 73   W. N. Cottingham, M. Lacombe, B. Loiseau, J. M. Richard, and R. Vinh Mau, *Phys. Rev. D* **8**:800 (1973).

CS 66   A. H. Cromer and M. I. Sobel, *Phys. Rev.* **152**:1351 (1966).

CSA 69   D. M. Clement, F. J. D. Serduke, and I. R. Afnan, *Nucl. Phys.* **A139**:407 (1969).

DBP 74   J. L. Durand, O. M. Bilaniuk, and C. Perrin, *Nucl. Phys.* **A224**:77 (1974).

Dey+ 69   J. Dey, J. P. Elliott, A. D. Jackson, H. A. Mavromatis, E. A. Sanderson, and B. Singh, *Nucl. Phys.* **A134**:385 (1969).

DM 68a   D. Drechsel and L. C. Maximon, *Ann. Phys. (N.Y.)* **49**:403 (1968).

DM 68b   D. Drechsel and L. C. Maximon, *Phys. Lett.* **26B**:477 (1968).

DMW 69   D. Drechsel, L. C. Maximon, and R. E. Warner, *Phys. Rev.* **181**:1720 (1969).

DOB 69   G. Dahll, E. Østgaard, and B. Brandow, *Nucl. Phys.* **A124**:481 (1969).

Dol 74   P. Doleschall, *Nucl. Phys.* **A220**:491 (1974).

Edg+ 74   J. A. Edgington, V. J. Howard, I. M. Blair, B. E. Bonner, F. P. Brady, and M. W. McNaughton, *Nucl. Phys.* **A218**:151 (1974).

EHB 69   K. Erkelenz, K. Holinde, and K. Bleuler, *Nucl. Phys.* **A139**:308 (1969).

EJ 68   J. P. Elliott and A. D. Jackson, *Nucl. Phys.* **A121**:279 (1968).

Eks 60   H. Ekstein, *Phys. Rev.* **117**:1590 (1960).

Ell+ 68   J. P. Elliott, A. D. Jackson, H. A. Mavromatis, E. A. Sanderson, and B. Singh, *Nucl. Phys.* **A121**:241 (1968).

EMS 67   J. P. Elliott, H. A. Mavromatis, and E. A. Sanderson, *Phys. Lett.* **24B**:358 (1967).

Ern+ 73   D. J. Ernst, C. M. Shakin, R. M. Thaler, and D. L. Weiss, *Phys. Rev. C* **8**:2056 (1973).

EST 73   D. J. Ernst, C. M. Shakin, and R. M. Thaler, *Phys. Rev. C* **8**:46 (1973).

EST 74   D. J. Ernst, C. M. Shakin, and R. M. Thaler, *Phys. Rev. C* **9**:1780 (1974).

Fad 61   L. D. Faddeev, *Sov. Phys.—JETP* **12**:1014 (1961).

Fad 65   L. D. Faddeev, *Mathematical Aspects of the Three-Body Problem in Quantum Scattering Theory,* Davey, New York (1965).

Fel 67   G. Felsner, *Phys. Lett.* **25B**:290 (1967).

Fie 69a   H. Fiedeldey, *Nucl. Phys.* **A135**:353 (1969).

Fie 69b   H. Fiedeldey, *Phys. Lett.* **30B**:603 (1969).

Fie 70   H. Fiedeldey, *Nucl. Phys.* **A156**:242 (1970).

Fie 71   H. Fiedeldey, *Phys. Lett.* **35B**:195 (1971).

Fie 72   H. Fiedeldey, in: *Proc. Int. Conf. on Few Particle Problems in Nuclear Interaction* (I. Slaus, S. A. Moszkowski, R. P. Haddock, and W. T. H. van Oers, eds.), Los Angeles (1972), p. 407.

Fie 74   H. Fiedeldey, *Lett. Nuovo Cimento* **9**:301 (1974).

Fil 28   L. N. G. Filon, *Proc. Roy. Soc. (Edinburgh)* **49**:38 (1928).

FK 67   H. Feshbach and A. K. Kerman, *Comment Nucl. Particle Phys.* **1**:132 (1967).

FK 68a   H. Feshbach and A. K. Kerman, *Comment Nucl. Particle Phys.* **2**:22 (1968).

FK 68b   H. Feshbach and A. K. Kerman, *Comment Nucl. Particle Phys.* **2**:78 (1968).

FM 72   H. Fiedeldey and N. J. McGurk, *Nucl. Phys.* **A189**:83 (1972).

FS 59   T. Fulton and P. Schwed, *Phys. Rev.* **115**:973 (1959).

Fud 68a   M. G. Fuda, *Phys. Rev.* **174**:1134 (1968).

Fud 68b   M. G. Fuda, *Nucl. Phys.* **A116**:83 (1968).

Fud 69a   M. G. Fuda, *Phys. Rev.* **178**:1682 (1969).

Fud 69b   M. G. Fuda, *Phys. Rev.* **186**:1078 (1969).

Fud 70   M. G. Fuda, *Phys. Rev. C* **1**:1910 (1970).

Fud 71   M. G. Fuda, *Phys. Rev. C* **3**:55 (1971).

Fud 72   M. G. Fuda, *Phys. Rev. C* **5**:275 (1972).

Fur+ 70   M. Furić, V. Valković, D. Miljanić, P. Tomaš, and B. Antolković, *Nucl. Phys.* **A156**:105 (1970).

FW 73   M. G. Fuda and J. S. Whiting, *Phys. Rev. C* **8**:1255 (1973).
GFA 70   J. E. Galonska, A. Faessler, and K. Appel, *Nucl. Phys.* **A155**:465 (1970).
GH 74   A. M. Green and P. Haapakoski, *Nucl. Phys.* **A221**:429 (1974).
GK 64   R. J. Griffiths and K. M. Knight, *Nucl. Phys.* **54**:56 (1964).
GL 51   I. M. Gel'fand and B. M. Levitan, *Izv. Akad. Nauk SSSR, Ser. Math.* **15**:309 (1951).
GMW 67   A. E. S. Green, M. H. McGregor, and R. Wilson, *Rev. Mod. Phys.* **39**:498 (1967).
GPT 70   P. Gogny, P. Pires, and R. de Tourreil, *Phys. Lett.* **32B**:591 (1970).
GR 64   G. C. Ghirardi and A. Rimini, *J. Math. Phys.* **5**:722 (1964).
Gre 62a   A. M. Green, *Phys. Lett.* **1**:136 (1962).
Gre 62b   A. M. Green, *Nucl. Phys.* **33**:218 (1962).
GS 67   A. E. S. Green and T. Sawada, *Rev. Mod. Phys.* **39**:594 (1967).
GS 68   D. Gutkowski and A. Scalia, *J. Math. Phys.* **9**:588 (1968).
GT 57   J. L. Gammel and R. M. Thaler, *Phys. Rev.* **107**:291, 1337 (1957).
Gue 66   J. Y. Guennéguès, *Nuovo Cimento* **42A**:549 (1966).
GW 64   M. L. Goldberger and K. M. Watson, *Collision Theory*, Wiley, New York (1964).
Haa 74   P. Haapakoski, *Physics Letters* **48B**:307 (1974).
Haf 70   M. I. Haftel, *Phys. Rev. Lett.* **25**:120 (1970).
Haf 73   M. I. Haftel, *Phys. Rev. C* **7**:80 (1973).
Hal 72   M. L. Halbert, Review of Experiments on Nucleon–Nucleon Bremsstrahlung, in: *The Two-Body Force in Nuclei* (S. M. Austin and G. M. Crawley, eds.), Plenum, New York (1972).
Har 58   D. R. Hartree, *Numerical Analysis*, Oxford (1958).
Har 67   M. Harada, *Prog. Theor. Phys.* **38**:353 (1967).
Har 70   E. Harms, *Phys. Rev. C* **1**:1667 (1970).
Har 72   E. Harms, *Phys. Lett.* **41B**:26 (1972).
Har 74   E. P. Harper, *Phys. Rev. C* **9**:2106 (1974).
HD 72   M. A. Hennell and L. M. Delves, *Phys. Lett.* **40B**:20 (1972).
Hel 72   L. Heller, Some Basic Questions in Nucleon–Nucleon Bremsstrahlung, in *The Two-Body Force in Nuclei* (S. M. Austin and G. M. Crawley, eds.), Plenum, New York (1972).
HH 69   T. F. Hammann and Q. Ho-Kim, *Nuovo Cimento* **64B**:356 (1969).
HHN 72   E. Hadjimichael, E. Harms, and V. Newton, *Phys. Lett.* **40B**:61 (1972).
HHO 68   J. W. Humberston, R. L. Hall, and T. A. Osborn, *Phys. Lett.* **27B**:195 (1968).
HJ 62   T. Hamada and I. D. Johnston, *Nucl. Phys.* **34**:382 (1962).
HJ 72   E. Hadjimichael and A. D. Jackson, *Nucl. Phys.* **A180**:217 (1972).
HKT 70   E. P. Harper, Y. E. Kim, and A. Tubis, *Phys. Rev. C* **2**:877 (1970).
HKT 72a   E. P. Harper, Y. E. Kim, and A. Tubis, *Phys. Rev. Lett.* **28**:1533 (1972).
HKT 72b   E. P. Harper, Y. E. Kim, and A. Tubis, *Phys. Rev. C* **6**:126 (1972).
HKT 72c   E. P. Harper, Y. E. Kim, and A. Tubis, *Phys. Rev. C* **6**:1601 (1972).
HL 66   M. M. Hoenig and E. L. Lomon, *Ann. Phys.* (*N.Y.*) **36**:363 (1966).
HL 69   E. Harms and J. S. Levinger, *Phys. Lett.* **30B**:449 (1969).
HL 71   E. Harms and L. Laroze, *Nucl. Phys.* **A160**:449 (1971).
HLS 72   M. I. Haftel, E. Lambert, and P. U. Sauer, *Nucl. Phys.* **A192**:225 (1972).
HN 70   E. Harms and V. Newton, *Phys. Rev. C* **2**:1214 (1970).

Hod 69	R. J. W. Hodgson, *Can. J. Phys.* **47**:499 (1969).
Hoe 73	M. M. Hoenig, *Nucl. Phys.* **A206**:169 (1973).
HR 71	L. Heller and M. Rich, *Bull. Am. Phys. Soc.* **16**:559 (1971).
HR 73	E. J. Heller and W. P. Reinhardt, *Phys. Rev.* **A7**:365 (1973).
HT 70	M. I. Haftel and F. Tabakin, *Nucl. Phys.* **A158**:1 (1970).
HT 71	M. I. Haftel and F. Tabakin, *Phys. Rev. C* **3**:921 (1971).
HTT 66	M. Harada, R. Tamagaki, and H. Tanaka, *Prog. Theor. Phys.* **36**:1003 (1966).
Ing 68	L. Ingber, *Phys. Rev.* **174**:1250 (1968).
IP 70	L. Ingber and R. M. Potenza, *Phys. Rev. C* **1**:112 (1970).
JK 52	R. Jost and W. Kohn, *Phys. Rev.* **87**:977 (1952).
JK 53	R. Jost and W. Kohn, *Kgl. Dansk. Videnskab. Selskab. Mat. Fys. Medd.* **27**(9) (1953).
JLS 70	A. D. Jackson, A. Lande, and P. U. Sauer, *Nucl. Phys.* **A156**:1 (1970).
Jov+ 71	J. V. Jovanovich, L. G. Greeniaus, J. McKeown, T. W. Millar, D. G. Peterson, W. F. Prickett, K. F. Suen, and J. C. Thompson, *Phys. Rev. Lett.* **26**:277 (1971).
JP 51	R. Jost and A. Pais, *Phys. Rev.* **82**:840 (1951).
JRS 73	M. Jain, J. G. Rogers, and D. P. Saylor, *Phys. Rev. Lett.* **31**:838 (1973).
JT 70	A. D. Jackson and J. A. Tjon, *Phys. Lett.* **32B**:9 (1970).
Kar 72	B. R. Karlsson, *Phys. Rev.* **D6**:1662 (1972).
KEW 68	L. P. Kok, G. Erens, and R. Van Wageningen, *Nucl. Phys.* **A122**:684 (1968).
KF 61	K. L. Kowalski and D. Feldman, *J. Math. Phys.* **2**:499 (1961).
KF 63	K. L. Kowalski and D. Feldman, *J. Math. Phys.* **4**:507 (1963).
Kim 69	Y. E. Kim, *J. Math. Phys.* **10**:1491 (1969).
Kis 69	S. Kistler, *Z. Physik* **223**:447 (1969).
Knu+ 73	L. D. Knutson, E. J. Stephenson, N. Rohrig, and W. Haeberli, *Phys. Rev. Lett.* **31**:392 (1973).
KNV 74	M. W. Kermode, Y. Nogami, and W. Van Dijk, *Can. J. Phys.* **53**:207 (1975).
Kol 67	D. S. Koltun, *Phys. Rev. Lett.* **19**:910 (1967).
Koo+ 69	E. Ley Koo, M. de Llano, D. V. Grillot, and H. McManus, *Nucl. Phys.* **A133**:610 (1969).
Kow 65	K. L. Kowalski, *Phys. Rev. Lett.* **15**:798 (1965).
Kow 72	K. L. Kowalski, *Nucl. Phys.* **A190**:645 (1972).
Kow+ 71	K. L. Kowalski, J. E. Monahan, C. M. Shakin, and R. M. Thaler, *Phys. Rev. C* **3**:1146 (1971).
KPY 68	D. Kiang, M. A. Preston, and P. C. Yip, *Phys. Rev.* **170**:907 (1968).
KS 69	M. W. Kermode and D. W. L. Sprung, *Nucl. Phys.* **A135**:535 (1969).
KSS 71	V. F. Kharchenko, S. A. Shadchin, and S. A. Storozhenko, *Phys. Lett.* **37B**:131 (1971).
KT 70a	Y. E. Kim and A. Tubis, *Phys. Rev. C* **1**:414 (1970).
KT 70b	Y. E. Kim and A. Tubis, *Phys. Rev. C* **2**:2118 (1970).
KT 72	Y. E. Kim and A. Tubis, *Phys. Rev. C* **7**:1710 (1972).
KT 73a	Y. E. Kim and A. Tubis, *Phys. Rev. Lett.* **31**:952 (1973).
KT 73b	W. M. Kloet and J. A. Tjon, *Nucl. Phys.* **A210**:380 (1973).
KT 74a	W. M. Kloet and J. A. Tjon, *Phys. Lett.* **49B**:419 (1974).
KT 74b	W. M. Kloet and J. A. Tjon, Preprint (1974).

Las+ 62   K. E. Lassila, M. H. Hull, H. M. Ruppel, F. A. MacDonald, and G. Breit, *Phys. Rev.* **126**:881 (1962).

Law 68   J. Law, *Nuovo Cimento* **58B**:258 (1968).

LC 71   M. K. Liou and K. S. Cho, *Nucl. Phys.* **A160**:417 (1971).

Lee 69   H. C. Lee, Thesis, McGill University (1969), unpublished.

Lev 73   J. S. Levinger, *J. Math. Phys.* **14**:1314 (1973).

LF 67   E. L. Lomon and H. Feshbach, *Rev. Mod. Phys.* **39**:611 (1967).

LG 73   A. Laverne and C. Gignoux, *Nucl. Phys.* **A203**:597 (1973).

L'Hu 74   M. L'Huillier, Thèse d'etat, Université de Paris VII, March 1974.

Lim 73   T. K. Lim, *Phys. Rev. Lett.* **30**:709 (1973).

Lio 70   M. K. Liou, *Phys. Rev. C* **2**:131 (1970).

LK 67   R. P. Lynch and T. T. S. Kuo, *Nucl. Phys.* **A95**:561 (1967).

LLS 69   J. S. Levinger, A. H. Lu, and R. Stagat, *Phys. Rev.* **179**:926 (1969).

LMS 73   J. P. Lavine, S. K. Mukhopadhyay, and G. J. Stephenson, Jr., *Phys. Rev. C* **7**:968 (1973).

LN 67   B. A. Loiseau and Y. Nogami, *Nucl. Phys.* **B2**:470 (1967).

LO 73   J. S. Levinger and J. O'Donoghue, *Bull. Am. Phys. Soc.* **17**:440 (1973).

Lov 64   C. Lovelace, *Phys. Rev.* **135**:B1225 (1964).

LR 68   J. G. Lucas and M. L. Rustgi, *Nucl. Phys.* **A112**:503 (1968).

LS 68   R. Laughlin and B. L. Scott, *Phys. Rev.* **171**:1196 (1968).

LS 72   M. K. Liou and M. I. Sobel, *Ann. of Phys.* **72**:323 (1972).

LS 74   J. P. Lavine and G. J. Stephenson, Jr., *Phys. Rev. C* **9**:2095 (1974).

Mar 50   V. A. Marchenko, *Dokl. Akad. Nauk SSSR* **72**:47 (1950).

MC 69   J. H. McGuire and A. H. Cromer, *Phys. Rev.* **184**:1018 (1969).

McC+ 70   J. S. McCarthy, I. Sick, R. R. Whitney, and M. R. Yearian, *Phys. Rev. Lett.* **25**:884 (1970).

McG 70a   J. H. McGuire, *Phys. Rev. C* **1**:371 (1970).

McG 70b   J. H. McGuire, *Phys. Lett.* **32B**:73 (1970).

McG 70c   J. H. McGuire, Preprint, LAMPF Proposal (1970).

McG+ 74   N. J. McGurk, H. Fiedeldey, H. deGroot, and H. J. Boersma, *Phys. Lett.* **49B**:13 (1974).

MCS 69   J. H. McGuire, A. H. Cromer, and M. I. Sobel, *Phys. Rev.* **179**:948 (1969).

Mee 61   K. Meetz, *J. Math. Phys.* **3**:690 (1961).

Mil+ 69   M. D. Miller, M. S. Sher, P. Signell, N. R. Yoder, and D. Marker, *Phys. Lett.* **30B**:157 (1969).

Mit 62   A. N. Mitra, *Nucl. Phys.* **32**:529 (1962).

Mit 69   A. N. Mitra, The Nuclear Three Body Problem, in: *Advances in Nuclear Physics*, Vol. 3 (M. Baranger and E. Vogt, eds.), Plenum, New York (1969).

MLM 71   V. S. Mathur, A. V. Lagu, and C. Maheshwari, *Nucl. Phys.* **A178**:365 (1971).

MN 59   A. N. Mitra and V. L. Narasimham, *Nucl. Phys.* **14**:407 (1959).

Mon 68   T. R. Mongan, *Phys. Rev.* **175**:1260 (1968).

Mon 69a   T. R. Mongan, *Phys. Rev.* **178**:1597 (1969).

Mon 69b   T. R. Mongan, *Phys. Rev.* **180**:1514 (1969).

Mon 69c   T. R. Mongan, *Nuovo Cimento* **63B**:539 (1969).

Mon 69d   T. R. Mongan, *Phys. Rev.* **184**:1888 (1969).

Mor 72   M. J. Moravcsik, *Rep. Prog. Phys.* **53**:587 (1972).

Mor+ 69   M. Morlet, R. Frascaria, B. Geoffrion, N. Marty, B. Tatischeff, and A. Willis, *Nucl. Phys.* **A129**:177 (1969).

Mor+ 72   M. Morlet, R. Frascaria, N. Marty, and A. Willis, *Nucl. Phys.* **A191**:385 (1972).

MP 71a   J. H. McGuire and W. A. Pearce, *Nucl. Phys.* **A162**:561 (1971).

MP 71b   J. H. McGuire and W. A. Pearce, *Nucl. Phys.* **A162**:573 (1971).

MR 66   P. Mittelstaedt and M. Ristig, *Z. Physik* **193**:349 (1966).

MS 69a   H. A. Mavromatis and B. Singh, *Phys. Lett.* **29B**:282 (1969).

MS 69b   D. Marker and P. Signell, *Phys. Rev.* **185**:1286 (1969).

MST 71a   J. E. Monahan, C. M. Shakin, and R. M. Thaler, *Phys. Rev. C* **4**:43 (1971).

MST 71b   J. E. Monahan, C. M. Shakin, and R. M. Thaler, *Phys. Rev. Lett.* **27**:518 (1971).

MST 72   J. E. Monahan, C. M. Shakin, and R. M. Thaler, *Phys. Rev. C* **5**:59 (1972).

MT 69a   R. A. Malfliet and J. A. Tjon, *Nucl. Phys.* **A127**:161 (1969).

MT 69b   R. A. Malfliet and J. A. Tjon, *Phys. Lett.* **30B**:293 (1969).

MT 70   R. A. Malfliet and J. A. Tjon, *Ann. Phys. (N.Y.)* **61**:425 (1970).

MT 71   I. E. McCarthy and P. C. Tandy, *Nucl. Phys.* **A178**:1 (1971).

Naq 64   J. H. Naqvi, *Nucl. Phys.* **58**:289 (1964).

Neg 70   J. W. Negele, *Phys. Rev. C* **1**:1260 (1970).

Nes+ 68   C. W. Nestor, Jr., K. T. R. Davies, S. J. Krieger, and M. Baranger, *Nucl. Phys.* **A113**:14 (1968).

New 55   R. G. Newton, *Phys. Rev.* **100**:412 (1955).

New 57   R. G. Newton, *Phys. Rev.* **107**:1025 (1957).

New 66   R. G. Newton, *Scattering Theory of Waves and Particles*, McGraw-Hill, New York (1966).

NF 57   R. G. Newton and T. Fulton, *Phys. Rev.* **107**:1103 (1957).

NJ 55   R. G. Newton and R. Jost, *Nuovo Cimento* **1**:590 (1955).

Noy 65   H. P. Noyes, *Phys. Rev. Lett.* **15**:538 (1965).

Nym 67   E. M. Nyman, *Phys. Lett.* **25B**:135 (1967).

Nym 68   E. M. Nyman, *Phys. Rev.* **170**:1628 (1968).

Nym 74   E. M. Nyman, *Phys. Rep.* **9**:179 (1974).

OIS 74   S. Oryu, T. Ishihara, and S. Shioyama, *Prog. Theor. Phys.* **51**:1626 (1974).

Omn 61   R. Omnes, *Nuovo Cimento* **21**:524 (1961).

Ory 71   S. Oryu, *Prog. Theor. Phys.* **45**:386 (1971).

Osb 69   T. A. Osborn, *Nucl. Phys.* **A138**:305 (1969).

Osb 71   T. A. Osborn, *Phys. Rev. D* **3**:395 (1971).

Osb 73a   T. A. Osborn, *J. Math. Phys* **14**:373 (1973).

Osb 73b   T. A. Osborn, *Nucl. Phys.* **A211**:211 (1973).

PBS 71   G. Purkayastha, S. N. Banerjee, and N. C. Sil, *Nucl. Phys.* **A167**:376 (1971).

Pet 70   L. Petris, *Nucl. Phys.* **A148**:583 (1970).

PGD 67   W. A. Pearce, W. A. Gale, and I. M. Duck, *Nucl. Phys.* **B3**:241 (1967).

Pie 74   S. C. Pieper, *Phys. Rev. C* **9**:883 (1974).

PL 70   M. H. Partovi and E. L. Lomon, *Phys. Rev. D* **2**:1999 (1970).

PL 72   H. S. Picker and J. P. Lavine, *Phys. Rev. C* **6**:1542 (1972).

PRS 71   H. S. Picker, E. F. Redish, and G. J. Stephenson, Jr., *Phys. Rev. C* **4**:287 (1971).

PRS 72   H. S. Picker, E. F. Redish, and G. J. Stephenson, Jr., in: *Proc. Int. Conf. on Few Particle Problems in the Nuclear Interaction* (I. Slaus, S. A. Moszkowski, R. P. Haddock, and W. T. H. van Oers, eds.), North-Holland, Amsterdam (1972).

PRS 73   H. S. Picker, E. F. Redish, and G. J. Stephenson, Jr., *Phys. Rev. C* **8**:2495 (1973).
PS 73   H. C. Pradhan and Y. Singh, *Can. J. Phys.* **51**:343 (1973).
PS 75   H. S. Picker and G. J. Stephenson, *Nucl. Phys.* **A241**:443 (1975).
PSV 72   H. C. Pradhan, P. U. Sauer, and J. P. Vary, *Phys. Rev. C* **6**:407 (1972).
Pug+ 69   H. G. Pugh, J. W. Watson, D. A. Goldberg, P. G. Roos, D. I. Bonbright, and R. A. J. Riddle, *Phys. Rev. Lett.* **22**:408 (1969).
Red 73   E. F. Redish, *Phys. Rev. Lett.* **31**:617 (1973).
Rei 67   A. S. Reiner, *Nuovo Cimento* **51A**:1 (1967).
Rei 68   R. V. Reid, Jr., *Ann. Phys.* (*N.Y.*) **50**:411 (1968).
Rei 74   M. J. Reiner, *Phys. Rev. Lett.* **32**:236 (1974).
Rid 57   E. C. Ridley, *Proc. Cambridge Phil. Soc.* **53**:442 (1957).
Ris 67   M. Ristig, *Z. Physik* **199**:325 (1967).
RK 68   M. Ristig and S. Kistler, *Z. Physik* **215**:419 (1968).
RM 71   P. Ripa and E. Maqueda, *Nucl. Phys.* **A166**:534 (1971).
Rot 62   M. Rotenberg, *Ann. Phys.* (*N.Y.*) **19**:262 (1962).
RS 64   M. Razavy and D. W. L. Sprung, *Phys. Rev.* **133**:B300 (1964).
RS 71   S. S. Raghavan and B. K. Srivastava, *Can. J. Phys.* **49**:2211 (1971).
RSL 70   E. F. Redish, G. J. Stephenson, Jr., and G. M. Lerner, *Phys. Rev. C* **2**:1665 (1970).
RSP 72   E. F. Redish, G. J. Stephenson, Jr., and H. S. Picker, *Phys. Rev. C* **5**:707 (1972).
Sas 63   T. Sasakawa, *Prog. Theor. Phys. Suppl.* **27**:1 (1963).
Sau 70   P. U. Sauer, *Nucl. Phys.* **A150**:467 (1970).
Sau 71   P. U. Sauer, *Nucl. Phys.* **A170**:497 (1971).
Sau 73   P. U. Sauer, *Ann. Phys.* (*N.Y.*) **80**:242 (1973).
Sau 74a   P. U. Sauer, *Phys. Rev. Lett.* **32**:626 (1974).
Sau 74b   P. U. Sauer, Contribution to the Int. Conf. on the Few Body Problem in Nuclear and Particle Physics, Laval University, Quebec City (1974).
SBS 69   M. K. Srivastava, P. K. Banerjee, and D. W. L. Sprung, *Phys. Lett.* **29B**:635 (1969).
SBS 70   M. K. Srivastava, P. K. Banerjee, and D. W. L. Sprung, *Phys. Lett.* **31B**:499 (1970).
SG 73   I. H. Sloan and J. D. Gray, *Phys. Lett.* **44B**:354 (1973).
SH 68   H. Sugawara and F. von Hippel, *Phys. Rev.* **172**:1764 (1968).
SH 71   P. Signell and J. Holdeman, Jr., *Phys. Rev. Lett.* **27**:1393 (1971).
Sie 70   P. J. Siemens, *Nucl. Phys.* **A141**:225 (1970).
Sig 69a   P. Signell, The Nuclear Potential, in: *Advances in Nuclear Physics*, Vol. 2 (M. Baranger and E. Vogt, eds.), Plenum, New York (1969).
Sig 69b   P. Signell, in: *Proc. Int. Conf. on Light Nuclei, Few Body Problems and Nuclear Forces* (Brela, Yugoslavia, 1967), Gordon and Breach, New York (1969).
SJS 69   M. K. Srivastava, A. M. Jopko, and D. W. L. Sprung, *Can. J. Phys.* **47**:2459 (1969).
SL 74   B. Siebert and J. S. Levinger, *Z. Physik* **266**:95 (1974).
Sla+ 66   I. Slaus, J. W. Verba, J. R. Richardson, R. F. Carlson, L. S. August, and E. L. Petersen, *Phys. Lett.* **23**:358 (1966).
SLH 72   B. Siebert, J. S. Levinger, and E. Harms, *Nucl. Phys.* **A197**:33 (1972).

SM 68a	P. Signell and D. Marker, *Phys. Lett.* **26B**:559 (1968).
SM 68b	P. Signell and D. Marker, *Phys. Lett.* **28B**:79 (1968).
Sob 68	M. I. Sobel, *J. Phys. A* **1**:610 (1968).
Spr 70	D. W. L. Sprung, *Proc. Herceg Novi Summer School on Nuclear Physics* (1970).
Spr 72	D. W. L. Sprung, Nuclear Matter Calculations, in: *Advances in Nuclear Physics*, Vol. 5 (M. Baranger and E. Vogt, eds.), Plenum, New York (1972).
Spr+ 70	D. W. L. Sprung, P. K. Banerjee, A. M. Jopko, and M. K. Srivastava, *Nucl. Phys.* **A144**:245 (1970).
Sri 70a	M. K. Srivastava, *Nucl. Phys.* **A144**:236 (1970).
Sri 70b	M. K. Srivastava, *Nucl. Phys.* **A157**:61 (1970).
Sri 70c	M. K. Srivastava, Thesis, McMaster University (1970), unpublished.
Sri 70d	M. K. Srivastava, *Phys. Lett.* **33B**:341 (1970).
Sri 71	M. K. Srivastava, *Nucl. Phys.* **A178**:332 (1971).
Sri 73a	M. K. Srivastava, *Nucl. Phys.* **A202**:145 (1973).
Sri 73b	M. K. Srivastava, *Ind. J. Pure Appl. Phys.* **11**:780 (1973).
Sri 74	M. K. Srivastava, *Nucl. Phys.* **A221**:183 (1974).
SS 69	D. W. L. Sprung and M. K. Srivastava, *Nucl. Phys.* **A139**:605 (1969).
SS 70	M. K. Srivastava and D. W. L. Sprung, *Nucl. Phys.* **A149**:113 (1970).
SS 72	A. P. S. Sirohi and M. K. Srivastava, *Nucl. Phys.* **A179**:524 (1972).
SS 73a	A. P. S. Sirohi and M. K. Srivastava, *Nucl Phys.* **A201**:66 (1973).
SS 73b	M. K. Srivastava and A. P. S. Sirohi, *Nucl. Phys.* **A207**:527 (1973).
SS 74	M. Strayer and P. U. Sauer, *Nucl. Phys.* **A231**:1 (1974).
SSB 70	M. K. Srivastava, Y. Singh, and R. K. Bhaduri, *Phys. Lett.* **32B**:333 (1970).
ST 73a	P. U. Sauer and J. A. Tjon, *Nucl. Phys.* **A216**:541 (1973).
ST 73b	M. Sotona and E. Truhlík, *Phys. Lett.* **43B**:362 (1973).
Ste 72	M. S. Stern, *J. Phys. A: Gen. Phys.* **5**:426 (1972).
Ste+ 72	G. J. Stephenson, Jr., E. F. Redish, G. M. Lerner, and M. I. Haftel, *Phys. Rev. C* **6**:1559 (1972).
Str 68	G. L. Strobel, *Nucl. Phys.* **A116**:465 (1968).
STS 68	F. Sannes, J. Trischuk, and D. G. Stairs, *Phys. Rev. Lett.* **21**:1474 (1968).
SW 65	A. Scotti and D. Y. Wong, *Phys. Rev.* **138**:B145 (1965).
SW 71	Y. Singh and C. S. Warke, *Can. J. Phys.* **49**:1029 (1971).
SW 74	P. U. Sauer and H. Walliser, Contribution to the Int. Conf. on the Few Body Problem in Nuclear and Particle Physics, Laval University, Quebec City (1974).
SWB 72	Y. Singh, C. S. Warke, and R. K. Bhaduri, *Can. J. Phys.* **50**:2574 (1972).
TA 71	A. W. Thomas and I. R. Afnan, *Phys. Rev. Letters* **26**:906 (1971).
Tab 64	F. Tabakin, *Ann. Phys.* (*N.Y.*) **30**:51 (1964).
Tab 65	F. Tabakin, *Phys. Rev.* **137**:B75 (1965).
Tab 68	F. Tabakin, *Phys. Rev.* **174**:1208 (1968).
Tab 69	F. Tabakin, *Phys. Rev.* **177**:1443 (1969).
Tab 72	F. Tabakin, Off-Energy-Shell Effects in Many-Nucleon Systems, in: *The Two-Body Force in Nuclei* (S. M. Austin and G. M. Crawley, eds.), Plenum, New York (1972), p. 119.
Tak 72	T. Takemiya, *Prog. Theor. Phys.* **48**:1547 (1972).
Tak 73	T. Takemiya, *Prog. Theor. Phys.* **49**:1602 (1973).
Tan 66a	S. Tani, *Ann. Phys.* (*N.Y.*) **37**:411 (1966).

Tan 66b   S. Tani, *Ann. Phys.* (*N.Y.*) **37**:451 (1966).
Tan 68   S. Tani, *Phys. Rev.* **174**:2054 (1968).
TD 66   F. Tabakin and K. T. R. Davies, *Phys. Rev.* **150**:793 (1966).
TGC 70   J. A. Tjon, B. F. Gibson, and J. S. O'Connell, *Phys. Rev. Lett.* **25**:540 (1970).
TGG 71   R. H. Thompson, A. Gersten, and A. E. S. Green, *Phys. Rev. D* **3**:2069 (1971).
Tjo 70   J. A. Tjon, *Phys. Rev. D* **1**:2109 (1970).
TRS 74   R. de Tourreil, B. Rouben, and D. W. L. Sprung, *Nucl. Phys.*, **A242**:445 (1975).
TS 73   R. de Tourreil and D. W. L. Sprung, *Nucl. Phys.* **A201**:193 (1973).
Und 70   J. Underhill, *J. Math. Phys.* **11**:1409 (1970).
Val+ 71   V. Valković, P. Rendić, V. A. Otte, W. von Witsch, and G. C. Phillips, *Nucl. Phys.* **A166**:547 (1971).
Vor 65   Yu. V. Vorobyev, *Method of Moments in Applied Mathematics*, Gordon and Breach, New York (1965).
VR 70   W. Van Dijk and M. Razavy, *Nucl. Phys.* **A159**:161 (1970).
VR 73   W. Van Dijk and M. Razavy, *Nucl. Phys.* **A204**:412 (1973).
Wag+ 71   R. Van Wageningen, B. L. G. Bakker, J. Bruinsma, G. Erens, and J. H. Stuivenberg, Budapest Symposium on the Nuclear Three-Body Problem (1971).
Wal 71   H. R. J. Walters, *J. Phys.* **B6**:437 (1971).
Wal 72   J. M. Wallace, *Phys. Rev. C* **5**:609 (1972).
Wat+ 71   J. W. Watson, H. G. Pugh, P. G. Roos, D. A. Goldberg, R. A. J. Riddle, and D. I. Bonbright, *Nucl. Phys.* **A172**:513 (1971).
WB 71   C. S. Warke and R. K. Bhaduri, *Nucl. Phys.* **A162**:289 (1971).
WE 71   R. Van Wageningen and G. Erens, *Phys. Lett.* **34B**:184 (1971).
Wei 63   S. Weinberg, *Phys. Rev.* **131**:440 (1963).
Wil+ 72   A. Willis, V. Comparat, R. Frascaria, N. Marty, M. Morlet, and N. Willis, *Phys. Rev. Lett.* **28**:1063 (1972).
WK 75   W. T. Weng and T. T. S. Kuo, Stony Brook preprint.
WN 67   K. M. Watson and J. Nuttall, *Topics in Several Particles Dynamics*, Holden-Day, San Francisco (1967).
WS 72   C. W. Wong and T. Sawada, *Ann. Phys.* (*N.Y.*) **72**:107 (1972).
WZ 67   D. Y. Wong and G. Zambotti, *Phys. Rev.* **154**:1540 (1967).
Yam 54   Y. Yamaguchi, *Phys. Rev.* **95**:1628 (1954).
YJ 71   S. N. Yang and A. D. Jackson, *Phys. Lett.* **36B**:1 (1971).
Zoh 73   O. Zohni, *Phys. Rev. C* **8**:1164 (1973).

Chapter 3

THEORETICAL AND EXPERIMENTAL DETERMINATION OF NUCLEAR CHARGE DISTRIBUTIONS*

J. L. Friar

Department of Physics
Brown University
Providence, Rhode Island

and

J. W. Negele[‡]

Laboratory for Nuclear Science and Department of Physics
Massachusetts Institute of Technology
Cambridge, Massachusetts

---

1. INTRODUCTION AND DEFINITION OF THE PROBLEM

1.1. Outline

Because of their purely electromagnetic interaction, charged leptons provide the most precise probe of the nuclear many-body wave function presently available. Although, in principle, precise scattering data for e^+, e^-, μ^+,

* This work supported in part through funds provided by the Atomic Energy Commission under Contracts AT(11-1)3069 and AT(11-1)3235.
‡ Alfred P. Sloan Foundation Research Fellow.

and μ^- and bound-state energy levels for e^- and μ^- could provide a wealth
of complementary information, in practice the most useful information
presently arises from muonic X-rays in medium and heavy nuclei and
from electron scattering data. The challenge to which this present review
is addressed is to utilize the available information from muonic X-rays
and elastic electron scattering as a tool to critically test our present theoretical understanding of the nuclear wave function.

The goal of critically testing nuclear theory provides severe restrictions
on the scope of this work. Since we are not interested in simply fitting data
with phenomenological shapes for charge densities or transition densities,
we must restrict our attention to those cases for which a meaningful microscopic theory of the nuclear wave function exists. With the exception of
the deuteron, ^{3}He, and ^{3}H, which are not treated in this work, this restriction
limits our consideration to ground states of closed-shell spherical nuclei.
Thus we shall deliberately ignore a wealth of interesting inelastic scattering
data and results on other nuclei, and consider only elastic electron scattering
and muonic X-ray transitions for spherical nuclei.

The first step in reducing the analysis of electron scattering and muonic
data to manageable form is to reduce the problem to the solution of bound
states and scattering states of the Dirac equation in a static Coulomb potential. In the remainder of this section, it is shown how the general problem
can systematically be separated into the static Coulomb contribution plus
a hierarchy of small correction terms which can be approximately evaluated.
Removal of these corrections from the experimental data yields corrected
data, with associated experimental and theoretical uncertainties, which
should be reproduced by solving the Dirac equation in the static Coulomb
field of the nuclear charge distribution.

The most direct means of testing theoretical wave functions is to directly
compare the theoretical cross sections and muonic energies calculated by
solving the Dirac equation with the experimental data corrected for all
nonstatic effects. A detailed analysis for all the available theoretical densities
of ^{208}Pb is carried out in Section 2. In order to assess the physical significance
of the salient features of these charge density distributions, the underlying
nuclear theories are reviewed and the physical origin of these features is
discussed.

The direct comparison of Section 2 leaves several important problems
unsolved, which are addressed in Section 3. One problem is to specify in a
precise and unambiguous way exactly what constraint is imposed on the
theoretical charge distribution by each experimental datum. Thus we will
attempt to replace vague notions about measuring radii, the surface thick-

ness, radial moments, and "wiggles" by a precise formulation of linear constraints. In addition, we will address the question of how to display in coordinate space the changes required in a theoretical charge distribution to bring it into agreement with experiment, as well as the statistical errors and error correlations implied by the experimental errors.

The results of analyses to determine the linear constraints specified by the data and to determine density distributions in coordinate space are reviewed in Section 4. Finally, the conclusions reached in this review are summarized in Section 5 and recommendations for future experimental and theoretical investigations are presented.

1.2. The Static Coulomb Interaction

The electromagnetic interaction between charged particles is most easily developed and understood using time-dependent perturbation theory. In this framework, the force between particles is due to the exchange of virtual photons. In the simplest case, first Born approximation, a single photon is exchanged, which leads to an attractive force between two oppositely charged particles. Unfortunately, the picture becomes very complicated for multiple-photon exchanges, since in addition to the simple iteration of single-photon exchanges, two or more photons may propagate simultaneously. The presence of the negative energy sea of fermions will also play an important role in this case. However, for the present application there exists a limiting case where a simple picture of the interaction is restored.

The force due to one-photon exchange can be understood using classical arguments (LL 62). We are primarily interested in the interaction of a light lepton with a very heavy, slow-moving, spin-0^+, extended nucleus. It is thus necessary to treat the lepton in a relativistic fashion, while the nucleus may be treated in a nonrelativistic or semirelativistic way. The primary interaction between such objects is the static Coulomb interaction. It is always possible to separate the effect of any photon exchange into two parts by the use of Coulomb gauge. One part sums the effect of the retarded interaction between the charges and the effect of the component of the currents of the two particles along the momentum of the exchanged photon. The other part is the interaction due to the remaining "transverse" parts of the currents. The former part, on inspection, is seen to be the *static* Coulomb interaction, and the latter part now contains all the retardation and magnetic effects. In the static limit of the heavy particle (i.e., taking its mass m_t to be infinite) the magnetic effects vanish because the convection current vanishes and because a spin-0 object has no intrinsic magnetization. One is left with simply the static Coulomb interaction between the particles (Fri 73a). Similar considerations apply to multiple-photon exchanges, except that it is also necessary to consider the effect of the negative energy sea of leptons on the interaction, which leads to particle–antiparticle "pair" intermediate states.

If one selectively neglects the effect of vacuum polarization (virtual electron–positron pair creation, for example), the Lamb shift (absorption of a photon by the lepton that emitted it), and the effect of virtual excitation of the nucleus by the photons (the nucleus remains in its ground state), a simple result can still be obtained in the limit $m_t \to \infty$. It was shown many years ago that the $m_t \to \infty$ limit for two interacting point charges reduces to the problem of solving for the wave function of the light particle in the presence of a static Coulomb field using the Dirac or Klein–Gordon equation (Des 55, Fri 73b). If one has an extended heavy particle, the mere fact that this object has size means that it has a spectrum of excited states (EH 73). If we restrict ourselves to those contributions to the interaction in which the nucleus remains in the ground state and neglect certain effects of mesons inside the nucleus, we simply modify the Coulomb potential and the previous theorem holds in the infinite-mass limit for an extended, spinless, heavy particle. This decomposition of the problem into a basic static Coulomb problem and corrections has many attractive features. We will see that the corrections are quite small, and this offers us the opportunity of using wave functions that are exact for the static Coulomb problem and systematically calculating the corrections in powers of the appropriate small parameter (e.g., $1/m_t$ for the magnetic effects). We shall include in the category of corrections those effects that arise from motion within a fixed nucleus, although in principle they are part of the static charge distribution.

1.3. Corrections to the Static Coulomb Interaction

We shall divide the problem into several categories. Of primary importance, obviously, is the nucleon distribution, which is difficult to calculate theoretically and which involves theoretical uncertainties yielding effects considerably larger than some of the correction terms mentioned above. A detailed discussion of this problem will be presented in Section 2. Of secondary importance is the fact that the charge distribution of the nucleus is partly due to the charge distribution of the nucleons themselves as well as to the distribution of nucleons in the nucleus. This problem is intimately related to relativistic corrections to the interaction of the virtual photons with the moving nucleons in the nucleus, since even though the nucleus itself may be fixed by the $m_t \to \infty$ limit, the individual nucleons can move. In addition, our usual description of a nucleus as a collection of nucleons interacting with each other by means of *static* nucleon–nucleon potentials is only an approximation. For example, if at any time a picture is taken of the nucleus, we will observe, in addition to the A nucleons, mesons which are being exchanged between nucleons. The charge in the nucleus is carried by these exchanged mesons as well as by the Z protons. In addition, the nucleon is the lowest member of a spectrum of excited states. One expects to find small admixtures in the nucleus of these excited baryonic states (internal isobars). An example is the N*(1238), which has recently received considerable attention. The photon, when it probes the nucleus, can interact with the mesonic charge or with the internal isobars. It is also possible for mesons or isobars to be created (virtually) in the nucleus by the photon and these created particles interact with the rest of the nucleus. All of these effects occur for a stationary nucleus and contribute to the effective nuclear charge distribution.

Although our previous category of effects included virtual states of the meson and baryon spectrum, we have chosen to treat the virtual nuclear excited states separately. The reason for this distinction is the relatively large amount of energy needed to create a meson or isobar and the relatively small probability for doing so. The nuclear excited states, in contrast, are relatively easy to excite and the process requires little energy. In dealing with bound states of the lepton, the effects of the nuclear excited states are usually referred to as polarization corrections, while they are generally called dispersion corrections for the scattering problem.

Another category includes those effects whose origin lies in quantum electrodynamics: vacuum polarization and the Lamb shift for the bound-state problem and real radiation of photons for the scattering problem. It is in this area that some of the most innovative work in twentieth century physics has been done and where difficult problems remain for the future.

The effects that depend on the nucleus mass are most conveniently divided into two groups. Corrections of order $1/m_t$ arise from magnetic interactions and from the fact that the moving nucleus carries kinetic energy. Such corrections can be reliably described as recoil or center-of-mass effects. There are also effects of order $(1/m_t)^2$, which arise because of the effect of relativity on the wave function of the moving nucleus. Such phenomena as Lorentz contraction and the Thomas precession can be expected to play a role. Our treatment will show that the relativistic effects primarily modify the effective charge density of the nucleus, while the recoil effects modify the basic interaction between the photon and the nucleus.

Most electron scattering is performed at energies above 50 MeV, so that it is an excellent and convenient approximation to neglect the electron mass. Unless we explicitly comment, this approximation will be used for scattering. Because the radius of the lowest state of a hydrogenic atom is given approximately by the Bohr radius $r_0 \sim \hbar c / Z \alpha m c^2$ for a nucleus of charge Ze, a lepton of mass m will overlap the nuclear charge distribution only for large values of Zmc^2. Atomic electrons are scarcely affected by the nuclear charge distribution, and even muons spend a small fraction of their time inside small-Z nuclei. The Bohr radius of ^{12}C, for example, is approximately 40 fm, which is much larger than the charge radius of about 2.5 fm. Present levels of experimental accuracy do not justify use of muonic atoms as a tool for probing very light elements and we will implicitly assume in any further discussion of atoms that we are dealing with muonic atoms of medium- or high-Z nuclei.

Our present review of corrections will treat selected topics of current interest and does not represent an exhaustive coverage of either electron scattering or muonic atoms. The reader is referred to the complementary reviews of deForest and Walecka (DW 66), Barrett (Bar 74), Überall (Ube 71), and Wu and Wilets (WW 69).

1.3.1. Nucleon Form Factors

The effective interaction between a single photon (virtual or otherwise) and a nucleon in the nucleus is most conveniently described in terms of the matrix element of the nucleon charge and current operators. To a good approximation, which is discussed in Section 1.3.2, we can neglect the effect of the other nucleons on the electromagnetic interaction of any one nucleon. Thus our nucleus is composed of nucleons that can be regarded as "free" with their momenta specified by the nuclear wave function. Upon interacting with a photon, the momentum of a given nucleon is changed, as is the momentum of the entire nucleus. We wish to treat each nucleon relativistically and then expand the charge-current operator matrix elements in powers of (v/c). Because the nucleus is a weakly bound system, the potential energy and the kinetic energy are of roughly equal magnitude, and the kinetic energy operator contains one inverse power of the nucleon mass m, which we denote order $(1/m)$. Because momentum divided by mass is velocity, an expansion in $(1/m)$ is the same as a (v/c) expansion.

The matrix element of the charge-current operator for a nucleon can be written in terms of invariants in the usual manner (Ros 50, BD 65). We write

$$\langle \mathbf{p}'\lambda' \mid \hat{J}_\mu(0) \mid \mathbf{p}\lambda \rangle = \bar{u}_{\lambda'}(\mathbf{p}')[F_1\gamma_\mu + (i\varkappa/2m)F_2\sigma_{\mu\nu}q^\nu]u_\lambda(\mathbf{p}) \tag{1}$$

where $\hat{J}_\mu(0)$ is the four-vector current operator at the point $\mathbf{x} = 0$, $t = 0$; $\mathbf{p}$ and $\mathbf{p}'$ are the initial and final momenta of the nucleon; λ and λ' are the initial and final helicities of the nucleon; F_1 and F_2 are functions of q^2, the invariant momentum transfer $(q_\mu = p_\mu' - p_\mu)$; $\varkappa$ is the nucleon anomalous magnetic moment; and γ_μ and $\sigma_{\mu\nu}$ are Dirac matrices. Our metric is such that $p^2 = -m^2$. The invariant form factors are F_1, the Dirac form factor, and F_2, the Pauli form factor, which are normalized to unity at $q^2 = 0$.

The spinors u_λ have the representation

$$u_\lambda(\mathbf{p}) = \left[\frac{E+m}{2E}\right]^{1/2} \left(\begin{array}{c} 1 \\ \dfrac{\boldsymbol{\sigma}\cdot\mathbf{p}}{E+m} \end{array}\right)\chi_\lambda \tag{2}$$

where $E = (\mathbf{p}^2 + m^2)^{1/2}$ and χ_λ is the usual two-component spin wave function. We have normalized u according to the nonrelativistic convention rather than the more common covariant convention (BD 65), which would replace $2E$ by $2m$ in the square root. The overall normalization is not in question, Lorentz invariance of the total charge being sufficient to specify this. Rather than carry additional normalization factors in the phase space, for example, such factors are included in Eq. (2).

Expansion of the space components of Eq. (1) $(\mu = 1, 2, 3)$ to order $(1/m)$ leads to the usual nonrelativistic current operator, which includes both the convection current and the spin magnetization current. The $\mu = 0$ component expanded to order $(1/m^2)$ yields (FW 50, BD 65)

$$\langle \mathbf{p}'\lambda' \mid \hat{\varrho}(0) \mid \mathbf{p}\lambda \rangle \cong \chi_{\lambda'}^\dagger \left[F_1 - \left(\frac{q^2}{8m^2} + \frac{i\mathbf{q}\cdot\boldsymbol{\sigma}\times\mathbf{p}}{4m^2}\right)(F_1 + 2\varkappa F_2)\right]\chi_\lambda \tag{3}$$

The spin-independent parts are the Darwin–Foldy terms and are due to Zitterbewegung, a relativistic effect (Fol 58). It is conventional to define a new set of form factors which include part of the Darwin–Foldy terms in their definition. These new form factors are the Sachs (Sac 62) charge and magnetic form factors G_E and G_M, respectively:

$$G_E = F_1 - \varkappa q^2 F_2/4m^2, \qquad G_M = F_1 + \varkappa F_2 \tag{4}$$

The magnetic form factor becomes $1 + \varkappa$ at $q^2 = 0$ and this is the total nucleon magnetic moment μ. In terms of these definitions we have

$$\langle \mathbf{p}'\lambda' \mid \hat{\varrho}(0) \mid \mathbf{p}\lambda \rangle = \chi_{\lambda'}^{\dagger}\varrho(\mathbf{p}, \mathbf{q})\chi_{\lambda} \tag{5}$$

$$\varrho(\mathbf{p}, \mathbf{q}) = (1 - q^2/8m^2)G_E - (i\mathbf{q} \cdot \boldsymbol{\sigma} \times \mathbf{p}/4m^2)(2G_M - G_E) \tag{6}$$

The spin-dependent term is the spin–orbit interaction. It depends on the nucleon velocity within the nucleus and vanishes for a nucleon at rest or if $\mathbf{p}$ is collinear with $\mathbf{q}$, the momentum transfer. This term therefore does not contribute to electron scattering from free protons in the lab frame (MV 62, Fri 73a). The spin–orbit interaction has an interesting physical origin and consequences (Osb 68). A moving magnetic dipole generates, according to special relativity, a small electric dipole moment which can interact with an electric field. Such will be the case for a moving spin-$\frac{1}{2}$ object and this accounts for the $2G_M$ term. The electric field accelerates the nucleon and the accelerated frame of reference attached to the particle appears to rotate. The time rate of change of the rotation is the Thomas precession and it is necessary to correct the Hamiltonian for this rotational motion. This accounts for the G_E term in the spin–orbit contribution to ϱ.

Both the Darwin terms and the spin–orbit terms depend on $\varkappa$, the anomalous magnetic moment, and this moment will contribute to the effective charge density of neutrons as well as protons. The proton form factors $G_E{}^p$ and $G_M{}^p$ are the ones that are fit to electron–proton (e–p) scattering data using Eqs. (7) and (9) below, and are widely used.

The differential cross section for elastic electron scattering in first-Born approximation in the lab frame can be written in the form (DW 66, Fri 73a)

$$d\sigma/d\Omega = \sigma_{\text{Mott}}\{A(q^2) + B(q^2)[\tfrac{1}{2} + (1 + q^2/4m_t^2)\tan^2(\theta/2)]\} \tag{7}$$

where m_t is the target mass, θ is the electron scattering angle, B is a form factor derived from matrix elements of the part of the current operator that is perpendicular to $\mathbf{q}$, and A is a form factor derived from the charge operator. In addition, for electrons with energy E we have

$$\sigma_{\text{Mott}} = \frac{\alpha^2 \cos^2(\theta/2)}{4E^2[\sin^4(\theta/2)]\{1 + 2E[\sin^2(\theta/2)]/m_t\}}$$

$$q^2 = \frac{4E^2 \sin^2(\theta/2)}{1 + 2E[\sin^2(\theta/2)]/m_t} \tag{8}$$

where A and B contain all factors of Z and α is the fine structure constant.

For e–p scattering, where $m_t = m$, one finds (Ros 50, Fri 73a)

$$A = \frac{G_E^2}{1 + q^2/4m^2}, \qquad B = \frac{-\dfrac{q^2}{2m^2}\, G_M^2}{1 + q^2/4m^2} \qquad (9)$$

The factor of $(1 + q^2/4m^2)^{-1/2}$ associated with G_E is the *same* factor which arises in Eq. (6) with G_E. One can define the nucleon form factors in a variety of ways and a possibly more convenient set would lump the extra Darwin factor into the definition (YLR 57),

$$\tilde{G}_E = G_E/(1 + q^2/4m^2)^{1/2}, \qquad \tilde{G}_M = G_M/(1 + q^2/4m^2)^{1/2} \qquad (10)$$

A very good approximation to the experimentally observed proton electric and magnetic form factors $G_E{}^p$ and $G_M{}^p$ and the neutron magnetic form factor $G_M{}^n$ is the dipole form (BKL 71)

$$G_M{}^p/\mu_p \cong G_M{}^n/\mu_n \cong G_E{}^p \cong (1 + q^2/A_1)^{-2}, \qquad A_1 \cong 0.71 \;(\text{GeV}/c)^2 \qquad (11)$$

This corresponds to an exponential charge distribution

$$\varrho_p(\mathbf{r}) = (A_1^{3/2}/8\pi)\exp(-r\sqrt{A_1}) \qquad (12)$$

with an rms radius of 0.81 fm. Also commonly used is the more convenient but much less accurate Gaussian parameterization

$$G_E \cong \exp(-\beta_p{}^2 q^2/6), \qquad \beta_p \cong 0.8 \text{ fm} \qquad (13)$$

which corresponds to a Gaussian charge density

$$\varrho_p(\mathbf{r}) = (2\pi\beta_p{}^2/3)^{-3/2}\exp(-3r^2/2\beta_p{}^2) \qquad (14)$$

On the other hand, the neutron form factors $G_E{}^n$ and $G_M{}^n$ cannot be measured directly and must be deduced from elastic and inelastic electron–deuteron scattering. The slope of $G_E{}^n$ is also known at $q^2 = 0$ from thermal neutron–atom scattering (KR 66)

$$\beta_n \equiv \frac{dG_E{}^n}{dq^2}(0) = 0.0193 \pm 0.0004 \text{ fm}^2 \qquad (15)$$

and this number is almost consistent with $dF_1{}^n/dq^2 = 0$. Note that we must have $F_1{}^n(0) \equiv 0$. One finds that the thermal neutron slope is $\sim 90\%$ of the $\varkappa$ term (Fol 58). The neutron root-mean-square radius can be determined from β_n and we find

$$\langle r^2 \rangle_n^{1/2} = -0.340 \pm 0.003 \text{ fm} \qquad (16a)$$

which can be compared to the corresponding proton rms radius

$$\langle r^2 \rangle_p^{1/2} = 0.80 \pm 0.02 \text{ fm} \tag{16b}$$

where we have averaged a number of different fits by Bilen'kaya *et al.* (BKL 71). The sign of β_n is consistent with negative charge at large distances and positive charge at short distances, a result which is intuitively appealing in view of the fact that the neutron can virtually dissociate according to $n \rightarrow p + \pi^-$. The size of the neutron rms radius indicates that neutrons may measurably affect the nuclear charge distribution and the form factors obtained by scattering electrons from the charge distribution.

Experimentally the neutron charge form factor is poorly known because so many theoretical corrections must be made to electron–deuteron scattering data before extracting the neutron form factor. A number of these corrections will be discussed subsequently, in Section 1.3.5. Consequently, a variety of forms may be used to parameterize the neutron form factor with roughly equal justification. Neglecting F_1^n and using the dipole form for G_M^n, we have, for example (Fol 58),

$$G_E^n = -\mu_n q^2 G_{\text{Dip}}/4m^2 \tag{17}$$

with G_{Dip} being given by Eq. (11). This is probably the upper limit for G_E^n. A somewhat better approximation is given by

$$G_E^n = \beta_n q^2/(1 + q^2/A_1)^3 \tag{18}$$

obtained from a quark model argument (Fri 72). Also useful is the difference of two dipole forms, with one parameter fixed to produce β_n. This form was used in Fig. 1 with various values of the one free parameter in order to determine reasonable estimates of the maximum and minimum values of the neutron charge form factor allowed by the data. Extraction of this form factor is difficult and the "data" shown in Fig. 1 are consistent with the theoretical assumptions made by Bertozzi *et al.* (Ber+ 72), from whom the figure is taken. The upper curve is quite close to the values given by Eq. (17), and the solid (best-fit) curve is quite close to the result of using Eq. (18). The sum and difference of the proton and neutron charge form factors, the isoscalar and isovector combinations, are also parameterized, and one widely used set is that of Janssens *et al.* (Jan+ 66). The only neutron data used in these fits is β_n.

Because the proton form factor falls off fairly rapidly, the neutron form factor is a nonnegligible fraction of the isoscalar or isovector form factors, and therefore contributes significantly to scattering. Using Eq. (7),

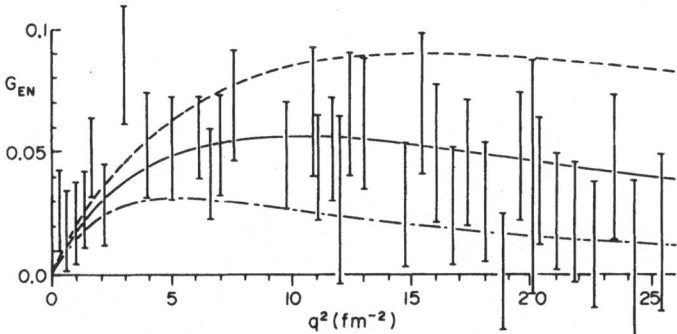

Fig. 1. Neutron form factor assuming an exponential neutron charge distribution compared with experimental data extracted by Bertozzi *et al.* (Ber+ 72). Solid, dashed, and dash-dot curves correspond to best fit, maximum, and minimum form factors determined by sight.

the cross section for electron scattering from a spin-0$^+$ nucleus in first Born approximation is given by

$$A(\mathbf{q}^2) = F^2(\mathbf{q}^2), \qquad B(\mathbf{q}^2) = 0 \tag{19}$$

$$F(\mathbf{q}^2) = \langle 0 \mid \sum_i \hat{e}_i \exp(i\mathbf{q} \cdot \mathbf{x}_i')$$
$$- i\mathbf{q} \cdot \sum_i \left(\frac{2\hat{\mu}_i - \hat{e}_i}{8m^2} \right) \{ \boldsymbol{\sigma}(i) \times \boldsymbol{\pi}_i, \exp(i\mathbf{q} \cdot \mathbf{x}_i') \} \mid 0 \rangle \tag{20}$$

with

$$\hat{e}_i = \tilde{G}_E^p \left(\frac{1 + \tau_3(i)}{2} \right) + \tilde{G}_E^n \left(\frac{1 - \tau_3(i)}{2} \right)$$
$$\hat{\mu}_i = \tilde{G}_M^p \left(\frac{1 + \tau_3(i)}{2} \right) + \tilde{G}_M^n \left(\frac{1 - \tau_3(i)}{2} \right) \tag{21}$$

$$\mathbf{x}_i' = \mathbf{x}_i - \mathbf{R}, \qquad \mathbf{R} = \sum_{i=1}^{A} \mathbf{x}_i/A, \qquad \boldsymbol{\pi}_i = \mathbf{p}_i - \mathbf{P}/A, \qquad \mathbf{P} = \sum_{i=1}^{A} \mathbf{p}_i \tag{22}$$

where $\mathbf{p}_i$ is the momentum of the ith particle, $\sigma(i)$ is (twice) its spin, $\tau_3(i)$ is the third component of its isospin operator, and $\mathbf{R}$ and $\mathbf{P}$ are the center-of-mass coordinate and total momentum operator, respectively. We have argued that $\tilde{G}_E$, rather than G_E, should be the charge form factor to be used in Eq. (21) and have assumed (without proof) that $\tilde{G}_M$, rather than G_M, should be used also. The neutrons and protons enter coherently in Eq. (20) and the charge of the neutrons has been shown (Ber+ 72) to produce a sizable change in the cross section for scattering from ^{40}Ca and ^{208}Pb, as demonstrated in Fig. 2. The two dashed curves in Fig. 1 (maximal

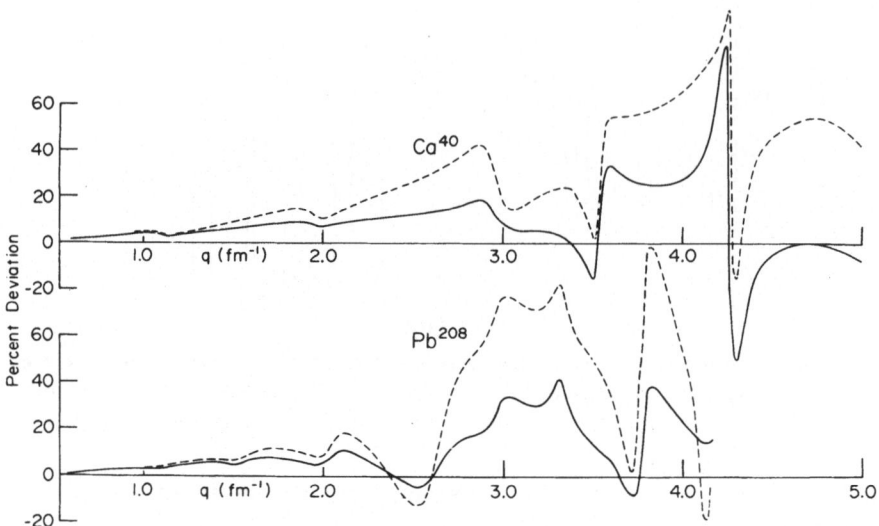

Fig. 2. Fractional change in electron scattering cross sections for ^{40}Ca and ^{208}Pb caused by including the neutron charge density. The dashed and solid curves denote the effect of the maximal and minimal form factors of Fig. 1, respectively.

and minimal $G_E{}^n$) correspond to the dashed and solid curves in Fig. 2. The effect of the neutrons is substantial, and indicates that a greater knowledge of the neutron charge form factor would be extremely useful.

If we are given nuclear proton and neutron distributions $\varrho_P(r)$ and $\varrho_N(r)$ obtained from nuclear wave functions, we can calculate the total charge distribution by inverting the Fourier transform in the first term in Eq. (20)

$$\varrho_P(\mathbf{r}) \equiv \left\langle 0 \left| \sum_i \left(\frac{1 + \tau_3(i)}{2} \right) \delta^3(\mathbf{r} - \mathbf{x}_i') \right| 0 \right\rangle$$

$$\varrho_N(\mathbf{r}) \equiv \left\langle 0 \left| \sum_i \left(\frac{1 - \tau_3(i)}{2} \right) \delta^3(\mathbf{r} - \mathbf{x}_i') \right| 0 \right\rangle$$

$$\varrho(\mathbf{r}) = \int d^3\mathbf{r}' \, (\varrho_p(|\,\mathbf{r} - \mathbf{r}'\,|)\varrho_P(\mathbf{r}') + \varrho_n(|\,\mathbf{r} - \mathbf{r}'\,|)\varrho_N(\mathbf{r}')) \qquad (23a)$$

For spherical nuclei, manipulating the angular $\mathbf{r}'$ integrals casts this expression into the computationally useful form

$$\varrho(r) = \frac{2\pi}{r} \int_0^\infty r' \, dr' \int_{|r-r'|}^{r+r'} y \, dy \, [\varrho_p(y)\varrho_P(r') + \varrho_n(y)\varrho_N(r')] \qquad (23b)$$

The resulting charge distribution $\varrho(r)$ can be used to calculate cross sections using the methods of Section 3.

The spin–orbit contribution can also be manipulated into a convenient form involving an effective charge density for 0^+ nuclei. For such a case the second term in Eq. (20) can be simplified using spherical symmetry

$$F_{so} = -\frac{q^2}{4m^2} \left\langle 0 \left| \sum_i (2\hat{\mu}_i - \hat{e}_i) \frac{j_1(qx_i')}{qx_i'} \, \boldsymbol{\sigma}(i) \cdot \mathbf{L}_i \right| 0 \right\rangle \tag{24}$$

with $\mathbf{L}_i = \mathbf{x}_i' \times \boldsymbol{\pi}_i$ being the angular momentum of the ith particle relative to the center of the nucleus. This form factor is the Fourier transform of the effective spin–orbit charge density for spherical nuclei, which is easily found to be

$$\varrho_{so}(r) = \frac{-1}{r^2} \frac{d}{dr} (r\varrho_{LS}(r))$$

$$\varrho_{LS} = \left\langle 0 \left| \sum_i \left(\frac{2\hat{\mu}_i - \hat{e}_i}{4m^2} \right) \delta^3(\mathbf{r} - \mathbf{x}_i') \boldsymbol{\sigma}(i) \cdot \mathbf{L}_i \right| 0 \right\rangle \tag{25}$$

Although Eq. (25) is completely general (for a spherical nucleus), it is convenient to specialize to j–j coupling. Since the total angular momentum of the ith particle can be written as

$$\hat{\mathbf{j}}_i = \mathbf{L}_i + \boldsymbol{\sigma}(i)/2 \tag{26}$$

we have

$$\boldsymbol{\sigma} \cdot \mathbf{L} = \begin{cases} l, & j = l + \tfrac{1}{2} \\ -(l+1), & j = l - \tfrac{1}{2} \end{cases} \tag{27}$$

Since $2(l+1)$ nucleons fit into a filled $j = l + \tfrac{1}{2}$ subshell and $2l$ nucleons fit into a $j = l - \tfrac{1}{2}$ subshell, a completely filled shell has no net spin–orbit density if the radial wave functions are the same for the two (j, l) subshells. The effect of the spin–orbit density will therefore be more pronounced in the nuclear surface, since the radial wave functions for the non-spin-saturated levels of highest l are concentrated in the surface.

The largest effect obviously is for filled $j = l + \tfrac{1}{2}$ subshells. Although the density contribution we are discussing is quite small compared to the entire density, it will not necessarily be small when comparing the densities of different isotopes, for example. Since this example is of some importance, let us specialize to the charge density of the neutrons in a nucleus with a partially filled $j = l + \tfrac{1}{2}$ shell. Denoting the density of each neutron in this shell by $\varrho_l(r)$, which is normalized to unity, and neglecting the very small contribution of the neutron charge form factor to the spin–orbit density as well as the magnetic moment form factor, we get

$$\varrho_{so}(r) = -\frac{l\mu_n N'}{2m^2}\frac{1}{r^2}\frac{d}{dr}(r\varrho_l(r)) \tag{28}$$

where N' is the number of neutrons in the l shell. This density should be added to the density in Eq. (23) in addition to any possible contribution of unfilled proton shells to the spin–orbit density.

It is worthwhile to examine the effect of the Darwin term and the spin–orbit term on the mean square radius of a nucleus. This can be obtained from Eqs. (20) and (24) by expanding these expressions for small $\mathbf{q}^2$. The coefficient of $-Z\mathbf{q}^2/6$ is the mean square radius $\langle r^2 \rangle$. Neglecting very small quantities $(1/m^4)$ as we did earlier and denoting $\mu_p' \equiv \mu_p - \frac{1}{2}$, we get

$$\langle r^2 \rangle = \langle r_p{}^2 \rangle + \langle r^2 \rangle_p + \frac{3}{4m^2} + \frac{N}{Z}\langle r^2 \rangle_n + \frac{Z'l'\mu_p' + N'l\mu_n}{Zm^2} \tag{29}$$

where $\langle r_p{}^2 \rangle$ is the mean square radius of the proton distribution in the nucleus, N and Z are the neutron and proton numbers, and Z' and N' are the numbers of protons and neutrons in partially filled $j = l + \frac{1}{2}$ shells with orbital quantum numbers l' and l, respectively. The third term is the Darwin–Foldy contribution. Since the first two terms are the largest, the rms radius can be written approximately as

$$\langle r^2 \rangle^{1/2} \cong r_c + \frac{3}{8m^2 r_c} + \frac{N}{2Z}\frac{\langle r^2 \rangle_n}{r_c} + \frac{Z'l'\mu_p' + N'l\mu_n}{2Zm^2 r_c}$$

$$r_c \equiv (\langle r_p{}^2 \rangle + \langle r^2 \rangle_p)^{1/2} \tag{30}$$

The neutron charge density changes the rms radius by approximately $-0.06N/Zr_c$ fm and the Darwin term by $0.02/r_c$ fm.

In order to illustrate these effects, let us consider the nuclei ^{40}Ca and ^{48}Ca (Ber+ 72), the former having completely filled shells and the latter having filled proton shells and a filled $f_{7/2}$ neutron shell. Thus we have $Z' = 0$, $N' = 8$, $N = 28$, $Z = 20$, and $l = 3$ for the latter nucleus and, in addition, $r_c \approx 3.5$ fm. The shift in radius due to the extra neutrons is -0.007 fm from the neutron's charge distribution and -0.014 fm from the spin–orbit term (note that μ_n is negative), giving a total shift of -0.021 fm. Analysis of electron scattering and muonic atom data yields a ^{48}Ca charge radius 0.009 fm *smaller* than that of ^{40}Ca. This means that the proton distribution radius in ^{48}Ca is actually larger than that of ^{40}Ca by about 0.012 fm. This increase in proton radius resolves the anomaly that originates from the incorrect assumption that the decrease in charge radius implies a decrease in the proton radius. The observed increase in the proton radius of ^{48}Ca is in qualitative agreement with Hartree–Fock predictions of 0.01–0.04 fm.

Inclusion of the neutron's electromagnetic structure tends to shrink the charge radius of a nucleus. This is seen in Fig. 3, taken from (Ber+ 72), which shows the various contributions to the ^{48}Ca charge density from the protons, core neutrons, $f_{7/2}$ neutrons, and the spin–orbit density. Note that the last three are negative for large r. It is quite possible for a nucleus to have a negative charge density at large distances if the neutrons stick out further than the protons, because the neutrons' charge density is negative at large distances. The spin–orbit contribution due to neutrons will also be negative, because at large distances the neutron wave function behaves like $e^{-\beta r}/r$.

The spin–orbit contribution to the density of ^{208}Pb has been calculated and is very small, because of a cancellation between the unsaturated proton and neutron shells (Ber+ 72). It is nevertheless important when considering isotopic shifts in the Pb isotopes where spin-unsaturated $3p_{1/2}$ and $2f_{5/2}$ neutron shells are depopulated. Although the low spin and two radial nodes of the $3p_{1/2}$ state result in a small effect for ^{208}Pb–^{206}Pb, the $2f_{5/2}$ contribution is sizable for ^{208}Pb–^{204}Pb. In addition, Campi *et al.* (CSM 74) have examined the effects of neutrons on the charge densities of the even isotopes of Sn and have found a substantial contribution to isotopic charge differences from this source. For future reference, the normalization of ϱ used in subsequent sections is changed to the more convenient convention $\int \varrho r^2 \, dr = Z$, rather than $Z/4\pi$.

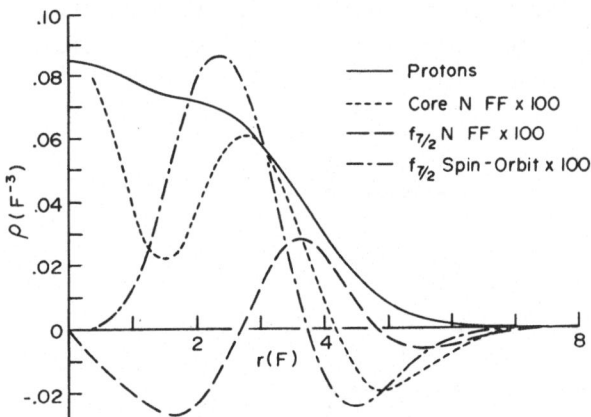

Fig. 3. Contributions to the total charge density of ^{48}Ca from protons, core neutrons, $f_{7/2}$ valence neutrons, and the spin–orbit density.

1.3.2. Meson and Isobar Effects

To a first approximation, a nucleus is a collection of nucleons inter-acting with one another by means of static potentials. The nucleons are not free, however, and a large number of virtual processes are possible. The physical nucleon is the lightest member of a family of baryons. Within the nucleus, therefore, the virtual nucleons spend part of their time as excited nucleons or isobars. Thus our usual description is inadequate and it is necessary to include physical processes such as those depicted in Figs. 4a and 4b in any treatment of the electromagnetic interactions of a nucleus. A convenient methodology for dealing with the extra degrees of freedom produced by the isobars is to extend the *nucleon* wave function to include any extra components, such as the N*'s shown in these figures. The state of the nucleon at any position is determined by its spin, isospin, and isobaric state. A detailed description of this approach and its many applications may be found in the review by Arenhövel and Weber (AW 72).

Williams, Arenhövel, and Miller (WAM 71, AM 74) calculated the effect of internal isobars on elastic electron–deuteron scattering. They found that the effect of the isobars on the charge form factor was roughly the same size as the effect of the neutron's charge form factor (the charge form factor F is proportional to the isoscalar form factor, $G_E^p + G_E^n$). The reason that these effects may be comparable is that the momentum components of the wave function associated with the isobars are different from those associated with the nucleons. They find that the dominant contribution is from Fig. 4b for the N*(1238) and that this piece of the wave function has a probability of about 1%. Isospin conservation for the deuteron is sufficient to rule out the N*(1238) contribution to Fig. 4a, although this nucleon resonance can contribute to the charge form factor

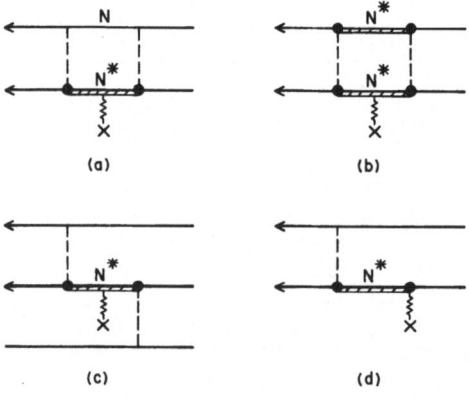

Fig. 4. Nucleon isobar contributions to nuclear electromagnetic properties.

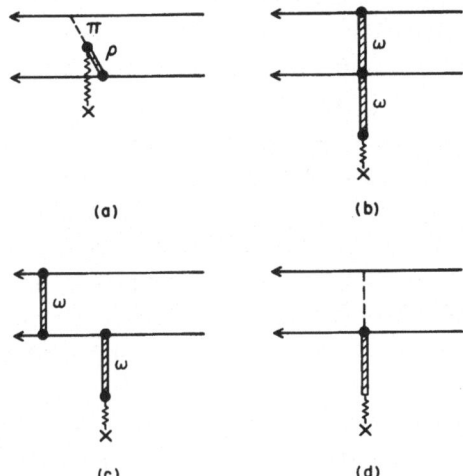

Fig. 5. Mesonic contribution to nuclear electromagnetic properties. Dashed lines denote pions, and shaded lines denote vector mesons.

of other nuclei. In heavier nuclei, genuine three-body effects related to the three-body force can be expected to contribute as shown in Fig. 4c. It is, of course, necessary to normalize the wave function so that we include the additional degrees of freedom. Recently, Kallio *et al.* (Kal+ 74) have investigated the effect of a 5% N*(1238) component on the charge form factor of ^{3}He. They find a relatively small effect, of the wrong sign to help resolve disagreements between experiment (McC+ 70) and theoretical calculations (KT 74a). The isobars decrease the rms radius slightly.

Apart from probing the details of what is already present in a nucleus, the photon is also capable of "polarizing" a nucleus by creating virtual particles. These are independent contributions to the charge form factor. An example of this is shown in Fig. 4d where the incident virtual photon creates an isobar, which subsequently decays by emitting a pion. The overall normalization is not changed by this process; in the static limit ($q^2 = 0$) the contribution of this diagram vanishes.

Mesons also play a role in the charge form factor. Figure 5a shows a process which is believed by some to play a significant role in the deuteron charge form factor at large momentum transfer q. The reason is that the ordinary charge form factor falls off very rapidly for large q. Processes like Fig. 5a, however, share the momentum of the photon between the two nucleons and thus depend on small-momentum components of the nuclear wave function, which are much larger than the large-momentum components. The contribution of this diagram to the charge form factor vanishes at $q^2 = 0$. Calculations of the contribution of this process to the deuteron charge form factor have been performed by Adler and Drell (AD 66), Adler (Adl 66), and Chemtob *et al.* (CMR 74).

A different viewpoint was adopted by Blankenbecler and Gunion (BG 71). Their proposal was that since vector dominance was the mechanism mediating the nucleon interaction with a photon, a vector meson attached to a photon could rescatter from one nucleon to another. The momentum transferred from the photon is again shared between nucleons and this process may be important at large momentum transfers when the ordinary charge form factor is very small. An example of this type of process is shown in Fig. 5b, where the vector meson is an ω. Depending on the isospin of the nucleus, it could also be a ϱ. For the deuteron case the initial vector meson must be isoscalar (ω), but the rescattered one can be isovector as well (ϱ). In fact, the latter process includes the process depicted in Fig. 5a, since the γ–π–ϱ vertex can be mediated by an ω and this process can be visualized as the decay of the ω-meson, which produces part of the nucleon form factor by means of the vector dominance mechanism. Thus, instead of the ω attaching to one nucleon, its decay products go to different nucleons, and this process is subsumed in the diagram of Fig. 5b. Blankenbecler and Gunion take the ω-nucleon scattering amplitude to be diffractive (Gaussian type of dependence on momentum transfer) and find that the effect of this rescattering process on the deuteron charge form factor may be large. One difficulty with this approach is that it is too inclusive. The process depicted in Fig. 5c is obviously a part of the process shown in Fig. 5b. Unfortunately, the former process is already included in the usual nonrelativistic charge form factor, the ω-exchange being part of the nuclear force and the other ω producing the nucleon isoscalar form factor. Modification of Fig. 5c to include only excited nucleon states (with additional mesons perhaps) between the two ω's illustrates the contributions that should be included. This may account for at least part of the reason why the calculation of Blankenbecler and Gunion (BG 71) led to an improperly normalized charge form factor [$F(q^2 = 0) \neq 1$].

Recognizing these defects, Lehman (Leh 71) tried to subtract the effect of Fig. 5c (the Born term) from the overall process. For a variety of reasons discussed by Lehman (Leh 71), this is not a completely satisfactory procedure. The model was then applied to ^{3}He and Lehman concluded that it probably could not explain the peculiar shape of the ^{3}He charge form factor (McC+ 70).

The vector mesons that we have been discussing are heavy and do not like to be exchanged over large distances. Because the nucleon–nucleon force is strongly repulsive at short distances, heavy meson exchange contributions to the charge density are suppressed. Unfortunately, this region of the nuclear potential is also the region of our greatest ignorance. Pions, on the other hand, are very light and can travel large distances. It has been part of the folklore for many years that pion exchanges cannot contribute to the deuteron's charge and magnetic form factors. According to this folklore, processes like Fig. 5d, where the dashed line denotes a pion, do not contribute. Recently, Kloet and Tjon (KT 74b) examined the effect of pion exchanges on the charge form factor of ^{3}He. They examined a diagram very much like Fig. 5c, except with an exchanged pion, and noted that since the nucleon line between the mesons is virtual, it contains negative energy or nucleon–antinucleon "pair" contributions, a subject explored in some detail by Chemtob and Rho (CR 71). Such a pair graph is shown in the last of the time-ordered diagrams of Fig. 6, where all dashed lines depict pions. They found that this graph alone made a substantial contribution to the ^{3}He charge form factor at large momentum transfer. They also calculated Fig. 6c and found that it vanishes if one neglects nucleon recoil upon absorbing or emitting the pion. The latter diagram is isovector and vanishes for the deuteron case (as it does for any isoscalar nucleus), which is the origin of the folklore mentioned earlier. There are other contributions to the form factor as well, which are shown in Figs. 6a and 6b. The effect of the meson exchange in Fig. 6a is largely to change the nuclear wave function, since the nuclear potential is generated by such exchanges. If, however, we calculate wave functions using static potentials, as is customary, we are neglecting something. If we take a picture of a nucleus, there is a chance that the picture will show not just a collection of nucleons, but one or more mesons, since the exchanges are not instantaneous. Since the total probability must be normalized to one, the probability of finding just nucleons and no pions must be less than one. Thus the

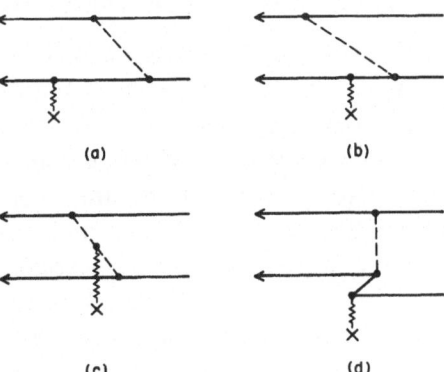

Fig. 6. Time-ordered graphs illustrating pionic contributions of lowest order to nuclear electromagnetic properties.

normalization of our "nucleon-only" wave function must be reduced. In perturbation theory, the normalization change arises in treating Fig. 6a and its mirror image. Other exchange processes are also possible; it should be a good approximation, in fact, to treat heavy meson exchanges as instantaneous and contributing to the static nucleon–nucleon potential while doing the pion part perturbatively, as indicated in Fig. 6.

The remaining graph is shown in Fig. 6b, which Chemtob and Rho (CR 71) call the recoil graph. It is *not* included in the usual wave function treatment of charge form factors. In effect, a photon, when it lands on a nucleon in a nucleus with no mesons "in the air," samples the nuclear momentum distribution specified by the initial and final states. In Fig. 6b, however, this distribution is disturbed through recoil by the emission of the pion and the momentum components of the wave function that the photon sees are different from those contained in the ground-state wave functions. The contribution of this diagram for $q^2 = 0$ exactly compensates for the change in normalization discussed above.

The latter two effects contain isoscalar contributions and do not vanish for the deuteron, as noted by Jackson *et al.* (JLR 75) and Friar (Fri 75a). The latter author also considered the effect of nucleon recoil on Figs. 6a–6c. Numerical estimates of the size of those pionic contributions to the deuteron charge form factor were made by Jackson *et al.* (JLR 75).

Our discussion has centered on "exotic" contributions to the nuclear charge density, and they will affect muonic atoms as well as electron scattering. The calculations which have been performed have emphasized the deuteron and the trinucleon systems, but investigations of the charge distributions of heavier nuclei will undoubtedly be made in the near future. This area of nuclear physics is currently one of the most exciting and is becoming increasingly active.

1.3.3. Dispersion and Polarization Corrections

When a projectile passes a nucleus or orbits around it, virtual excitation of the nucleus is possible. If a lepton scatters from the nucleus, the effects of the virtual processes on the scattering amplitude are referred to as dispersion corrections. If the lepton is orbiting in a bound state, the corrections to the energy levels are called polarization corrections. The treatment of these two cases is very similar. At this time the dispersion corrections have the largest theoretical uncertainty of all the corrections to electron scattering that we will discuss. Uncertainties in the polarization corrections to muonic atom energy levels are also a serious practical limitation on the analysis of muonic atom transition data.

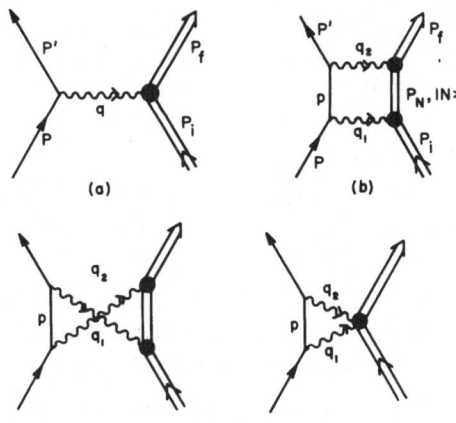

Fig. 7. One-photon and two-photon exchange contributions to the lepton–nucleus interaction. The four-momentum of initial, final, and intermediate states is shown.

In order to understand how these corrections arise and how they can be calculated, it suffices to work to second order in perturbation theory. We will treat the dispersion corrections at first, and later generalize our results to include polarization corrections. The appropriate Feynman diagrams are illustrated in Fig. 7. The single-photon exchange process shown in Fig. 7a was previously discussed. Figures 7b and 7c depict the contributions of the direct and crossed two-photon exchange processes, while Fig. 7d illustrates the so-called "seagull" contribution.

The latter term arises from the $A^2/2m$ term in the nonrelativistic Hamiltonian (FR 74). The other graphs reflect the various time-orderings of the electromagnetic interaction between the electron (single line) and the nucleus (double line). Although it is not difficult to include the effect of the motion of the nucleus on the scattering process, and we will do so in Section 1.3.4, the nucleus will be assumed to be fixed for the remainder of this section.

The interactions between the nucleus and the virtual photons, as well as the interaction of the electron with the photons, can be shown to be gauge invariant. We are therefore free to work in any gauge we choose, and Coulomb gauge is found to be the most convenient because the static Coulomb interaction is separated from the rest of the interaction. The sum of the contributions of Figs. 7b–7d can be conveniently separated into several physically distinct parts. One important part is the total Coulomb interaction between the electron and the nucleus. Other terms represent the transverse current–current and charge–current interactions of the electron and the nucleus, and these are partly magnetic in origin. The latter interactions have not been fully investigated, although preliminary investigations have been made (Lin 72, Lin 73, BD 71, BKK 73, KR 74). The Coulomb inter-

action has been presumed to be the dominant interaction and such information as is available supports this presumption. For this reason we
will confine our attention to this interaction.

The Coulomb interaction part of the two-photon exchange processes
is not simply the once-iterated static Coulomb interaction between the
nucleons in the nucleus and the electron. There is an additional interaction
term caused by presence of virtual electron–positron "pair" states in the
internal electron line in Figs. 7b–7d. Only excited intermediate nuclear
states contribute to the latter term and it has been shown to make a small
contribution to high-energy electron scattering by Friar and Rosen (FR 74).
Similar terms arise in the discussion of recoil corrections to both the bound-
state and scattering problems in Section 1.3.4.

Thus far, no mention has been made of the relative importance of the
intermediate nuclear states. Clearly any final nuclear state that can be
produced by the single-photon exchange process in Fig. 7a can contribute.
The dominant contribution comes from a single intermediate state, the
ground state, and the contribution of all the excited states can be regarded
as a correction. The former part involves the nuclear ground-state matrix
element of the Coulomb interaction between the electron and the nucleons.
This corresponds to the Coulomb potential

$$V_C(r) = -\frac{\alpha}{4\pi} \int d^3\mathbf{r}' \frac{\varrho(r')}{|\mathbf{r} - \mathbf{r}'|} \tag{31}$$

with $\varrho(r)$ given by Eqs. (23) and (25). The normalization of ϱ, however,
is $\int \varrho r^2\, dr = Z$. The factor of Z is crucial to the argument and arises because
the scattering is coherent when the nucleus remains in the ground state.
This produces a second Born amplitude proportional to $(Z\alpha)^2$. The remaining terms involve virtual breakup of the nucleus and are essentially in-
coherent, the resulting contribution to the second Born amplitude being of
order $Z\alpha^2$. Thus the intermediate ground-state term, the elastic Coulomb
part, dominates the intermediate excited-state part, the dispersion correction.
Detailed calculations bear out this argument (except possibly for the
deuteron).

Higher order diagrams in perturbation theory can be split up in the
same way, and the dominant part will be the higher order iterations of
V_C, with higher order (in α) dispersion corrections making up the rest.
Since the effect of V_C is so large, it is appropriate to sum its contribution in
all orders of perturbation theory, and make this the basic interaction between
the nucleus and the electron. We can do this by calculating the wave function

for an electron of energy E using the Dirac equation

$$(\boldsymbol{\alpha} \cdot \mathbf{p} + \beta m + V_C(r))\psi(\mathbf{r}) = E\psi(\mathbf{r}) \tag{32}$$

This forms the basis for calculating the electron scattering amplitude, as we mentioned in Section 1.2. In order to see how the dispersion corrections can be calculated, we define the effective charge operator $\hat{\varrho}(\mathbf{r})$ using Eq. (23),

$$\varrho(r) \equiv \langle 0 \mid \hat{\varrho}(\mathbf{r}) \mid 0 \rangle \tag{33}$$

and note that $\hat{\varrho}(\mathbf{r}) - \varrho(r)$ has vanishing expectation value between nuclear ground states. A transition Coulomb operator can be constructed using this density difference,

$$\Delta V_C = - \frac{\alpha}{4\pi} \int \frac{d^3\mathbf{r}'}{\mid \mathbf{r} - \mathbf{r}' \mid} (\hat{\varrho}(\mathbf{r}') - \varrho(r')) \tag{34}$$

Equation (32) can be extended to include ΔV_C, which is assumed to be small, and the scattering amplitude can then be calculated in perturbation theory to second order in ΔV_C. Using an arbitrary normalization, the complete scattering amplitude is found to be

$$f = f_0 + \langle f \mid \Delta V_C \, \hat{G}(E) \, \Delta V_C \mid i \rangle \tag{35}$$

where f_0 is the scattering amplitude due to V_C alone. The terms that are first order in ΔV_C vanish because this operator has a vanishing expectation value. The states $\mid f \rangle$ and $\mid i \rangle$ are products of different electron scattering states with the nuclear ground state, $\mid 0 \rangle$. The Green's function $\hat{G}$ has the spectral decomposition

$$\tilde{G} = \sum_{N,n} \frac{\mid Nn \rangle \langle Nn \mid}{E - E_N - E_n + i\varepsilon} \tag{36}$$

where the sum has been split into two parts: N referring to nuclear states and n to electron states (including negative energy states), which are calculated using Eq. (32). The difficulty in evaluating Eq. (35) lies in the coupling between electron and nuclear states seen explicitly in Eq. (36). Because the electron scattering experiments in which we are interested involve electrons with energies of the order of several hundred MeV, it is sensible to consider E_n, the nuclear excitation energy, small compared to E, the total electron–nucleus energy, in Eq. (36). In first approximation E_n can be ignored compared to E and the nuclear part of the summation in Eq. (36) can then be evaluated. This is known as the closure approximation. The validity of this approximation was challenged recently by deForest (Def 70). By essentially calculating the sum over N in Eq. (36) nuclear state by nuclear state in a simple model, the harmonic oscillator shell model, he could compare the closure approximation with an exact result. He also verified that keeping only a few nuclear states in the sum, a common practice in the past, was a very poor approximation. In order to improve on the closure approximation, Friar and Rosen (FR 74) expanded the denominator in Eq. (36), assuming E_N is small and approximately equal to $\bar{\omega}$, an effective excitation energy. The expansion procedure produced an improved closure approximation. Recently deForest and Friar (DF 75) verified, using deForest's model discussed above, that this corrected closure expansion is an excellent approximation to the exact model calculation.

Friar and Rosen (FR 72) and deForest also showed that small contributions to Eqs. (35) and (36) may become important because of cancellations. Using the closure approximation, the sum over N in the dispersion term yields a correlation function and this is exceedingly difficult to calculate accurately. Intricate cancellations may take place between pieces which are caused by different physical mechanisms, such as the Pauli principle and the strong short-range repulsion of nucleons. For light nuclei, probably the most important remaining task in dispersion calculations is the development and use of a "good" correlation function. For technical reasons, the electron sum in Eq. (36) is difficult to calculate unless plane wave electron states are used, even if we neglect E_N. For heavy nuclei this is undoubtedly a bad approximation and much work remains to be done in this area. Several recent attempts have used an eikonal approximation to Eq. (36) and this is a promising approach (BK 71, Ros 73). These calculations are probably too crude to be very useful at the present time, however.

Dispersion corrections are customarily calculated as amplitudes which must be added to the static Coulomb amplitude, as shown in Eq. (35). From these amplitudes cross sections can be calculated in the usual way. Figure 8 shows the percentage corrections to electron–^{12}C elastic scattering cross sections calculated by Friar and Rosen (FR 74). Since all the amplitudes are energy dependent, we show the results for two electron energies, 374.5 and 747.2 MeV, and have indicated with arrows the extent of the experimental data that are available at those energies. The region of rapid variation is caused by the diffraction minimum in the ^{12}C form factor, where the real part of the amplitude f_0 vanishes. The percentage dispersion correction is seen to vanish for small momentum transfer. The reason is that static Coulomb scattering amplitude diverges as $1/q^2$, while the dispersion correction is finite. Indeed, the imaginary part of the dispersion cor-

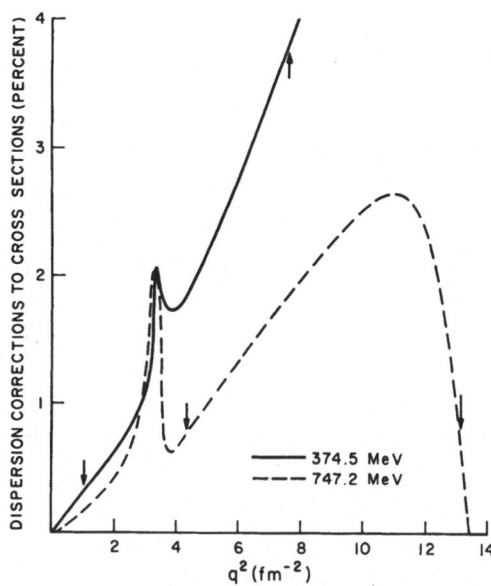

Fig. 8. Percentage change in ^{12}C cross section due to the dispersion corrections of Friar and Rosen (FR 74).

rection amplitude for $\mathbf{q}^2 = 0$ is proportional to the total cross section for excitation of the nucleus by the electron, and this is finite (BM 71). Thus the *fractional* correction vanishes.

A complete list of references and a critical discussion of most of the common approximation techniques can be found in (FR 74), together with a comprehensive set of numerical calculations [see also (FR 72) and (Def 70)].

Much of the formalism, as well as the formulas, remains the same when we consider the effect of the nuclear excited-state spectrum on the energy levels of an atom composed of a lepton and a nucleus. If we reinterpret f and f_0 as the energies of the atom with polarization correction (E) and without polarization corrections (E_i), Eqs. (35) and Eq. (36) can be used if we make $|f\rangle$ and $|i\rangle$ the same. In this case we can combine and rewrite these equations in a slightly different form

$$E - E_i = \sum_{N,n \neq i} \frac{|\langle Nn | \Delta V_{\mathrm{C}} | i \rangle|^2}{E_i - E_N - E_n} \tag{37}$$

where the sum over states excludes the initial state $|i\rangle$. The energy shift is negative for the ground state because the nucleus deforms in such a way as to lower the total energy by moving the nuclear charge closer to the orbiting lepton. This interpretation is also consistent with the sign of the fractional dispersion correction for small momentum transfers in Fig. 8.

Modern calculations of Eq. (37) have been performed (Col 69, Che 70, Ska 70, BN 68) for low-lying levels. It is clear from our interpretation of Eq. (37) that high-lying levels have a very small polarization correction because the wave functions barely overlap the nucleus. In fact, the effective polarization potential can be shown on general grounds (FS 70, EH 72) to behave like

$$V_{\text{pol}} = -\alpha \alpha_N / 2r^4 \tag{38}$$

for large distances from the deformable component (the nucleus in our case) and this falls off rapidly. The quantity α_N is the electrical polarizability of the nucleus. Other methods have also been used (Tan 71).

One simplification which helps in evaluating Eq. (37) is the use of nonrelativistic muon wave functions. This is clearly not an acceptable procedure in the electron scattering case. The approximation is meaningful because the muon mass is about ten times the binding energy of the most-bound state in Pb, for example. In lighter nuclei, the approximation is even better. This eliminates the sum over negative energy states mentioned following Eq. (36). All the difficulty which remains is caused by the double sum over nuclear states and muonic states. It has been found that serious errors can arise by neglecting energetic muon continuum states, and all the calculations mentioned earlier include the full muon spectrum of states in one way or another. This means that the nuclear physics must be approximated. Two primary techniques have been used: Sum over a small set of model nuclear states which, we hope, does justice to the nuclear physics, or use an effective nuclear energy $\bar{\omega}$ in place of E_n in the denominator of Eq. (37) and then use the closure approximation. The latter technique is the same as the technique used in calculating dispersion corrections. The results of the two different approaches are not very different, since the only quantities which seem to be important are the effective excitation energy and certain nuclear multipole matrix elements. Both techniques can build this information into the calculation if sufficient care is taken. This contrasts with the dispersion calculations, where more detailed nuclear physics is needed, as shown by deForest (Def 70).

Comparable results for the polarization corrections to the lowest four muonic states of Pb were obtained by Cole (Col 69), Chen (Che 70), Skardhamar (Ska 70), and Bethe and Negele (BN 68). The calculations to be discussed in Section 4 used Chen's results (Che 70) but extended some-

TABLE I

Contributions to the Binding Energies (in keV) of Various Levels of the ^{208}Pb Muonic Atom

	$1s_{1/2}$	$2s_{1/2}$	$2p_{1/2}$	$2p_{3/2}$
VP($\alpha Z\alpha$)	67.146	19.353	32.355	29.814
VP($\alpha^2 Z\alpha$)	0.55	0.15	0.25	0.23
VP($\alpha(Z\alpha)^3$)	-0.81	-0.30	-0.46	-0.44
VP($\alpha(Z\alpha)^5$)	-0.10	-0.04	-0.06	-0.05
VP($\alpha(Z\alpha)^7$)	-0.02	-0.01	-0.01	-0.01
VP($\pi^+\pi^-$)	0.02			
Total VP	66.79 ± 0.5	19.15 ± 0.3	32.08 ± 0.3	29.54 ± 0.3
Lamb shift	-2.99 ± 0.41	-0.72 ± 0.17	-0.35 ± 0.27	-0.69 ± 0.23
Recoil	0.38	0.09	0.11	0.09
Nuclear polarization	6.0 ± 1.0	1.2 ± 0.7	1.9 ± 0.3	1.9 ± 0.3

what his estimate of the possible error in his calculations. These numbers are shown in Table I.

The effect of a very small energy denominator in Eq. (37) on the energy of one level in the Pb muonic atom has been investigated by Shakin and Weiss (SW 73).

1.3.4. Recoil Corrections

Corrections of order $(1/m_l)$ to the motion of the lepton, which arise for a variety of physical reasons, are referred to as recoil corrections. There are additional terms of order $(1/m_l)^2$ which could also be denoted recoil corrections, but these are dimensionally the same as relativistic corrections to be considered in Section 1.3.5. The recoil corrections affect the form of the interaction, while the relativistic corrections, at least for spin-0^+ nuclei, affect the charge distribution itself. Our interest will be confined to those contributions in which the nucleus remains in the ground state during the interaction; recoil corrections to dispersion corrections are quite small (FR 74).

It is well known that calculations of scattering amplitudes are most easily accomplished in the center-of-mass frame. Nonrelativistically, the center of mass of an object moves uniformly if there is no external force, and the center of mass is a good frame for performing calculations. Calculations in the lab frame, for example, are complicated by the fact that the target, initially at rest, picks up energy in the collision. In addition, the electromagnetic interaction depends on the velocities of the particles, not just on the particle separations, and the description of the collision is more complicated in the lab frame. The cross section for electron–nucleus scattering in first Born approximation in the lab frame is

$$d\sigma/d\Omega = \sigma_{\text{Mott}} \mid F(\mathbf{q}^2) \mid^2 \tag{39}$$

for a spin-0^+ nucleus, where σ_{Mott} is given in Eq. (8). The recoil factor, $\{1 + 2E[\sin^2(\theta/2)]/m_t\}$, involves the nucleus mass m_t explicitly, and is the result of a complicated cancellation between a kinematical phase space factor and a factor which has a dynamical origin. In the center-of-mass frame, however, these factors exactly cancel and the recoil factor can be dropped. The resulting expression for the cross section is just what one would obtain for electron scattering from a *fixed* nucleus. The electron center-of-mass energy and center-of-mass scattering angle must be used in the expression for the cross section. The cancellation in the center-of-mass frame involves two kinematical terms and one dynamical term. It is to be hoped, therefore, that some such cancellation may also occur in higher order terms in the Born series. If this were the case, we could solve the Dirac equation for the lepton's wave function in the scattering problem using the center-of-mass variables, calculate the cross sections, and then transform them to any other frame of reference, such as the lab frame. Similar simplifications might be expected in the bound-state problem.

The electromagnetic interaction of two particles is most rigorously handled by means of the Bethe–Salpeter (BS) equation (SB 51). The scattering amplitude for two particles can be calculated using time-dependent perturbation theory in the usual way (BD 65) and this is completely equivalent to the scattering amplitude calculated perturbatively using the BS equation. Perturbation theory using a plane-wave basis is not appropriate for the bound-state problem, however. The BS equation is not easy to work with and in ladder approximation (neglecting crossed kernels) it has notorious diseases (BP 69). Many people therefore prefer to work with an approximation to this equation that has many useful properties. This approximate equation is the Breit equation (Bre 29), which, unlike the BS equation, has a structure similar to the Dirac equation. For a very heavy, extended, spin-0^+ particle of mass m_t interacting electromagnetically with a lepton of mass m, Grotch and Yennie (GY 69) write the Breit equation in the form

$$H\psi \equiv [\boldsymbol{\alpha} \cdot \mathbf{p} + \beta m + (\mathbf{p}^2/2m_t) + V_T(r)]\psi = E\psi \tag{40}$$

where

$$V_T(r) = V_C(r) + \{\boldsymbol{\alpha} \cdot \mathbf{p}, V_C(r)\}/2m_t + [\boldsymbol{\alpha} \cdot \mathbf{p}, [\mathbf{p}^2, w(r)]]/4m_t \quad (41)$$

and

$$w(r) = -(\alpha/4\pi) \int \varrho(r') \mid \mathbf{r} - \mathbf{r}' \mid d^3\mathbf{r}' \quad (42)$$

Working in the center-of-mass frame, $\mathbf{p}$ is the momentum of either particle, $\mathbf{p}^2/2m_t$ represents the kinetic energy of the heavy particle, and the potential terms other than the static Coulomb potential reflect the magnetic interaction between the two moving particles and the effects of retardation on the electric interaction. The equation has the obvious property that the $m_t \to \infty$ limit reduces to the Dirac equation with a static Coulomb potential, which we remarked in Section 1.2 was necessary. Because the Breit equation is only an approximation to the correct (BS) approach, the potential V_T is not uniquely specified but in practice must be specified in such a way that the correction terms, the difference of the exact and approximate results, are small. This has been discussed in detail (GY 69).

For a spherically symmetric nuclear charge distribution the Coulomb potential V_C and the additional function w can be simplified in terms of moments of the charge distribution:

$$V_C(r) = Q(r)/r + Q'(r) \quad (43a)$$

$$w(r) = rQ(r) + r^2Q'(r)/3 + R(r)/3r + R'(r) \quad (43b)$$

$$Q(r) = -\alpha \int_0^r \varrho(r')r'^2 \, dr' \quad (43c)$$

$$Q'(r) = -\alpha \int_r^\infty \varrho(r')r' \, dr' \quad (43d)$$

$$R(r) = -\alpha \int_0^r \varrho(r')r'^4 \, dr' \quad (43e)$$

$$R'(r) = -\alpha \int_r^\infty \varrho(r')r'^3 \, dr' \quad (43f)$$

For point particles, Q', R, and R' vanish, while Q is a constant (BB 48). Because m_t is so large, it is an excellent approximation to calculate the shift in energy of an atom or the change in the scattering amplitude, caused by the $1/m_t$ term, using perturbation theory and keeping only terms linear in $1/m_t$. We use as a basis the eigenfunctions of the unperturbed Hamiltonian H_0,

$$H_0 = \boldsymbol{\alpha} \cdot \mathbf{p} + \beta m + V_C(r) \tag{44}$$

and note that $H_0{}^2$ has the same basis. The square of H_0 can be used in place of $\mathbf{p}^2$ in the nuclear kinetic energy term

$$H = H_0 + \frac{H_0{}^2 - m^2 - 2\beta m V_C}{2m_t} + \frac{[H_0, [\mathbf{p}^2, w]]}{4m_t} - \frac{V_C{}^2 + [V_C, [\mathbf{p}^2, w]]/2}{2m_t} \tag{45}$$

For both the bound and scattering problems, first-order perturbation theory involves a matrix element of the $1/m_t$ terms between eigenstates of H_0 with the same energy if we are in the center-of-mass frame, and this eliminates the third term.

Taking expectation values with wave functions for bound states of energy E_0 and using the identity (FN 73b)

$$m\langle \beta V_C \rangle = E_0{}^2 - m^2 - E_0 \langle Q' \rangle \tag{46}$$

we obtain the following expression for the binding energy B ($B \equiv m - E$)

$$B = B_0 + (B_0{}^2 - 2mB_0)/2m_t$$
$$+ (\langle V_C{}^2 + [V_C, [\mathbf{p}^2, w]]/2 \rangle - 2(m - B_0)\langle Q' \rangle)/2m_t \tag{47}$$

The potential terms in the first bracket vanish for a point nucleus, as does Q'. Since binding energies are always small compared to rest masses, the two terms proportional to m/m_t are by far the largest contribution to Eq. (47). These terms simply represent the effect of a reduced mass $\mu \approx m - m^2/m_t$. Using μ instead of m in the Dirac equation changes the eigenfunctions and the eigenvalues slightly, and using the identity

$$m\langle \beta \rangle = E_0 - \langle Q' \rangle \tag{48}$$

we see that the m/m_t terms in Eq. (47) are eliminated completely and the binding energy shift becomes

$$B - B_0 = \frac{B_0{}^2}{2m_t} + \frac{\langle Q'^2 + Q(R/r^4 + 4Q'/r)/3 + 2B_0 Q' \rangle}{2m_t} \tag{49}$$

The bracketed term vanishes for a point nucleus, in agreement with the results of Breit and Brown (BB 48). Outside the charge distribution, Q'

vanishes and the function inside the brackets is just

$$\frac{(Z\alpha)^2 \langle r^2 \rangle / 6m_t}{r^4} \tag{50}$$

which is similar in structure to the polarization correction potential, Eq. (38), and is also attractive. For most states, the term $B_0{}^2/2m_t$ is the largest contribution to Eq. (49). For example, using a representative charge distribution for ^{208}Pb, this term is 0.29 keV for the $1s_{1/2}$ state, while the bracketted term is 0.09 keV, (FN 73b). For comparison, the ordinary reduced mass correction is about 2 keV out of a total binding energy of roughly 10 MeV. Identical numerical results were obtained by Barrett et al. (Bar+ 73) using Eq. (40) and a slightly different approach. Numerical results for the four lowest states of muonic ^{208}Pb are listed in Table I.

Similar results are expected in the scattering problem. As indicated earlier, we set $m = 0$ in Eq. (45) and note that

$$(H_0 + H_0{}^2/2m_t)\psi_0 = (\varepsilon + \varepsilon^2/2m_t)\psi_0 = E\psi_0 \tag{51}$$

where ε is the *electron* energy in the center-of-mass frame, E is the total energy of both particles, and ψ_0 is the eigenfunction of H_0 with energy ε. Thus we solve for the scattering in the center-of-mass frame, using the scattering solutions of the Dirac equation corresponding to the center-of-mass energy of the *electron*. The remaining term in Eq. (45) can be used to calculate the change in scattering amplitude Δf in the usual fashion (Fri 75b):

$$\Delta f = -\langle \psi_0^{(-)} \mid Q'^2 + Q(R/r^4 + 4Q'/r)/3 \mid \psi_0^{(+)} \rangle / 2m_t \tag{52}$$

The effect of recoil on scattering, determined by Δf, vanishes for a point nucleus. This equation also agrees with the calculation of Foldy et al. (FFY 59), who showed that neglecting terms of order $(Z\alpha)^2$ and of order $(1/m_t)^2$ leads to a cross section in the center-of-mass frame which is unchanged by recoil. Equation (52) represents the correction to this of order $((Z\alpha)^2/m_t)$. Because recoil effects are small for heavy nuclei and because wave function distortion effects are not large for light nuclei, it is probably not a bad approximation to calculate the matrix element in Eq. (52) using plane waves. Such a calculation for ^{16}O is shown in Fig. 9, using the best-fit charge distribution of Sick and McCarthy (SM 70) for $\varrho(r)$, and shows the corrections which result from adding Δf to the static Coulomb amplitude (Fri 75b). The scattering amplitude for the static Coulomb potential alone

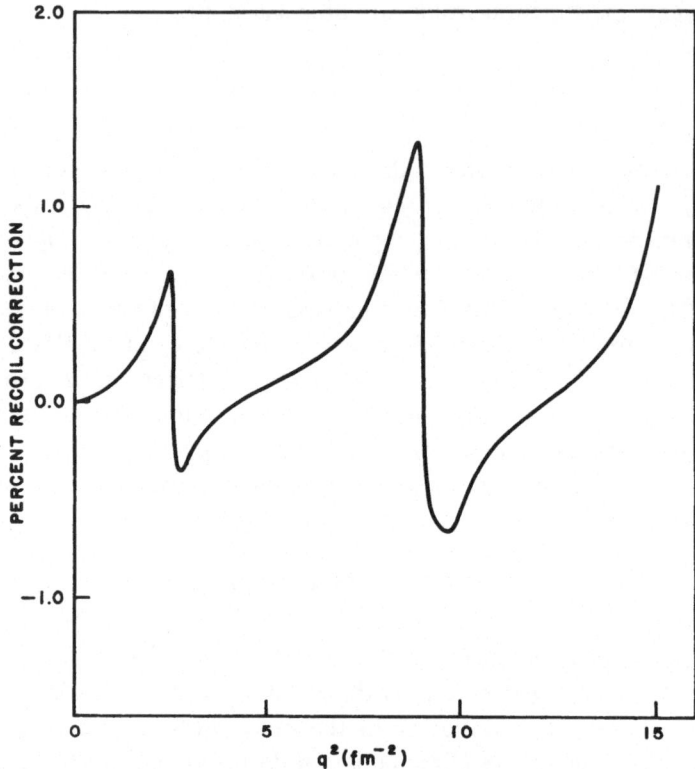

Fig. 9. Percent recoil corrections to ^{16}O elastic electron scattering cross sections obtained using a Breit equation treatment.

will be discussed in Section 3. It is apparent that these corrections of order $(1/m_t)$ are quite small.

It is clear that a unified description of bound and scattering recoil effects exists and the same is presumably true for those corrections *not* given by the Breit equation. The only calculation which exists for these additional corrections are for bound states of electronic atoms and involve an approximate treatment of the Bethe–Salpeter equation (Sal 52, FM 54) or the effective potential method (GY 69).

1.3.5. Relativistic Corrections

In Section 1.3.1 we discussed corrections to the nucleon charge operator that were of order $(v/c)^2$, i.e., relativistic corrections. There are also corrections of this order that arise not from the nucleon motion within the nucleus but from the motion of the nucleus itself. These corrections are generally small, except for extremely light nuclei, since they are inversely proportional to powers of the nucleus mass m_t. Most of the phenomena we are interested in are also classical, so our discussion will be semiclassical.

The usual nonrelativistic description of a nucleus fails for a number of reasons. The wave function of a uniformly moving nonrelativistic nucleus is simply the product of the velocity-independent internal wave function, which describes the structure of the nucleus, and a plane wave for the center-of-mass motion. Relativity tells us, however, that a moving object is contracted along the direction of its motion. It is therefore necessary to build the Lorentz contraction into the internal wave function of the nucleus.

Because an accelerated object is continuously changing its "instantaneous" rest frame, our description of quantities such as spin, which are most sensibly discussed in the rest frame, is continuously changing. In particular, two frames of reference differing only slightly in velocity relative to a third frame of reference appear to be rotated with respect to one another if their velocities (with respect to the third frame) are not collinear. This Wigner rotation becomes the Thomas precession for an accelerated system and primarily affects the total spin of a moving nucleus and the spin of moving nucleons within such a nucleus.

A third category of kinematical effects concerns the proper definition of the center-of-mass coordinate of an object. The center-of-mass coordinate of an extended object is defined as that combination of the coordinates of the individual particles comprising the object that moves according to Newton's laws when a force is applied. In particular, when a composite particle has no external forces acting upon it, the center-of-mass moves with uniform velocity. We expect that this definition changes when relativity is introduced, since the relationship between energy (mass) and velocity changes. If one examines those quantities that are conserved for a uniformly moving system, one finds that the nonrelativistic center-of-mass relationship should be changed to a center-of-energy relationship. Instead of summing the coordinates of each particle weighted by its mass in order to define the inertial "center" of an object, we should weight the coordinate by the total energy: rest mass plus kinetic and potential energy. This changes the nature of our description of a moving nucleus and must be taken into account (Hag 63, LL 62).

The actual calculation of the effect of these various aspects of Lorentz kinematics on the motion of a nucleus being accelerated by the electric field of a passing electron can be done in a variety of ways. Work has established (Osb 68, CC 70, CO 70, KF 70, BP 69, KF 74) the effect of relativity on a moving wave function. Since the usual description of a nucleus is by means of a nucleus-rest-frame wave function (a nonrelativistic internal wave function, but that is unimportant), it is necessary to relate the wave function in the moving frame to a rest-frame wave function. This is done by modifying the effective interaction between the nucleus and the external field caused by the electron so that this effective interaction can be used in conjunction with rest-frame nuclear wave functions to calculate transition probabilities. Thus the effect of such relativistic phenomena as Lorentz contraction is shifted from the wave function to the electromagnetic transition operator.

The most complete and elegant treatment of this problem has been made by Krajcik and Foldy (KF 70, KF 74) and the results of their fundamental work on the effective interaction were used by Friar (Fri 73a) to calculate changes in the electron scattering cross-section formulas, Eqs. (7) and (19), to order $(v/c)^2$. For spin-0^+ nuclei many of the corrections for elastic scattering vanish, particularly those due to the Thomas precession. For this special case (the only one we are interested in) the only effects in first Born approximation are modifications of the form factor $F(\mathbf{q}^2)$. These modifications are of two types. One type modifies the definition of the form factor slightly. Lorentz contraction contributes to this, as does, the Thomas precession, which generates an additional spin–orbit type of interaction. In addition, the difference of the center-of-energy and the center-of-mass coordinates also produces a change in F. Calculations of the latter two effects are somewhat difficult to perform and have been estimated for $^3\text{He}-^3\text{H}$; they have been shown to vanish for the deuteron. The effect of Lorentz contraction on the form factor is to alter slightly the argument of the form factor. That is, the form factor F should be evaluated at a slightly different effective momentum transfer. The other type of modification is more interesting since it confirms some nuclear physics folklore of long standing: the three-momentum transfer $\mathbf{q}^2$ used as the argument of F in Eqs. (7) and (19) should be replaced by the four-momentum transfer q^2, where

$$q^2 = \mathbf{q}^2 - \omega^2 \qquad (53)$$

and in the lab frame the energy loss ω is given by

$$\omega = q^2/2m_t \qquad (54)$$

As we indicated in Section 1.3.4, the center-of-mass frame is actually the most convenient to work in. In this frame, $\omega \equiv 0$. The procedure we use to calculate cross sections in the lab for comparison with data is to calculate cross sections in the center-of-mass system and transform the result to the lab frame. Thus the only relativistic modification of the form factor which depends on our choice of frames is *automatically* built into our basic procedure and requires no additional effort. This argument and all calculations to date have been based on the first Born approximation. Higher terms in the Born series are not treated exactly by this procedure but effects of order $(1/m_t)^2$ are probably much smaller in these terms than in the first Born term.

The effect of such relativistic modifications is most important for very light nuclei and most calculations have involved the deuteron or trinucleon system. For the deuteron, relativistic corrections were pioneered by Gross (Gro 65, Gro 66, CG 67) using the Bethe–Salpeter equation (SB 51) for the two nucleons. Gross found that there are five relativistic corrections: the Darwin–Foldy terms, the spin–orbit term, the Lorentz contraction factor, the use of the four-momentum transfer in the form factor, and, in addition to those we have already discussed, a term proportional to the nucleon–nucleon potential. The latter term, according to Gross (Gro 67), results from the fact that synchronized clocks on the two nucleons in the nucleus rest frame will not be synchronized in any other frame. Unfortunately, all the techniques except Gross' which we previously mentioned cannot uniquely determine the form of any relativistic corrections that are proportional to the nucleon–nucleon potential strength (KF 74). No calculations of the effect of the Gross potential term on electron scattering from nuclei heavier than the deuteron have been made. Recently, numerical calculations of relativistic corrections to the deuteron charge form factor were performed by Sprung and Rao (SR 75).

All the corrections mentioned in this section are most important at high momentum transfer. Few of the correction terms contribute to the mean-square radius, for example. These effects are also very small for heavy nuclei. For both these reasons, we expect the relativistic effects of order $(1/m_t)^2$ to be a negligible factor in muonic atoms, which generally involve medium to heavy nuclei and, as we will see in Section 4, are not substantially affected by high Fourier components of the charge density.

1.3.6. Quantum Electrodynamic Corrections

There are two important effects whose origin and calculation lie within the realm of quantum electrodynamics: vacuum polarization and the Lamb shift. The former phenomenon is extremely important in muonic atoms, more so than the Lamb shift. The vacuum polarization itself is due to the effect of the nuclear charge on the filled negative energy sea of electrons, muons, etc. If a positive point charge q_0 is introduced in the homogeneous sea, negative charge is drawn toward the positive charge, which results in a cloud of negative charge surrounding the positive charge. Since one only measures deviations of the charge from the uniform, negatively charged background, a large sphere placed around the point charge contains more negative charge than before, and the original "bare" charge q_0 has been reduced to the "physical" charge q ($q < q_0$). A test charge probing *inside* the cloud sees a greater amount of positive charge and therefore its binding is increased. Because the electric field of a nucleus with large Z is very strong, the polarization of the vacuum is fairly large. The radius of the cloud is roughly a Compton wavelength.

The basic Feynman diagram for the vacuum polarization process of order α is shown in Fig. 10a. The double lines correspond to leptons in the nuclear electrostatic potential, which has a coupling constant $Z\alpha$. Powers of α and $Z\alpha$ will be counted separately. The closed double loop is the exact lepton Green's function, which can be expanded in a power series in $Z\alpha$. By Furry's theorem only odd powers of $Z\alpha$ in this expansion contribute (BD 65), and three of these contributions are shown in Figs. 10b–10d. The dominant contribution of order $\alpha Z\alpha$ is shown in Fig. 10b, where the nuclear electrostatic potential originates at the cross. This process leads to a polarization charge density and a polarization potential called the Uehling potential (Ueh 35).

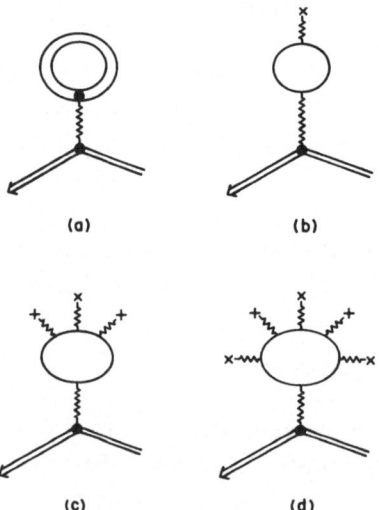

Fig. 10. Vacuum polarization diagrams of order α. The three contributions of order $\alpha(Z\alpha)$, $\alpha(Z\alpha)^3$, and $\alpha(Z\alpha)^5$ shown in (b), (c), and (d) are contained in the graph shown in (a).

The Uehling potential can be expressed in the following form for a point nucleus:

$$V_{\text{Ueh}} = -\frac{\alpha Z \alpha}{\pi r} \int_1^\infty dt \, e^{-2tmr} \left(\frac{2}{3t^2} + \frac{1}{3t^4} \right)(t^2 - 1)^{1/2} \tag{55}$$

As indicated earlier, the range of the vacuum polarization charge density is measured in units of $\hbar/mc$, the Compton wavelength. For pions and muons this is a very small distance and simplifies the treatment of the vacuum polarization corrections. Henceforth, we will only discuss the vacuum polarization caused by electrons. For distances much less than an electron Compton wavelength, an expansion of the integral can be made:

$$V_{\text{Ueh}} \cong \frac{\alpha Z \alpha}{\pi r} \left\{ \frac{2}{3} \left[\ln (mr) + \gamma \right] + \frac{5}{9} - \frac{\pi}{2} (mr) + (mr)^2 \right.$$
$$\left. - \frac{2\pi}{9} (mr)^3 + \frac{(mr)^4}{6} \left[\ln (mr) + \dot{\gamma} \right] + \cdots \right\} \tag{56}$$

where γ is Euler's constant. Additional terms in the expansion are given by Blomqvist (Blo 72). The charge distribution of the nucleus makes a substantial effect on the potential calculated from Fig. 10b, and since this diagram is linear in the nuclear electrostatic potential, it is only necessary to fold a nuclear charge distribution into the point Uehling potential (FW 62, Bar+ 68, Bar 68, Blo 72):

$$V_{\text{Ueh}}^{\text{finite}}(r) = \int V_{\text{Ueh}}(r')\varrho(|\, \mathbf{r} - \mathbf{r}' \,|) \, d^3\mathbf{r}' \Big/ \int \varrho(r') \, d^3\mathbf{r}' \tag{57}$$

In calculating the numerical values of the Uehling potential in Eq. (57), the expansion in Eq. (56) is extremely helpful, but care must be taken to include enough terms for high muonic levels. By adding the Uehling potential to the Coulomb potential in the Dirac equation and calculating energy eigenvalues for the muonic levels, we can obtain the shift in energy due to the lowest order vacuum polarization potential. This shift includes the iteration of the Uehling potential indicated in Fig. 11a. The latter contribution is small, being less than 0.1 keV for the $1s$ state in Pb (FN 73a, EV 74).

The polarization of the vacuum is not linear. The polarization depends on higher powers of the electric field of the nucleus, and two of the higher order diagrams are indicated in Figs. 10c and 10d. In an elegant calculation, Wichmann and Kroll (WK 56) calculated the complete vacuum polarization charge density indicated in Fig. 10a, for a point nucleus, and the various point charge results for the $Z\alpha$ expansion shown in Figs. 10b–10d. Values of the binding energy shift for Pb from the $\alpha(Z\alpha)^3$ potential calculated by

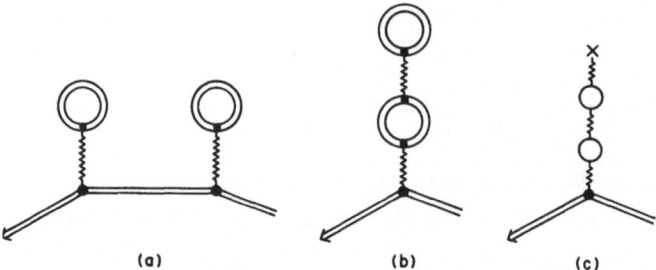

Fig. 11. Vacuum polarization diagrams of order α^2. Graph (c) is the lowest order in $Z\alpha$ part of (b).

Fricke (Fri 69) were used in the analysis of Friar and Negele (FN 73a), to be discussed later. Because of a long-standing discrepancy* between the theoretical and experimental values of the energy of certain high-lying levels in heavy element muonic atoms [reviewed by Engfer and Vuilleumier (EV 74) and Watson and Sundaresan (WS 74)], the vacuum polarization calculations were redone (Blo 72, SW 72, Bel 73). It was found that Fricke's $\alpha(Z\alpha)^3$ results had the wrong sign and were slightly too small in magnitude for these levels. The correct sign was used in the analysis of muonic atom data by Friar and Negele (FN 73a) but the magnitudes were incorrect. Recent calculations of the $\alpha(Z\alpha)^3$, $\alpha(Z\alpha)^5$, and $\alpha(Z\alpha)^7$ vacuum polarization contributions by Engfer and Vuilleumier (EV 74) are listed in Table I.

Interest has focused on the effect of the nuclear charge distribution on Fig. 10c. Rinker and Wilets (RW 73) performed a purely numerical evaluation of Fig. 10a. This calculation requires a difficult charge renormalization which must be performed numerically and their original numbers were inaccurate. In a series of similar calculations, Arafune (Ara 74), Brown *et al.* (BCM 74a), and Gyulassy (Gyu 74) have calculated the dependence of the vacuum polarization binding energy shifts on the nuclear radius for a variety of high-lying levels. Unfortunately, their method is not immediately applicable to low-lying levels. They used a partial wave expansion of the complete electron Green's function (the loop in Fig. 10a) and found that a few low partial waves determine the shift in the energy caused by the finite nuclear radius. Detailed discussions of the procedure have been given (Gyu 75, BCM 74b). Recent unpublished results by Rinker and Wilets agree with these results for high muonic orbitals, and also include all lower levels. An alternative method has been developed by Owen (Owe 73) and numerical calculations, which are quite difficult, have been promised.

* See Note Added in Proof, p. 376.

In addition to the vacuum polarization graphs of order α, there are contributions of order α^2 which have been calculated. In addition to the iteration of the Uehling result shown in Fig. 11a, there can be consecutive loops as shown in Fig. 11b. To lowest order in $Z\alpha$, this process is given by the contribution of Fig. 11c. An additional process of order α^2 is the radiative correction to the vacuum polarization loop indicated in Fig. 12a. An expansion to lowest order in $Z\alpha$ gives the three contributions shown in Figs. 12b–12d. Källen and Sabry (KS 55) calculated the last three graphs, and the polarization potential corresponding to these diagrams is given by Blomqvist (Blo 72) together with a convenient expansion for small values of the radius. The older results by Fricke (Fri 69) contained a counting error, including the diagram of Fig. 11c twice, as well as the previously mentioned error in the $\alpha(Z\alpha)^3$ term. Recent numerical calculations by Engfer and Vuilleumier (EV 74) are listed in Table I with the label $\alpha^2 Z\alpha$. The small contribution of $\pi^+\pi^-$ pairs to the Uehling potential is also included in Table I. Traditionally the effect of μ pairs is included in the numerical values of the Lamb shift, which we will discuss later.

The remaining α^2 contributions of the vacuum polarization type are shown in Figs. 13a and 13b. An expansion of the electron Green's functions in powers of $Z\alpha$ can be performed and diagrams of the type $\alpha^2(Z\alpha)^2$, one of which is shown in Fig. 13c, can be separated from the graph of order α^2 shown in Fig. 13d. The latter diagram depicts the vacuum polarization modification of the Lamb shift contribution. Recently Chen (Che 75) numerically calculated the contribution of the $\alpha^2(Z\alpha)^2$ diagrams for high-

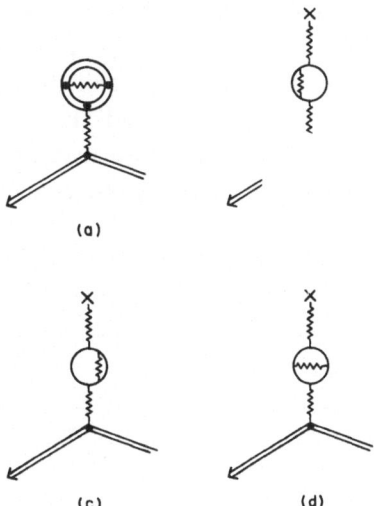

(a)

(c)   (d)

Fig. 12. Vacuum polarization diagrams of order α^2. Graphs (b), (c), and (d) are the lowest order in $Z\alpha$ parts of (a).

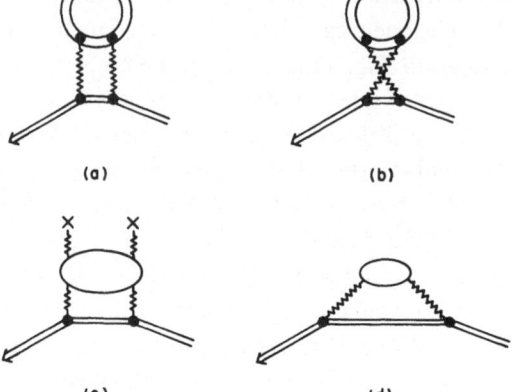

(a)   (b)

(c)   (d)

Fig. 13. Vacuum polarization diagrams of order α^2. Graph (c) is one of the $(Z\alpha)^2$ contributions contained in (a) and (b), while (d) is a correction to the Lamb shift.

lying states, assuming a point nucleus. He found binding energy shifts substantially greater than those expected on the basis of estimates (WR 75). The reasons for his large results are not known,* nor is the effect of this process on low-lying states. The uncertainty for these states is such that we assign generous error estimates to the total vacuum polarization contribution in Table I.

Considerable theoretical work on vacuum polarization is needed, particularly as it concerns the low-lying states. More accurate data for these states will not be of much use until known theoretical deficiencies can be eliminated. In particular, the physics of the $\alpha^2(Z\alpha)^2$ diagrams must be better understood, and the dependence of the $\alpha(Z\alpha)^3$ potential on the nuclear charge distribution for small values of the radius must be deduced.

The Lamb shift is a purely radiative effect, the lepton emitting and then absorbing a photon, as shown in Fig. 14a. The traditional method of evaluating the energy shift due to this phenomenon is to split the virtual photon into "soft" and "hard" parts (BD 65). The "hard" part is shown in Fig. 14b and a renormalization procedure must be carried out for this diagram. All but the $Z\alpha$ contribution has been neglected in the lepton's intermediate Green's function, since additional potential effects should be small for a hard photon. For the soft part, however, the effect of the potential is important and all potential insertions must be summed. A typical term is shown in Fig. 14c. Both the soft and hard parts have a logarithmic divergence, which cancels when the two terms are added. The soft part gives the famous Bethe logarithm (Bet 47). The lepton part of the hard diagram has the form of Eq. (1), which is no surprise, since the effect of

* See Note Added in Proof, p. 376.

recoil on the lepton emitting the photon is to smear out the lepton's charge distribution and to produce an anomalous magnetic moment, $\varkappa = \alpha/2\pi$. If we perform a nonrelativistic expansion similar to that in Eq. (3), we see that the anomalous magnetic moment contributes to the lepton charge distribution through Zitterbewegung and that there will be a spin–orbit interaction proportional to this anomalous magnetic moment. Putting everything together, the change in binding energy ΔB_{LS} due to the Lamb shift is given by (Bar+ 68)

$$\Delta B_{\mathrm{LS}} = - \frac{\alpha^2}{3\pi m^2} \langle \varrho(r) \rangle \left[\ln \frac{m}{2\bar\varepsilon} + \frac{11}{24} + \frac{3}{8} - \frac{1}{5} \right]$$
$$- \frac{\alpha}{4\pi m^2} \left\langle \frac{1}{r} \frac{dV_c}{dr} \, \boldsymbol{\sigma} \cdot \mathbf{L} \right\rangle \tag{58}$$

where $\bar\varepsilon$ is the average excitation energy defined by the Bethe logarithm, the 3/8 term and the spin–orbit term are due to the anomalous magnetic moment, the 1/5 is the contribution of μ pairs to the vacuum polarization, and the remaining pieces are caused by the nonmagnetic smearing out of the lepton charge. The quantity $\bar\varepsilon$ is not easy to calculate. Estimates have been made by Barrett *et al.* (Bar+ 68). By using sum rules and the inequality

$$- \ln \left(\sum_i \omega_i \frac{1}{E_i - E_0} \right) \leq \sum_i \omega_i \ln (E_i - E_0) \leq \ln \left[\sum_i \omega_i (E_i - E_0) \right]$$

to calculate upper and lower bounds to the logarithmic term, Bethe and Negele (BN 68) calculated improved estimates for the ground state. Barrett (Bar 68) made extensive calculations using this method.

As noted by Barrett (Bar+ 68), the nuclear charge distribution $\varrho(r)$ prevents the binding energy shifts ΔB_{LS} from being large. Typical numbers from Barrett (Bar 68) for the Pb shift are listed in Table I for the lowest few states. The part of the Lamb shift that reflects the smearing out of the

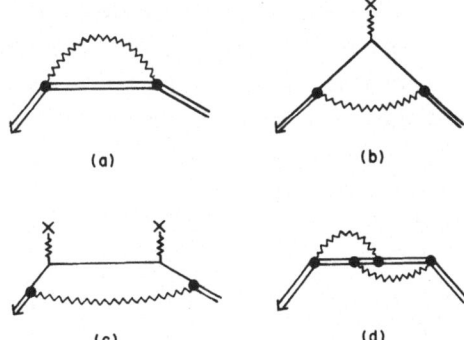

Fig. 14. Lamb shift diagrams of order α and α^2. Graphs (b) and (c) are part of (a).

(a)   (b)   (c)   (d)

lepton charge, and which always lessens binding, dominates the magnetic spin–orbit contribution for the lowest states. Although the magnetic spin–orbit part is rather small for these states, it is sufficiently large for some higher states to change the sign of the binding energy shift. McKee (McK 69) has suggested that adding an anomalous magnetic moment interaction term to the Dirac equation allows one to avoid the nonrelativistic approximation used in deriving Eq. (58).

Estimates by Barrett *et al.* (Bar+ 68) suggest that Lamb shift calculations using Eq. (58) may be as much as 30% in error for the $1s$ state, because of the various approximations which were made. It is possible to avoid the separation into hard and soft photons (with considerable increase in complexity), and this has been investigated by Erickson and Yennie (EY 65a, EY 65b). A method developed by Brown *et al.* (BLS 59), which utilizes a partial wave decomposition of the lepton Green's function, appears to be a likely candidate for calculating the Lamb shift in muonic atoms. Numerical calculations for the Hg electronic atom by Brown and Mayers (BM 59) appear to be in error, but the method has recently been used by Desiderio and Johnson (DJ 71) for a variety of electronic atoms. Recent work by Erickson (Eri 71) has confirmed the numbers of Desiderio and Johnson (DJ 71), and similar results have been obtained more recently by Mohr (Moh 74a, Moh 74b). An approximate calculation of the Lamb shift for the hypothetical $Z = 137$ atom has been performed by Labzovskii (Lab 71) using the method of Braun *et al.* (BDL 70).

Regrettably, no calculations using any of these methods exist for muonic atoms. This represents another area where considerably more work is needed. Diagrams such as Fig. 14d are uncalculated except for the hard photon pieces proportional to $Z\alpha$. The anomalous magnetic moment contributions to the Lamb shift from this and other diagrams of order α^2 are obviously not important. A detailed review of these calculations is given by Lautrup *et al.* (LPD 72).

All of the effects which we have discussed for the lepton bound-state problem exist for the scattering problem. In addition, the electron will radiate when scattered, as shown in Figs. 15a and 15b. Because of the radiation, elastically scattered electrons will not have a sharply defined

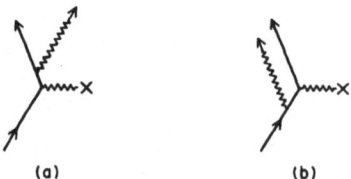

(a)   (b)

Fig. 15. Feynman diagrams indicating radiation of photons during electron scattering.

energy, but rather will have a distribution of energies depending on how much of the initial energy has been radiated. Fortunately many of the small corrections which we previously found necessary to consider because of the great precision of the muonic atom transition data are probably not as important for the scattering problem, because the data are not as precise ($\Delta\sigma/\sigma \sim 10^{-2}$ as opposed to $\Delta E/E \sim 10^{-4}$). The radiative correction problem is nevertheless extremely complex. Because of the certainty of energy loss by the electron, the extraction of a cross-section number from the data which would represent the strength of the scattering in the absence of radiative effects is a complicated procedure. Published cross sections have had the effects of real radiation, self-radiative effects, and vacuum polarization removed. The reader is referred to the reviews of Mo and Tsai (MT 69) and Maximon (Max 69).

1.3.7. Center-of-Mass Corrections

One of the corrections which must be applied to most theoretical calculations of charge densities is based on expediency rather than physical necessity. When calculating wave functions for a nucleus, it is extremely convenient to work with basis wave functions fixed in space. That is, the nuclear wave functions are not translation invariant as fundamental principles dictate. They will, in fact, depend not only on the differences of coordinates of different nucleons, but on the coordinate **R** of the center of mass of the nucleus. Since the amplitudes for electron scattering from a nucleus are derived on the assumption that the target wave function is translation invariant, the noninvariant wave functions ψ must be modified to remove the **R** dependence. Lipkin (Lip 58) proposed a way of producing an invariant wave function ψ'

$$\psi' \equiv \int d^3\mathbf{R}\, G(\mathbf{R})\psi \tag{59}$$

where the integral leaves relative coordinates untouched. The function G is arbitrary. Recently Friar (Fri 71) suggested the alternative method of projecting the **R** dependence out of the density matrix,

$$\psi'\psi'^* \equiv \int d^3\mathbf{R}\, G(\mathbf{R})\psi\psi^* \tag{60}$$

This method is somewhat more convenient because two commonly used methods, the fixed center-of-mass method and the Gartenhaus–Schwartz method (GS 57), correspond to $G(\mathbf{R}) = \lambda\, \delta^3(\mathbf{R})$ and $G(\mathbf{R}) = $ const, respectively. For heavy nuclei the correction is not very large, so that any correction is probably reasonably satisfactory. One special case, the harmonic oscillator shell model wave function, has a particularly simple center-of-mass correction, which is unique (i.e., independent of G). For this special case, the corrected form factor F' can be expressed in terms of the uncorrected form factor F as (TB 58)

$$F'(\mathbf{q}^2) = [\exp(b^2\mathbf{q}^2/4A)]F(\mathbf{q}^2) \tag{61}$$

where b is the usual oscillator parameter and A is the nucleon number. Many people, in fact, use the above correction for arbitrary shell model wave functions. This is done by including the exponential in Eq. (61) with the nucleon charge form factor in Eq. (20), and the correction is therefore folded into the charge distribution in Eq. (23). This procedure is rather crude, but if the overall correction is not large, we are better off using it than doing nothing.

1.3.8. Experimental Considerations

There are a number of corrections of a purely experimental origin which must be dealt with in analyzing electron scattering data (or any scattering data). Because the number of scattered particles is usually small, experiments are a compromise between counting statistics and the introduction of significant correction factors. Figure 16 shows a typical experimental situation with a target of thickness t and an exaggerated spectrometer opening corresponding to an electron scattering angle θ. This opening has a finite solid angle, and electrons passing through the center of the aperture, which defines the angle θ in the scattering plane, are scattered through a different angle in general than electrons that pass through a different part of the aperture. The target in many experimental facilities will be oriented at an angle of $\theta/2$ with respect to the incident beam, so that the incident and exit beams for scattering angle θ make equal angles with respect to the target plane. This arrangement guarantees that the thickness of material through which the beam passes is always the same, $t/\cos(\theta/2)$, regardless of the depth in the target at which the interaction takes place. Because the Coulomb interaction strongly favors small-angle scatterings, it is probable that in passing through the target many very small-angle scatterings will occur in addition to any large-angle scattering. For simplicity one usually assumes that the small-angle scatterings are distributed in a Gaussian fashion. In practice it is necessary to parameterize the cross section in order to deal with both these effects, which require averaging the cross sections

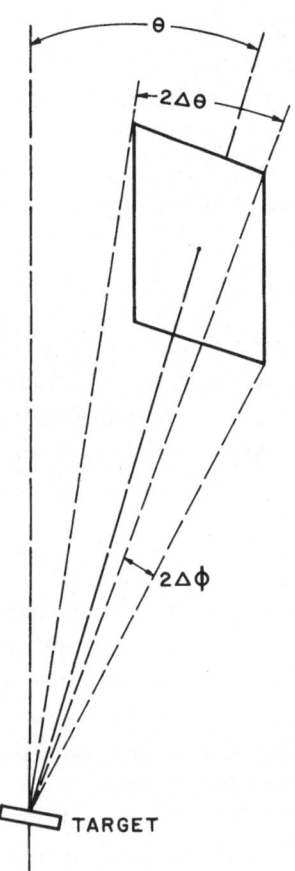

Fig. 16. Target–beam–spectrometer geometry. The large rectangle is the spectrometer opening and the two long-dashed lines indicate the initial and final electron directions.

over a range of scattering angles. Because most cross sections fall off very rapidly, it is not a reliable procedure to parameterize them with polynomials. One should rather parameterize the logarithm of the cross section. Typically, it is assumed that for scattering through an angle θ' which is close to an angle θ we can write

$$\sigma(\theta') \cong \sigma(\theta) \exp[\alpha^{1/2}(\theta' - \theta) + \beta(\theta' - \theta)^2] \qquad (62)$$

This expression should be averaged over all the directions a scattered electron can travel and pass through the slit. Expressed in terms of the horizontal and vertical acceptance angles $\Delta\theta$ and $\Delta\phi$ in Fig. 16, the average is

$$\bar{\sigma}(\theta) \cong \sigma(\theta)\left[1 + \alpha\frac{\cot\theta}{6}(\Delta\phi)^2 + \frac{(\Delta\theta)^2}{3}\left(\beta + \frac{\alpha}{2}\right) + \frac{(\Delta\theta)^4}{10}\left(\beta^2 + \beta\alpha + \frac{\alpha^2}{12}\right)\right]$$

$$\equiv \sigma(\theta)C_{\text{fold}} \qquad (63)$$

where terms of higher order ($\Delta\theta^6$, $\Delta\phi^2\,\Delta\theta^2$, etc.) have been neglected. This expansion should be quite sufficient, since the $\Delta\theta$ terms are considerably larger than the $\Delta\phi$ terms except for very small angles θ. The spectrometer folding correction C_{fold} can be evaluated by calculating cross sections $\sigma(\theta)$ at the angle θ and also at angles $\theta \pm \delta$, where δ is small (usually less than $1°$). The cross sections at these angles are denoted $\sigma_\pm(\theta)$. Using Eq. (62) we find

$$\alpha = \tfrac{1}{4}[\ln(\sigma_-/\sigma_+)]^2/\delta^2, \qquad \beta = \tfrac{1}{2}[\ln(\sigma_+\sigma_-/\sigma^2)]/\delta^2 \tag{64}$$

A similar procedure can be used to determine the effect of multiple scattering on the cross section in terms of a multiple scattering correction factor C_{ms}, which is given by Friar and Negele (FN 73a) and Rossi and Greisen (RG 41). This correction depends on the effective target thickness. Putting both corrections together, we have

$$\bar\sigma(\theta) \cong \sigma(\theta)C_{fold}C_{ms} \tag{65}$$

The effect of the horizontal part of the folding (in the scattering plane) is to average the cross section over scattering angles both larger and smaller than θ. The vertical correction, on the other hand, reflects the fact that an electron passing above or below the scattering plane has an *increased* scattering angle compared to one at a corresponding angle in the scattering plane. The vertical correction is most important for small angles.

The corrections we have considered in this section are those that must be applied to calculations of cross sections, where comparison with data is to be made. There are undoubtedly other small geometrical corrections that we have not considered. The experimental considerations that arise in muonic atom work are discussed in the excellent review by Wu and Wilets (WW 69).

2. DIRECT COMPARISON OF THEORETICAL CHARGE DENSITIES WITH EXPERIMENT

Given that a primary intent in studying elastic electron scattering and muonic X-rays is to critically test our present theoretical understanding of nuclear charge distributions, a thorough analysis of presently available* calculations of ^{208}Pb has been performed. The nucleus ^{208}Pb was chosen

---

* We have included all published Hartree–Fock calculations known to us for which the original authors were willing to provide tabulated charge densities.

for several reasons: (1) The criteria for validity of the most common finite nucleus approximations are best satisfied in a heavy, doubly closed-shell, spherical nucleus. (2) A heavy nucleus is more sensitive to the saturation properties of the two-body interaction and many-body approximations than a light nucleus, and thus provides a more stringent test of the theory. (3) The variety of data available for ^{208}Pb, including muonic X-ray data and elastic electron scattering data at five incident energies, provides accurate and complementary tests of the theoretical density distributions. (4) Finally, as a purely practical consideration, ^{208}Pb is the most commonly calculated heavy nucleus, and thus offers the optimal opportunity for critical comparison of the largest number of present theories.

2.1. Theories of the Ground-State Density Distribution

In order to meaningfully compare various theoretical charge densities, it is necessary to first enumerate the theoretical and phenomenological content of each theory. This present review does not undertake an exhaustive critique of the theories under consideration; rather it attempts to summarize those aspects that are directly relevant to understanding differences in theoretical charge distributions and distinguishing pure theoretical predictions from phenomenological parameterizations. The theories are presented in five separate categories, ranging from fundamental to purely phenomenological, and within each category are arranged in chronological order of publication. The salient parameters of the various theories are summarized in Table II and the densities are compared in Fig. 17.

2.1.1. Renormalized Brueckner–Hartree–Fock

Renormalized Brueckner–Hartree–Fock theory (RBHF) is the most fundamental attempt to calculate the structure of heavy, finite nuclei in the context of nonrelativistic many-body perturbation theory starting from a realistic two-body potential with a repulsive core. Because of the presence of the repulsive core, perturbation theory is rearranged in terms of the reaction matrix, denoted by G or K in the literature, which is the ladder sum of all possible successive interactions between two nucleons with normally unoccupied intermediate states:

$$G(W) = v - vQ[H_0 + W]^{-1}QG(W) \qquad (66)$$

where Q projects onto unoccupied two-particle states, H_0 defines the basis

TABLE II

Summary of Constraints Used to Determine Adjustable Parameters and the Nuclear Matter Properties of Various Theories

The abbreviations for theories are defined in the text in Section 2.1. Integers in the columns for finite nucleus properties denote the number of spherical closed-shell nuclei used in fitting when there are too many to conveniently enumerate. Underlined nuclear matter properties were introduced as predetermined constraints rather than calculated from the parameters fit to properties of finite nuclei.

Theory	Number of adjustable parameters	Finite nucleus properties, constraining parameters			Nuclear matter properties				
		BE	$\langle r^2 \rangle$	S.P.E.	BE/A, MeV	k_F, fm^{-1}	$\varkappa$, MeV	E_{sym}, MeV	m^*/m
RBHF	0	—	—	—	11.2	1.44	134	32	—
DDHF	2	—	^{40}Ca	—	15.7^a	1.31^a	—	28	—
DME	2	^{208}Pb	^{208}Pb	—	16.6	1.34	255	—	0.63
CS	2	5	5	—	16.5	1.35	190	36	—
FN-β	8	—	—	—	16.0	1.36	205	—	—
CK-1	4	^{40}Ca	^{40}Ca	—	15.68	1.37	190	35	—
CK-2	4	^{40}Ca	^{40}Ca	—	15.68	1.32	235	35	—
SKM-I	5	^{16}O, ^{208}Pb	^{16}O, ^{208}Pb	—	16	1.32	370	29.3	0.91

SKM-II	5	^{16}O, ^{208}Pb	^{16}O, ^{208}Pb	—	16	1.30	342	34.1	0.58
SKM-III	5	7	7	—	15.87	1.29	356	28.2	0.76
SKM-V	5	7	7	—	16.06	1.32	306	32.7	0.38
MDI-2	3	^{208}Pb, ^{16}O	^{208}Pb	—	16.46	1.33	310	31.9	0.51
MDI-4	4	—	^{208}Pb	—	16	1.368	200	32.8	0.49
NESTOR	20	—	—	—	15.5	1.42	311	42	0.40
BL-F1	9	4	4	4	16.15	1.325	280	34	0.90
BL-F2	9	4	4	4	16.0	1.32	280	34	0.67
SRP	6	5	5	^{208}Pb	17.2	1.36	295	30	—
CTMW	8	3	3	3	16.4	1.36	150	33	—
OBEP II	0	—	—	—	11.2	1.47	120	—	0.63
RH-1	4	^{40}Ca, ^{48}Ca	^{40}Ca, ^{48}Ca	—	16	—	—	—	—
RH-2	4	^{40}Ca	^{40}Ca	^{40}Ca	16	1.26	—	—	—

[a] DDHF parameter resulting from a slightly inconsistent treatment of starting energies in nuclear matter and finite systems. A consistent treatment yields values close to those tabulated for DME.

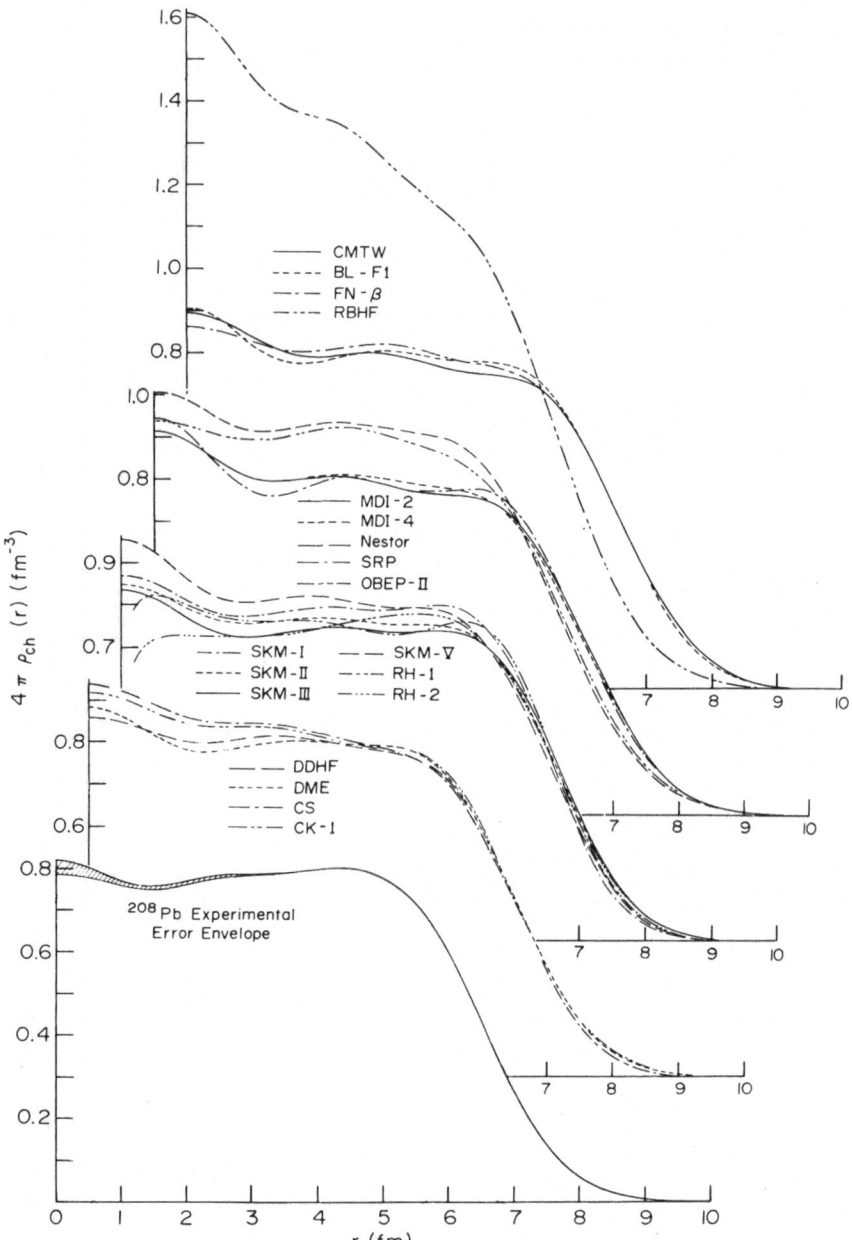

Fig. 17. ^{208}Pb charge density distributions. The abbreviations for the theories are defined in the text in Section 2.1 and the definition and limitations of the experimental envelope are discussed in Section 3.

in which the perturbation theory is performed, and W is the available energy for the two interacting nucleons.

Calculations at Oak Ridge (DM 71, DMS 72) use essentially exact G-matrices in a large oscillator basis, so that the primary approximations concern the manner in which the infinite perturbation series is truncated. Although the sum of many orders of perturbation theory should be independent of the choice of the single-particle basis, or, equivalently, H_0 in Eq. (66), the lowest order result is extremely sensitive to the choice of basis, and, especially, the particle–hole matrix elements of H_0. In order to cancel the largest low-order corrections to the one-body density, it is desirable to define the particle–hole matrix elements of H_0 by the sum of the diagrams of Figs. 18a–18c as shown by Negele (Neg 74b) and Davies *et al.* (Dav+ 74).

The primary problem in realistic calculations of the density is the lack of saturation, that is, the nucleus collapses to a density which is much higher than observed experimentally and the radius correspondingly shrinks to a value much smaller than determined experimentally. Inclusion of only the diagram of Fig. 18a, referred to as the Brueckner Hartree–Fock approximation (BHF), yields an average interior charge density roughly twice the observed value (DMS 72). The RBHF approximation improves saturation significantly, decreasing the interior density by the order of 20% (DMS 72) by including the diagram of Fig. 18b as well as Fig. 18a. For comparison with other theories, we have selected the RBHF density of Davies and McCarthy (DM 71), which differs only with respect to relatively minor technical details from the result of Davies *et al.* (DMS 72). In the subsequent comparison with experimental data, this RBHF density will serve as a "benchmark" to show just how far present theory with no additional approximations can go in describing the charge density of ^{208}Pb.

Although the RBHF density is still much higher than given by experiment and the binding energy per particle is only 2.5 MeV, there is reason for optimism that inclusion of additional corrections will bring both the density and energy into much closer agreement with experiment. As discussed subsequently in connection with the local density approxima-

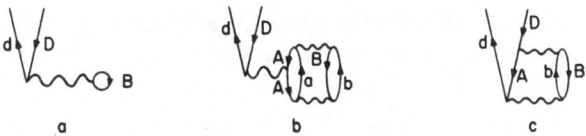

Fig. 18. Goldstone graphs which may be included in the definition of the particle–hole single-particle potential.

tion, approximate inclusion of the diagram of Fig. 18c produces a further significant decrease in the interior density. Higher order corrections from three- and four-body clusters, three-body forces, and relativistic corrections increase the binding energy of nuclear matter by roughly the amount required to reproduce the semiempirical mass formula value of 16 MeV per particle, as explained in review articles by Bethe (Bet 71) and Sprung (Spr 72). Since the semiempirical mass formula is fit by these estimates, it is expected that these effects will increase the binding energy of ^{208}Pb to roughly the experimental value. Finally, Green and Haapakoski (GH 74) have shown that explicit treatment of the Δ_{33} resonance decreases the saturation density of nuclear matter sufficiently that it is reasonable to expect ^{208}Pb to saturate properly when this effect is included. All of these arguments, then, suggest that there are additional corrections of the required order of magnitude to bring about agreement with the experimental energy and interior density (or rms radius) of ^{208}Pb. In the absence of exact calculations of these corrections in finite nuclei, it is reasonable to phenomenologically parameterize them, and attempt to systematically reproduce properties of nuclei throughout the periodic table. This leads to the next group of theories.

2.1.2. Effective G-Matrices with Phenomenological Adjustments

The density-dependent Hartree–Fock theory (DDHF) of Negele (Neg 70) differs from RBHF in three essential respects:

1. The local density approximation replaces the exact G-matrix by the nuclear matter G-matrix, which is a function of relative coordinate, density, and available energy calculated at the density occurring at the center of mass of two interacting particles. This has been shown (Neg 74b, Dav+ 74) to be essentially exact in the nuclear interior and extremely accurate in the nuclear surface.

2. Using the local density approximation, the variational definition of the single-particle potential approximately includes the diagrams of Figs. 18a–18c, which, as shown by Negele (Neg 74) and Davies *et al.* (Dav+ 74), greatly improves saturation. In the case of ^{40}Ca treated by Davies *et al.*, the inclusion of the diagram of Fig. 18c reduced the RBHF central density by 15%. The corresponding effect is even more dramatic in ^{208}Pb, where inclusion of the diagram of Fig. 18c in DDHF with the unadjusted Reid G-matrix reduces the charge density ϱ_{ch} at the origin from the RBHF value

of $4\pi\varrho_{ch} = 1.6\,\text{fm}^{-3}$ shown in Fig. 17 to $4\pi\varrho_{ch} = 0.99\,\text{fm}^{-3}$, constituting a 38% decrease in density. To avoid confusion in the graph in Fig. 17, this density is not plotted. It is very similar to the curve labeled Nestor, which is discussed later, agreeing within 1% at the origin, 3 fm, and 5 fm, and oscillating about the Nestor result elsewhere due to its larger density fluctuations. In light nuclei, the diagram of Fig. 18c may be calculated straightforwardly without recourse to the local density approximation, and its importance in producing saturation has been demonstrated (Str 71, Rao 72, SBK 73, TFM 73, TFM 74, Coo 75, RPS 74). None of these other techniques, however, is presently applicable to ^{208}Pb.

3. A two-parameter, zero-range, density-dependent interaction is added to the realistic G-matrix to parameterize the effect of the omitted higher order corrections. The parameters were adjusted to yield the binding energy of nuclear matter and the observed charge radius of ^{40}Ca. This adjustment further decreases the interior density of ^{208}Pb from $4\pi\varrho_{ch} = 0.99\,\text{fm}^{-3}$ to the value of $0.86\,\text{fm}^{-3}$ shown in Fig. 17 and yields a charge radius of 5.495 fm, which is in excellent agreement with experiment. Although systematics throughout the periodic table are beyond the scope of this review, it is a nontrivial consistency check on the parameterization of the higher order corrections that adjustment to fit ^{40}Ca also simultaneously reproduces ^{208}Pb. In all of the theories discussed in this section, a spin–orbit potential of the Thomas form is included with a constant either derived theoretically or fit experimental data, and we shall not consider the spin–orbit term when discussing adjustments to the forces or when counting phenomenological parameters.

The density matrix expansion (DME) of Negele and Vautherin (NV 72) is an approximation to the DDHF theory described above incorporating further approximations for both computational simplicity and to understand the physics of the simple phenomenological interactions discussed in Section 2.1.3 below. As the name implies, the essential feature of the approximation is to expand the nuclear density matrix $\varrho(\mathbf{x}, \mathbf{x}') = \sum_a \psi_a^*(\mathbf{x})\psi_a(\mathbf{x}')$ as a series of products of functions of the relative coordinate $\mathbf{r} = \mathbf{x} - \mathbf{x}'$ and functions of the center-of-mass coordinate $\mathbf{R} = \frac{1}{2}(\mathbf{x} + \mathbf{x}')$. The leading term is simply the nuclear matter Slater mixed density at the density $\varrho(\mathbf{R})$ and the first correction involves second derivatives of the density matrix, which can be expressed in terms of $\nabla^2\varrho(\mathbf{R})$ and $\tau(\mathbf{R}) \equiv \sum_a |\nabla\psi_a(\mathbf{R})|^2$. Truncation of all higher order terms and integration with respect to $\mathbf{r}$ of the product of the adjusted two-body G-matrix times the functions of relative coordinate generated by the expansion yields an extremely simple form for the energy density:

$$
\begin{aligned}
H(\mathbf{R}) = {}& (\hbar^2/2m)[\tau_n + \tau_p] + A[\varrho_p, \varrho_n] + B[\varrho_p, \varrho_n]\tau_p \\
& + B[\varrho_n, \varrho_p]\tau_n + C[\varrho_n, \varrho_p] \mid \nabla\varrho_n \mid^2 + C[\varrho_p, \varrho_n] \mid \nabla\varrho_p \mid^2 \\
& + D[\varrho_n, \varrho_p]\nabla\varrho_n \cdot \nabla\varrho_p
\end{aligned}
\tag{67}
$$

where ϱ_n and ϱ_p denote the neutron and proton density at $\mathbf{R}$; τ_n and τ_p denote $\tau(\mathbf{R})$ for neutrons and protons, respectively; and the functionals A, B, C, and D are determined by specific integrals of the G-matrix. The energies obtained with the DME differ insignificantly from those obtained in DDHF, and, as seen in Fig. 17, the density of ^{208}Pb is identical in all respects except the magnitude of quantum density oscillations. Since DME is an approximation to DDHF, the oscillations obtained in DDHF are theoretically more reliable, and the fact that DME happens to agree better with experiment must be regarded at present as fortuitous. Because the reduction of the energy density to a functional of ϱ and τ of the form in Eq. (67) introduces theoretical errors in the magnitude of oscillations, because many of the phenomenological interactions considered subsequently are expressed as similar functionals of ϱ and τ, and because the oscillations show up rather strongly in present experiments, it is important to understand precisely how this inaccuracy originates, and this will be explained in detail in Section 2.2.4.

The Hartree–Fock calculations of Campi and Sprung (CS 72), hereafter denoted as CS, utilize the three essential features of DDHF: the local density approximation, the variational definition of the single-particle potential, and the parameterization of higher order corrections to obtain the proper binding energy and saturation. There are many technical differences between DDHF and CS, the most significant of which appears to be the definition of the effective local approximation to the G-matrix. The DDHF interaction is defined as a weighted average of the exact G-matrix, where the weighting is proportional to the relevant nuclear matter phase space, so that the effective interaction is exact in nuclear matter and approaches the bare interactions at large relative coordinate. The G–0 interaction of Sprung and Banerjee (SB 71) used by CS is defined as a sum of Gaussians, the coefficients of which have been adjusted to reproduce diagonal G-matrix elements in momentum space and to yield qualitative agreement with the DDHF interaction at intermediate and long ranges. In addition, the CS interaction in partial waves beyond $j = 3$ has been approximately fitted to matrix elements obtained in the phase shift approximation, yielding slightly more attraction at high density than the

DDHF interaction, which reverts to OPEP for $j > 3$. Also, the parameterization of the density dependence of the CS interaction yields a lower compression modulus of nuclear matter than that of DDHF.

The theory of Fái and Németh (FN 73c), hereafter referred to as FN, is very similar to CS, with minor differences in the parameterization of the force. The effective interaction is again expanded as a sum of Gaussians, and the density dependence of Sprung and Banerjee (SB 71) is used. The coefficients of the Gaussians are determined by fitting the nuclear-matter-weighted effective interaction of Siemens (Sie 70), which was used in DDHF. Whereas the CS interaction was adjusted to obtain the desired binding energy and saturation of nuclear matter by adjusting the coefficients of the shortest range Gaussian at two different densities, the FN interaction is adjusted by adding an eight-parameter density-dependent delta-function force. Following the formulation of Nemeth and Ripka (NR 72), this zero-range force is defined to reproduce a predetermined eight-parameter nuclear matter saturation curve. For the β-force considered in the present review, three constraints on the eight parameters were imposed by the binding energy, saturation density, and compression modulus tabulated in Table II. The criteria considered in determining the other five constraints are not explained in the literature, rendering it difficult to understand the origin of differences in the results of CS and FN.

The density-dependent interactions of Coon and Köhler (CK 74), denoted CK-1 and CK-2, are very similar in principle to the preceding adjusted realistic interactions, although the derivation and parameterization superficially appear very different. All of the previous interactions were expressed as distinct local interactions in each spin–isospin channel, so that the interaction between like or unlike nucleons could be expressed as simple linear combinations of singlet–even, triplet–even, singlet–odd, and triplet–odd interactions. Coon and Köhler, on the other hand, write like and unlike interactions averaged over spin for spin-saturated systems which are manifestly nonlocal of the form

$$v(\mathbf{r}', \mathbf{r}'') = (f(\mathbf{r}) + g(\mathbf{r})\mathbf{\nabla}_x{}^2)\,\delta(\mathbf{x}) \tag{68}$$

where $\mathbf{r} = (\mathbf{r}' + \mathbf{r}'')/2$, $\mathbf{x} = \mathbf{r}' - \mathbf{r}''$, and f and g depend on energy and density as well as on $\mathbf{r}$. In momentum space, this becomes

$$\langle \mathbf{k} \mid v \mid \mathbf{k}' \rangle = \tilde{f}(\mathbf{k} - \mathbf{k}') + \left(\frac{\mathbf{k} + \mathbf{k}'}{2}\right)^2 \tilde{g}(\mathbf{k} - \mathbf{k}') \tag{69}$$

where $\tilde{f}$ and $\tilde{g}$ are the Fourier transforms of f and g. Fitting to nuclear matter matrix elements determines $\tilde{f}(0)$ and $\tilde{g}(0)$ as functions of density and the energy of the interacting particles. Expressing any of the previous interactions in terms of even- and odd-state projectors, where P_x denotes the space exchange operator, yields

$$v_{\text{even}}(r)\left(\frac{1 + P_x}{2}\right) + v_{\text{odd}}(r)\left(\frac{1 - P_x}{2}\right)$$

in coordinate space and

$$\langle \mathbf{k} \mid v \mid \mathbf{k'} \rangle = \tfrac{1}{2}[\tilde{V}_{\text{even}}(\mathbf{k} - \mathbf{k'}) + \tilde{V}_{\text{odd}}(\mathbf{k} - \mathbf{k'})]$$
$$+ \tfrac{1}{2}[\tilde{V}_{\text{even}}(\mathbf{k} + \mathbf{k'}) - \tilde{V}_{\text{odd}}(\mathbf{k} + \mathbf{k'})] \tag{70}$$

in momentum space. Thus, for diagonal matrix elements, which are the only matrix elements that arise in nuclear matter, (69) is a special case of (70) in which $\tilde{V}_{\text{even}} - \tilde{V}_{\text{odd}}$ is restricted to be a quadratic function of $(\mathbf{k} + \mathbf{k'})$. As shown by Coon and Köhler, such a quadratic approximation yields a reasonable representation of the nuclear matter matrix elements.

The $\mathbf{r}$ dependence of the functions f and g of Eq. (68) cannot be determined from diagonal nuclear matter matrix elements, and is therefore parameterized by CK in terms of a Yukawa with a range close to one pion range determined by fitting to the properties of ^{40}Ca. In principle, we know of no way to justify this parameterization of the long-range nonlocality appearing in Eq. (68).* For example, OPEP may be trivially expressed in the form (70) where $\tilde{V}_{\text{even}}(\mathbf{k} + \mathbf{k'}) - \tilde{V}_{\text{odd}}(\mathbf{k} + \mathbf{k'})$ is simply proportional to $(\mu^2 + (\mathbf{k} + \mathbf{k'})^2)^{-1}$. This is quite different from the form $(\mathbf{k} + \mathbf{k'})^2(\mu^2 + (\mathbf{k} - \mathbf{k'})^2)^{-1}$ obtained from Eq. (69) and the one-pion-range Yukawa utilized by CK. In practice the crude parameterization of the long-range nonlocality probably does not affect results severely, so we have included this interaction with the other realistic ones.

The two interactions CK-1 and CK-2 differ in the definition of the intermediate-state energies used in the G-matrix. The conventional choice

---

* In Eq. 10 of (Koh 65) a term similar to this form is obtained by truncation of an expansion of an S-wave potential defined to be an exponential beyond a specified radius and zero inside that radius. Angle-averaging this expression yields the form appearing in Eq. (68). Note, however, that at large r, although $f(r) \sim e^{-ur}/r$, $g(r) \sim e^{-ur}/r^2$, contrary to the form assumed by CK. The truncation of the expansion, the angle average, the assumption that g is proportional to f, and the omission of odd-state long-range forces are approximations of unsubstantiated validity.

of purely kinetic energies used in all the previous interactions was also used in CK-1. In CK-2, the particle states just above the Fermi surface were defined to have a self-consistent Hartree–Fock potential energy as well as the usual kinetic energy. Both interactions were adjusted to yield the same nuclear matter binding energy, nuclear matter symmetry energy, and the observed energy and radius of ^{40}Ca. As far as charge distributions are concerned, the two interactions are almost indistinguishable and the differences arising from different treatments of the excited-state spectrum are certainly less than the errors introduced by other approximations.

2.1.3. Simple Phenomenological Interactions

Recently, several exceedingly simple phenomenological interactions have been developed which, when utilized in HF calculations, reproduce the same systematic features as the realistic interactions described in the previous section. The most remarkable feature of these interactions is that all of the gross properties of nuclei can be reproduced using only three to five empirical constants.

Vautherin and Brink (VB 72) used the Skyrme interaction (Sky 56), which for our purposes can conveniently be written as follows:

$$t_0(1 + x_0 P_\sigma)\, \delta(\mathbf{r}_1 - \mathbf{r}_2) + \tfrac{1}{2}t_1\, [\delta(\mathbf{r}_1 - \mathbf{r}_2)k^2 - k'^2\, \delta(\mathbf{r}_1 - \mathbf{r}_2)]$$

$$+ t_2 \mathbf{k}' \cdot \delta(\mathbf{r}_1 - \mathbf{r}_2)\mathbf{k} + \tfrac{1}{6}t_3(1 + P_\sigma)\, \delta(\mathbf{r}_1 - \mathbf{r}_2)\, \varrho\!\left(\frac{\mathbf{r}_1 + \mathbf{r}_2}{2}\right) \quad (71)$$

where $\mathbf{k}$ denotes $(\boldsymbol{\nabla}_1 - \boldsymbol{\nabla}_2)/2i$ acting to the right, $\mathbf{k}'$ denotes $-(\boldsymbol{\nabla}_1 - \boldsymbol{\nabla}_2)/2i$ acting to the left, P_σ is the spin exchange operator, and, as before, we omit the spin–orbit potential. Although the form of Eq. (71) was originally motivated by a short-range expansion of a two-body force plus a three-body force, the coefficients cannot be obtained from a short-range expansion of a realistic interaction. Either they must be derived by a method such as the density matrix expansion (NV 72) or they must be determined empirically from nuclear masses and radii. When the five coefficients are determined from nuclear masses and radii, essentially four constraints are placed on the coefficients. In the language of the semiempirical mass formula, these are the Coulomb energy (which is determined by the A dependence of the radius), the volume energy, the surface energy, and the symmetry energy. Hence, one combination of the five parameters is essentially undetermined, which we may choose to be the range of non-locality of the Hartree–Fock potential which is specified by the effective mass m^*/m. Thus, Skyrme forces have been obtained (VB 72, Bei+ 75) ranging from the almost local SKM I force with $m^*/m = 0.91$ to the exceedingly nonlocal SKM V force with $m^*/m = 0.38$, where for comparison the Reid potential yields approximately $m^*/m = 0.6$. The nuclear matter properties of these forces are tabulated in Table II and the charge densities are graphed in Fig. 17.

The energy density with the Skyrme force can be simply expressed in the form of Eq. (67), where

$$A(\varrho_n, \varrho_p) = \tfrac{1}{4}t_0(1 - x_0)(\varrho_n^2 + \varrho_p^2) + t_0(1 + \tfrac{1}{2}x_0)\varrho_n\varrho_p$$
$$\qquad + \tfrac{1}{4}t_3\varrho_n\varrho_p(\varrho_n + \varrho_p)$$
$$B(\varrho_n, \varrho_p) = (\tfrac{3}{8}t_2 + \tfrac{1}{8}t_1)\varrho_n + \tfrac{1}{4}(t_1 + t_2)\varrho_p \qquad (72)$$
$$C(\varrho_n, \varrho_p) = (3/32)(t_1 - t_2)$$
$$D(\varrho_n, \varrho_p) = (3t_1 - t_2)/8$$

The magnitude of the quantum density fluctuations is therefore subject to the same uncertainty as in the DME, and we note from Fig. 17 that the Skyrme oscillations tend to be enhanced relative to the more realistic fluctuations of DDHF, CS, and FN.

Ehlers and Moszkowski (EM 72) used the modified delta interaction of Moszkowski (Mos 70), which is closely related to Eq. (71) but involves fewer parameters. Their force, which is referred to as MDI-2, sets $x_0 = t_2 = 0$ and replaces $(1 + P_\sigma)\varrho((\mathbf{r}_1 - \mathbf{r}_2)/2)$ by $\varrho^{2/3}((\mathbf{r}_1 + \mathbf{r}_2)/2)$. Since t_2 is always relatively weak in the Skyrme force, its omission does not change the force qualitatively, and Ehlers and Moszkowski showed that there is no significant difference between the ϱ and $\varrho^{2/3}$ density dependences. *A priori* one might expect the drastic difference in the spin dependence of MDI-2 due to the omission of the P_σ terms to significantly alter the symmetry energy. However, for reasons we regard as essentially fortuitous, the spin dependences of the t_0 and t_3 Skyrme terms cancel to a very high degree, so that the symmetry energies of both theories are comparable.

A slight variation of this interaction, denoted MDI-4, by Faessler *et al.* (Fae+ 75), introduces two density-dependent terms with different powers of the density in order to obtain a lower compression modulus $\varkappa$ in nuclear matter. As discussed in Section 2.2.2, we believe it is theoretically incorrect to adjust $\varkappa$ to reproduce the small compression modulus obtained in lowest order Brueckner theory in nuclear matter, and there is no firm experimental evidence supporting this lower value of $\varkappa$. Nevertheless it is interesting that this change in $\varkappa$ has a relatively minor effect on the charge distribution in Fig. 17, which will be discussed in Section 2.2.2.

As remarked above, one of the most significant results of investigations with these simplified forces is that the gross behavior of nuclear density distributions is governed by a very small number of parameters characterizing certain averages of the nuclear potential. Thus a primary question in considering the charge distributions in this study is whether any additional meaningful information beyond the gross properties already embodied in the simple phenomenological interactions described in this section is included by going to more complicated parameterizations or realistic interactions.

2.1.4. More Complicated Phenomenological Interactions

Interactions were classified in this section instead of either Section 2.1.2 or Section 2.1.3 for one of two reasons. Either the intermediate- and long-range parts of the interaction were not generated from a G-matrix calculated from a potential fit to nucleon–nucleon phase shifts, or the parameterization of the interaction was considerably more complicated than the Skyrme form in Eq. (71).

The earliest effective interaction reviewed in this work that was adjusted to yield realistic nuclear matter properties was force no. 1 of Nestor et al. (Nes+ 68), hereafter referred to as the Nestor force. This force was parameterized in terms of a Gaussian plus k^2 times another Gaussian in each spin–isospin channel, a tensor force which in the Hartree–Fock approximation contributes only in spin-unsaturated shells, and a spin–orbit term we will not consider. The 20 parameters were determined so as to match the values of the density and binding energy of nuclear matter tabulated in Table II, to roughly reproduce the scattering phase shifts in each partial wave, and to minimize the second-order corrections to nuclear matter. It is clear that the parameters of this force could now be readjusted to reproduce the properties of nuclei throughout the periodic table, as for the previous forces. With the hindsight provided by the more recent effective interactions, this would be desirable because there is no theoretical justification for constraining the coefficients of the effective interaction by the two-body scattering phase shifts. The unrealistically high interior density of ^{208}Pb in Fig. 17 evidently arises from the unusually high value of k_F which was selected in defining the force parameters.

The energy density formalism of Beiner and Lombard (BL 74) uses an energy density expression of the form of Eq. (67), with the following parameterization of the functionals:

$$A(\varrho_n, \varrho_p) = -b_1\varrho^2 + b_2\varrho^{7/3} + b_3\varrho^{8/3}$$
$$+ (s_1 + s_2\varrho^{1/3} + s_3\varrho^{2/3})(\varrho_n - \varrho_p)^2$$
$$B(\varrho_n, \varrho_p) = a_0\varrho \tag{73}$$
$$C(\varrho_n, \varrho_p) = \eta_0 + \eta_1$$
$$D(\varrho_n, \varrho_p) = 2(\eta_0 - \eta_1)$$

where $\varrho \equiv \varrho_n + \varrho_p$.

The nine parameters b_i, s_i, a_0, and η_i are constrained by G-matrix calculations only to the extent that the saturation curve of neutron matter is reproduced, and are otherwise fitted to the observed properties of finite nuclei. As with the Skyrme force, the properties of finite nuclei do not determine the effective mass, i.e., nonlocality, and two different interactions are defined, BL-F1 and BL-F2, with values of m^*/m of 0.90 and 0.67, respectively. Comparison of (73) with (72) indicates that BL introduce one additional parameter in B and three additional parameters in A. Since finite nuclei sample such a limited region of the ϱ_n-ϱ_p energy surface and are insensitive to the details of the very low-density region of this surface, we believe that the additional parameters do not modify the theory in any essential way beyond allowing a slightly better fit to masses due to the increased number of parameters.

Saunier et al. (SRP 74) define an interaction which combines several of the features of the Nestor force and the Skyrme force. In even partial waves, the force is defined as a realistic one-boson exchange potential plus an adjustable momentum-dependent function times a short-range Gaussian in each spin channel plus the density-dependent t_3 term of Eq. (71). In odd partial waves, the t_2 term of Eq. (71) is used, and the entire force contains a total of six free parameters, which are determined from the properties of finite nuclei. One should note that the presence of a realistic long-range potential does not automatically guarantee the damping of density fluctuations. Indeed, as observed in Fig. 17, the momentum-dependent phenomenological terms greatly enhance the fluctuations relative to those generated by any of the preceding interactions with realistic tails.

The K-matrix model calculations of Cusson et al. (Cus+ 74) differ from all the preceding theories in that the equations for the single-particle wave functions are not obtained variationally from a Hamiltonian density expressed in terms of density- and momentum-dependent effective interaction. There is, of course, no reason why the lowest order approximation in a many-body theory needs to be variational, and, in practice, the main effect of the nonvariational prescription of CTMW is that it allows a parameterization with greater freedom to separately fit experimental binding energies and experimental single-particle energies. One should also note, however, that there is no theoretical reason why any single-particle eigenvalues, other than those of the last bound neutron and proton, should necessarily be related to single-particle energies obtained from nuclear reactions. The form of the single-particle potential used in this work is very similar to that which would be obtained by variation using a density- and momentum-dependent potential. There is a purely density-dependent term, which corresponds to a zero-range density-dependent potential, a nonlocal density-dependent term, and a nonlocal term expressed as the product of a function of momentum times the convolution of $\varrho(r)$ with a Gaussian, which corresponds to including the Hartree term generated by a velocity-dependent Gaussian potential and omitting the exchange term. In addition to the spin–orbit interaction, there are eight parameters, which are determined by the nuclear matter properties tabulated in Table II and the energies of finite nuclei.

2.1.5. Theories Derived from One-Boson Exchange

The choice of whether to choose a purely phenomenological potential, such as the Reid potential, or a one-boson exchange potential (OBEP) is presently a matter of taste. Although the OBEP form is enticing because it seems to offer the promise of theoretically motivated off-shell behavior, it is subject to the obvious criticism that there is no justification for the omission of multiple meson exchanges in strong interaction physics. One obvious manifestation of this problem is the fact that OBEP potentials must add additional particles to mock up the strong intermediate-range attraction that arises from two-pion exchange, and it is not at all obvious that the off-shell behavior obtained by adjusting OBEP parameters to fit experimental phase shifts is any more fundamental than that arising from some other Ansatz. Thus we present calculations of ^{208}Pb based on OBEP potentials as exploration of different, though not necessarily more fundamental, off-shell behavior.

Machleidt et al. (MHN 74) use the OBEP II potential of Erkelenz et al. (EHM 74). The effective G-matrix obtained from this potential (MEH 74) is approximated using the method of Fái and Németh, and is used to perform a density-dependent Hartree–Fock calculation. Thus the calculation differs from DDHF, DME, CS, and FN-β both due to the replacement of the Reid G-matrix by the OBEP G-matrix and due to the fact that no phenomenological adjustment is added. As is the case with the Reid G-matrix with no adjustment, the resulting ^{208}Pb nucleus is underbound by approximately 4 MeV per particle and the interior density is unphysically high. It is very difficult to compare unadjusted Reid and unadjusted OBEP calculations, because minor technical differences may give effects as large as those arising from the difference in potentials. The gross features of DME calculation using the Reid potential without adjustment and the OBEP II calculation are certainly comparable: at $r = 0$ fm, OBEP is 5% lower, at $r = 2$ fm it is 3% higher, and at $r = 5$ fm it is 3% lower. Thus we conclude that there are no major differences arising from different off-shell behavior and that when the potentials have been adjusted to yield the proper energy and density in finite nuclei, ambiguities arising from the adjustment will dominate those originating from the use of different bare potentials.

A conceptually different approach using OBEP by Miller treats the nucleons in a self-consistent relativistic Hartree approximation. In two-body scattering, the nonrelativistic reduction of an interaction arising from scalar and vector meson exchange leads to a significant nonlocality, i.e., velocity dependence. Instead of formulating the many-nucleon problem in terms of this velocity-dependent potential, one might alternatively consider using a nuclear wave function expressed as a product of relativistic four-component nucleon single-particle wave functions, as is done in Hartree–Fock–Dirac calculations of atoms. This alternative approach was carried out for a simple OBEP potential with adjustable parameters by Miller and Green (MG 72).

The extent to which the theory contains parameters adjusted to the properties of finite nuclei, it is at least partly phenomenological, as are the theories in Sections 2.1.1–2.1.3. The primary theoretical omissions are the exchange terms and the effect of successive interactions corresponding to the G-matrix summation required by the repulsive core generated by the vector mesons. Subsequent calculations (Mil 74a) showed that the exchange terms in light nuclei do not qualitatively change the solutions, and it is plausible that adjustable constants corresponding to the nonlocal terms in a nonrelativistic potential are capable of accounting for the additional nonlocality which should arise from the exchange term. Adjusting the strengths of the vector meson coupling constants very roughly reproduces the effect of summing ladder diagrams, since in both cases the effect of the very short-range repulsion is significantly reduced, and calculations appear to be rather insensitive to the details of the precise spatial distribution of the short-range effective interaction or the form of its velocity dependence.

One difference which is potentially relevant to electron scattering is that the relativistic corrections of order v^4/c^4 are significantly larger in this relativistic theory than in nonrelativistic theories (see Section 1.3). This arises from the fact that the small Dirac components are significant, essentially because they see an extremely deep potential which goes like the sum of the magnitudes of the vector and scalar potentials, whereas the large components (and all nonrelativistic theories) see a potential which is much weaker, being the difference between the magnitudes of the scalar and vector potentials. The form factor correction is most pronounced at the origin, and whereas the point proton densities in the relativistic Hartree calculations have the usual s-state lumps at the origin, these are partially or completely removed from the charge density by the nucleon form factor corrections, as seen in Fig. 17.

The original calculations of Miller and Green (MG 72), denoted RH-1, used two vector mesons and two scalar mesons with four adjustable parameters (coupling constants and cutoff). This interaction gave a good fit to nuclear masses, and the resulting density is similar to the nonrelativistic densities except for the diminished central maximum and its extremely steep surface. A more recent interaction (Mil 74b) using three vector mesons and one scalar meson and having four adjustable parameters was developed to obtain more accurate charge densities and energies for ^{40}Ca. The resulting density of ^{208}Pb calculated with this interaction, denoted RH-2, is notable for the fact that it is the only distribution in this review with a central depression. Only part of this depression, the absence of an s-state lump, is associated with the nucleon form factor corrections mentioned above. The overall tendency toward a lower central density is also reflected in the point proton density and originates, we believe, from the incorrect balance of Coulomb energy and symmetry energy in this potential. Since the parameters of the potential were determined by properties of ^{40}Ca, there is no reason why the symmetry energy should be exact, and indeed ^{208}Pb comes out overbound by approximately 1 MeV per particle, indicating too small a symmetry energy. The connection between symmetry energy and spatial density distributions will be explored in the next section.

2.2. Physical Origin of Salient Features of Density Distributions

Having summarized the basic theoretical and phenomenological input to the variety of calculations which have been performed on ^{208}Pb, it is desirable to attempt to understand the physical origin of the essential features of the calculated distributions in Fig. 17. In addition to the asymptotic decay of the charge density, which is trivially governed by the removal energy of the last proton, the four most salient features are the average interior density, the average slope of the interior density, the surface thickness, and the magnitude of quantum density fluctuations. Each of these features relates to electron scattering in a characteristically different way. Although it is difficult to establish the precise origin of each feature, we can understand at least qualitatively how differences in the underlying theories in Section 2.1 manifest themselves in the final features of density distributions.

2.2.1. Average Interior Density

The average interior density, or alternatively the charge radius $\langle r^2 \rangle^{1/2}$, is related to the saturation problem discussed previously. RBHF theory, which uses the unadjusted Reid potential and the single-particle potential defined by the diagrams of Figs. 18a and 18b, yields an unphysically high interior density. Including the diagram of Fig. 18c greatly reduces the interior density, yielding the OBEP-II curve in Fig. 17 for the OBEP G-matrix and a very similar curve for the unadjusted Reid G-matrix. To obtain a lower density which agrees with experiment, the effective interaction must be adjusted to lower the saturation density of nuclear matter.

There is no exact relation between the saturation density of nuclear matter and the average interior density of ^{208}Pb, partially because there is no precise definition of the average interior density. Nevertheless, there is a very strong trend which can be observed easily in Fig. 17 using the k_F values from Table II. The Nestor force, with $k_F = 1.42$, yields by far the highest density of all the adjusted forces. The interior densities obtained with the Skyrme forces decrease directly with the values of k_F: The densities of SKM-V and SKM-I with $k_F = 1.32\,\text{fm}^{-1}$ are above SKM-II with $k_F = 1.30$, which is above SKM-III with $k_F = 1.29$. Similar behavior is evident with other forces, indicating that k_F must be in the range $k_F \sim 1.31$–$1.37\,\text{fm}^{-1}$ in order to obtain a realistic result. Since the adjustment is completely phenomenological and cannot as yet be calculated theoretically, it follows that the average interior densities are not determined theoretically and that comparisons of average densities are not meaningful tests of genuine differences between theories. For this reason, in subsequent comparisons with electron scattering data, we will also consider scaled versions of each theory, in which the theoretical shape is scaled to yield exactly the observed rms radius.

2.2.2. Average Slope of the Interior Density

Definition of the smooth behavior of the average density which is modulated by quantum density fluctuations is even more difficult than the average density. Fortunately, for our present purposes, a very loose and subjective definition of the average slope suffices. For each density, we imagine drawing a straight line such that between 0 and 5 fm the calculated density fluctuates equally above and below this line. A better definition would be to weight the deviation of the density from the straight line by r^2, to avoid undue influence by the central s-state maximum, but this is not necessary for the distributions considered here. Either way, the straight line is supposed to represent the average behavior of the density distribution and thus reflect gross structure implied by the underlying nuclear theory. In order to clearly separate saturation effects from all subsequent effects, we will omit the three nonsaturating cases of Fig. 17, RBHF, Nestor, and OBEP-II, from all further discussion in Section 2.2.

Although Thomas–Fermi theory is certainly inadequate to describe the nuclear surface, it provides a convenient language in which to discuss the slope of the interior density. First, consider the total nucleon density, which is the sum of neutron and proton densities, in the absence of the Coulomb force. Since the surface energy is repulsive, corresponding to a surface tension, the Thomas–Fermi density should be a nonincreasing function of the distance away from the origin. In the limit of an extremely "stiff" energy functional of the density, the average slope would be zero; for "softer" energy functionals, the density would decrease with distance from the origin yielding a negative slope. Because of the nuclear symmetry energy, both neutrons and protons will have roughly the same slope as the total density.

The stiffness or softness of the energy functional arises from the density and momentum dependence of the effective G-matrix or phenomenological interaction. Clearly, if the momentum or density dependence is large enough, the interior density will be very flat, and if it is sufficiently small, there will be a significant negative slope. A convenient parameter characterizing the stiffness of the energy functional is the compression modulus $\varkappa = -k_F^2\, \partial^2(\mathrm{BE}/A)/\partial k_F^2$ specifying the curvature of the nuclear matter saturation curve about the equilibrium density. The compression modulus for the unadjusted Reid potential including only two-body clusters in Brueckner theory is 134 MeV. However, in order to shift the binding energy from 11 to 16 MeV while maintaining the same or lower saturation density, higher order corrections should significantly increase the second derivative at the minimum. If one adds to the Reid saturation curve a reasonable parameterization of higher order corrections such that $k_F = 1.34$, BE$/A = 16.6$ MeV, and the very low-density part of the saturation curve is unchanged, $\varkappa$ increases to roughly 250 MeV. Thus we regard the lowest order result of $\varkappa = 134$ MeV as misleading and theoretically prefer a higher value of the order of 250 MeV. From the above argument, one would expect theories with low $\varkappa$, such as CMTW with 150 MeV, CK-1 with $\varkappa = 190$ MeV, and CS with 190 MeV, to yield slopes which are significantly more negative than calculations using forces with high $\varkappa$, such as any of the Skyrme forces, which have $\varkappa > 300$ MeV. This expectation is, in fact, clearly borne out in Fig. 17.

The above argument was oversimplified because of the omission of the Coulomb energy. The Coulomb interaction is the only mechanism which can cause a positive slope in the average proton density. If one considers a calculation without the Coulomb force and then treats it as a perturbation, there are two main effects. The repulsive one-body Coulomb potential is less repulsive near the surface than in the interior, so that protons move from the center to the edge and the decay of single-particle wave functions beyond their classical turning points is much faster, removing probability from the extreme tails. Thus there is a depletion of probability in the center and extreme surface and a corresponding increase around 5 fm, resulting in a proton density slope which is less negative or more positive than the original slope without the Coulomb force.

The tendency of the Coulomb force to decrease the interior proton density and increase the density near the nuclear surface is opposed by the symmetry energy, which acts to maintain a constant ratio of neutrons to protons throughout the nucleus. Even in ^{208}Pb, the Coulomb energy is a sufficiently weak driving term that a significant redistribution of proton density from the center to the edge is prevented by the prohibitive loss in symmetry energy that would result for a force that reproduced the symmetry energy term in the semiempirical mass formula. A distribution with a positive slope for both neutrons and protons is prevented by the large value of the compression modulus, which dominates the Coulomb driving term. Thus we would argue that for all interactions with a symmetry energy greater than 28 MeV, the Coulomb driving term is overwhelmed by the symmetry and incompressibility effects and results in no significant tendency toward a positive proton slope. Indeed, this expectation is fulfilled by all the curves in Fig. 17 except the RH-2 relativistic Hartree theory, for which we did not know the exact symmetry energy for tabulation in Table II.

The special case of RH-2 may be understood qualitatively in terms of symmetry energy in spite of the lack of a numerical value for the nuclear matter result. We first note that the balance between Coulomb and symmetry energy may be observed in the Skyrme force calculations if one looks in detail. The slopes for SKM-I and SKM-III, with $E_{sym} = 29.3$ and 28.2 MeV, respectively, are visually less negative than for SKM-II and SKM-IV with $E_{sym} = 34.1$ and 32.7, respectively. All of these forces reproduce binding energies of ^{40}Ca, ^{98}Ca, and ^{208}Pb to within 0.3 MeV per particle. In contrast, although the RH-2 interaction overbinds ^{40}Ca by only 0.16 MeV, per particle, it overbinds ^{48}Ca and ^{208}Pb by 0.47 and 0.89 MeV per particle. The significant overbinding of ^{48}Ca and ^{208}Pb relative to ^{40}Ca indicates much too small a symmetry energy, which clearly must be well below the range 28–34 covered by the Skyrme interactions, which yielded maximum deviations of 0.3 MeV. Thus we believe it is quite plausible that the incorrect balance of Coulomb energy against an unrealistically weak symmetry energy accounts for the positive slope in this case. In further support of this interpretation, the RH-1 theory, which has no positive slope, fits the binding energies of ^{40}Ca, ^{48}Ca, and ^{208}Pb to within 0.3 MeV per particle, indicating that there is no intrinsic feature of the relativistic theory producing a positive slope as long as the gross energy balance is correct. Finally, recall that the anomalous behavior of the density within 0.5 fm of the origin is a relativistic form factor effect, and as a fluctuation does not really effect our definition of the average slope.

The average slope of the interior density has been discussed at length because we believe it is really determined by the underlying theory, and offers the possibility, if measured unambiguously, of helping discriminate between alternative theories. Although many of the present theories, when applied to the superheavy nucleus $Z = 114$, yield distinct positive slopes because the Coulomb driving term with $Z = 114$ is comparable to symmetry and compressibility energies, we believe that the absence of a positive slope for $Z = 82$ is a very firm prediction of all theories incorporating the energy balance specified by the semiempirical mass formula. Thus the interpretation of experimental evidence suggesting the presence of a "central depression" or "wine bottle effect," which are the popular terms for what we have described as a positive average interior slope, will be extremely significant in the subsequent analysis of experimental data.

2.2.3. Surface Thickness

The surface thickness may be characterized in many ways. For our present discussion of ^{208}Pb distributions with rms radii in rough agreement with experiment, it suffices to approximate the slope at the half-density radius by

$$\varrho'_{HD} \equiv [\varrho(6.2) + \varrho(7.0)]/0.8$$

which is conveniently calculable from available density tabulations as a measure of the surface thickness. Values of $4\pi\varrho'_{HD}$ are tabulated in Table III (Section 2.3.1) and for purposes of comparison with the Fermi-function thickness parameter, $4\pi\varrho'_{HD} = 0.371$ fm^{-4} corresponds to $t = 2.33$ fm.

The surface thickness is governed essentially by the range of the effective nucleon–nucleon interaction, which manifests itself in two characteristically different ways. The first way is through the diffuseness of the Hartree potential obtained by performing a convolution of the effective interaction with the density. As the range of the nuclear interaction becomes longer, the Hartree potential becomes more diffuse, yielding a more diffuse density. A particularly clear way of visualizing the effect of a convolution is to consider a semiinfinite slab of nuclear matter with a surface shape described by a Fermi function which is symmetric about the half-density point. Then convolution leaves the half-density point unchanged and produces a distribution which is still symmetric about this point with a larger diffuseness.

The second way in which the range of the effective interaction affects the surface thickness is through its influence on the range of the nonlocality of the exchange term in the Hartree–Fock potential. A δ-function potential yields a local potential and increasing the range of the nuclear interaction increases the range of nonlocality. The effect of nonlocality on the surface thickness is qualitatively understood in terms of the Perey effect, in which the scattering wave functions in the interior of a nonlocal potential are suppressed relative to those in a corresponding local potential. For normalized bound states, suppression in the interior is compensated by enhancement in the tail, yielding a more diffuse surface for a nonlocal potential, and this effect has been investigated quantitatively by Negele (Neg 74b).

It is important to note that the effective interaction ranges appearing in the direct and exchange radial integrals after summing over spin are different, since the direct term involves positive statistical factors times the sum of the even and odd state forces, whereas the exchange term involves statistical factors times the difference between even and odd state forces. Thus the two effects affecting the surface thickness are independent, and, in principle, can be investigated separately. In the language of the DME functional, Eq. (67), the finite-range Hartree potential appears in the C and D terms. For a symmetric nucleus where $\varrho_n = \varrho_p = \frac{1}{2}\varrho$, these terms reduce to $[\frac{1}{2}C(\varrho/2, \varrho/2) + \frac{1}{4}D(\varrho/2, \varrho/2)] \mid \nabla\varrho \mid^2$. The relation of this term to the finite range is also obvious from considering a Taylor expansion of a short-range force since in this case the coefficient of the $|\nabla\varrho|^2$ term varies directly with the range. For the Skyrme form in Eq. (74), the coefficient of $|\nabla\varrho|^2$ is $(1/64)(9t_1 - 5t_2)$. Comparison of the values of $9t_1 - 5t_2$ for all the Skyrme and MDI forces and the DME with the corresponding values of ϱ'_{HD} verifies our expectation that the surface thickness increases with increasing values of $(9t_1 - 5t_2)$.

The range of nonlocality is best expressed in terms of the effective mass m^* appearing in the Schrödinger equation resulting from Eq. (67). For equal densities of neutrons and protons, m^* is defined as

$$\frac{\hbar^2}{2m^*} = \frac{\hbar^2}{2m} + B\left(\frac{\varrho}{2}, \frac{\varrho}{2}\right)$$

The functional B is proportional to integrals over the effective interaction appearing in the exchange term, so that increasing the range of this interaction corresponds to increased nonlocality and yields a decrease in m^*. For the Skyrme force, using Eq. (74), $B = (1/16)(3t_1 + 5t_2)\varrho$. Again, comparing $3t_1 + 5t_2$ for all the Skyrme, MDI, and DME forces with the calculated values of ϱ'_{HD} indicates that the surface thickness increases for decreasing values of m^*, corresponding to increasing nonlocality and increasing values of $3t_1 + 5t_2$, as expected.

The only unresolved question in this otherwise satisfactory understanding of the surface thickness is our inability to separate the roles of the direct and exchange effects. For the realistic interactions, this separation is not required since we believe that we know theoretically the correct balance of direct and exchange effects, and the zero-range adjustment has no effect on the finite-range arguments. It is indeed reassuring that all of the adjusted realistic interactions in Section 2.1.2 yield values of $4\pi\varrho'_{HD}$ between 0.32 and 0.35 fm^{-4}, nicely bracketing the experimental value of 0.34, and we regard this agreement as a significant accomplishment of these theories. Phenomenological interactions yielding significantly different values of $4\pi\varrho'_{HD}$, such as SKM-I and RH-1, should be rejected as unphysical, even though they yield reasonable nuclear energy systematics.

2.2.4. Quantum Density Fluctuations

Quantum density fluctuations cannot, in principle, be separated from the shape of the nuclear surface. Indeed, the nuclear surface is the most significant quantum density fluctuation in the nucleus: Its size is comparable to the characteristic wavelength π/k_F and it is highly quantal, as opposed to statistical or semiclassical, in the sense that all the single-particle wave functions are decaying precisely in phase. The fact that we are incapable of defining the proper reference density in the surface with respect to which fluctuations should be measured should not obscure the fact that the factors influencing the interior fluctuations also govern the details of the shape of the surface of the charge distribution, to which electron scattering is particularly sensitive. Fluctuations in the interior are easily defined in terms of the difference between the straight line defining the average slope of the interior density in Section 2.2.2 and the calculated interior density, and we shall measure the percentage fluctuation by the ratio of the maximum deviation of the actual interior density from this line to the average interior density.

From Fig. 17, we note that calculated densities yield fluctuations ranging from 11% for SRP to 3.8% for FN-β and DDHF, to be compared with the range 4.4–1.9% suggested by experiment. To understand the origin of this wide variation in fluctuations predicted by various theories, it is instructive to consider the difference between the DDHF and DME fluctuations. Recall that the DME was an approximation to DDHF in which the full finite-range interaction was replaced by the Hamiltonian density functional Eq. (67) involving at most second derivatives of nuclear wave functions. Thus the enhancement of the fluctuations in the DME relative to DDHF is associated with errors inherent in the form of energy density in Eq. (67) and therefore should also be expected in all of the phenomenological interactions utilizing similar parameterizations.

Consider, for simplicity, the Hartree potential generated by a finite-range attractive central potential $v(\mathbf{x})$:

$$V(\mathbf{r}) = \int \varrho(\mathbf{r} + \mathbf{x})v(\mathbf{x})\, d^3\mathbf{x} \tag{74}$$

If the density ϱ has fluctuations of wavelength π/k_F and the range of v is comparable to π/k_F, then $V(r)$ will be much smoother than ϱ because the fluctuations will be averaged out. Thus, with a realistic finite-range force, the self-consistent potential will have virtually no fluctuations, and the density calculated in this well should have fluctuations characteristic of the solution of the Schrödinger equation in a smooth well. If, instead of doing the convolution in (68), we perform a Taylor series expansion of $\varrho(\mathbf{r} + \mathbf{x})$ to second order and integrate, we obtain the following result:

$$\tilde{V}(\mathbf{r}) = A\varrho(\mathbf{r}) + B\,\nabla^2\varrho(\mathbf{r}) \tag{75}$$

where $A = \int v(\mathbf{x})\, d^3\mathbf{x}$ and $B = \frac{1}{2} \int v(\mathbf{x})x^2\, d^3\mathbf{x}$. The first term corresponds to a zero-range potential and thus reproduces the fluctuations of the density in the Hartree potential. Now, consider how the self-consistent density for a mirror nucleus with a zero-range potential differs from that of the original finite-range potential. Let the first iteration for the zero-range force start with the final finite-range density. The resulting Hartree potential will not be as smooth as in the finite-range case, but will be deeper at the maxima of the density and shallower at the minima. When the Schrödinger equation is solved in this fluctuating potential, the density will increase in the deep parts and decrease in the shallow parts, thereby amplifying the original (realistic) fluctuations. The net result is that keeping only the zero-range term A in Eq. (75) yields unrealistically large fluctuations in a mirror nucleus.

The second term in Eq. (75) tends to cancel the unrealistic fluctuations in the Hartree potential introduced by the first term. At a relative maximum of the density, $\nabla^2\varrho$ is negative, so that $B\,\nabla^2\varrho$ is positive, making the potential shallower, whereas at a relative minimum it makes the potential deeper. Unfortunately, for an arbitrary force, there is no reason to expect the correction to be of the right magnitude, since the two-term series will not converge for a long-range force. If the correction is too small, the resulting quantum density fluctuations will be unrealistically large, as argued above; if the correction is too large, it will overcompensate the effect of the first term, and the resulting fluctuations will be damped rather than amplified and will thus be unrealistically small.

To extend the above mirror nucleus argument to ^{208}Pb, the effect of the excess neutrons must be considered. Since the attraction between neutrons and protons is much stronger than between like nucleons, the dominant contribution to the proton single-particle well arises from the neutron density. It turns out that the oscillations in the neutron density in ^{208}Pb are exactly out of phase with the oscillations in proton density, yielding, incidentally, an extremely smooth total density. Thus the previous arguments are modified by a simple sign change. That is, if this second derivative correction term is too small, the proton single-particle potential tends to damp the proton fluctuations, and if the B term overcorrects for this effect of finite range, the proton fluctuations are artificially enhanced. The DME approximation to DDHF corresponds to the case in which the derivative term overcorrects. As seen in Fig. 17 it enhances fluctuations in ^{208}Pb and, as shown by Negele and Vautherin (NV 72), damps fluctuations for mirror nuclei.

We have already noted that, in principle, the shape of the surface is essentially related to quantum density fluctuations. This is also evident from the fact that integration by parts of the $|\nabla\varrho|^2$ terms in Eq. (67) before variation yields a single-particle potential containing $\nabla^2\varrho$, as in Eq. (75), so that the C and D terms of Eq. (67) govern the magnitude of fluctuations as well as the surface thickness. The only essentially new feature occurring in the case of interior fluctuations is the isospin dependence of the C and D functionals. In the nuclear surface the protons and neutrons are decaying in phase, whereas in the interior they are out of phase. Thus, although phenomenological interactions adjusted to reproduce nuclear masses may be expected to yield a reasonable surface thickness to obtain the right surface energy, there is no reason why the isospin dependence of the gradient terms should be well determined. This explains the relatively large variations in fluctuations obtained in different phenomenological theories.

The only theoretically reliable method of obtaining realistic density fluctuations is to retain the finite range of an effective interaction having a long-range part which is determined from a realistic potential rather than phenomenologically. It is reassuring that the DDHF, CS, and FN-β interactions, which approach the unadjusted Reid G-matrix at large distance, all yield fluctuations of the order of 3.8%, which is consistent with the range 1.9–4.4% suggested by experiment. Thus we regard the prediction of 3.8% fluctuations as another substantive prediction of realistic HF calculations that appears to be consistent with present experimental data.

2.3. Comparison with Experimental Results

Having surveyed the theoretical calculations of ^{208}Pb charge density distributions and the origin of the various features of these distributions, we now present a direct comparison of the elastic electron scattering cross sections and muonic X-ray transition energies predicted by these theories with experiment. Such a direct comparison with experimental data has definite advantages and limitations. One advantage is that calculation of χ^2 provides a convenient quantitative means of discriminating among various theories. Also, working directly with experimental data often makes systematic errors easier to isolate, such as inconsistencies in normalizations, filling in of diffraction minima, or energy dependence in form factors. On the other hand, however, such direct comparisons yield little insight as to what features of the charge distribution should be changed to improve agreement with experiment. This is particularly important given our previous conclusion that certain features of the charge distribution are well-determined theoretically, whereas others are virtually undetermined. Hence, in Section 3 we will address the problem of displaying in coordinate space the changes required in a theoretical charge distribution to bring it into agreement with experiment, as well as displaying in coordinate space the errors and error correlations implied by the statistical experimental errors. The result of such an analysis, which is described in detail in Section 4, is included in Fig. 17 and denoted as the experimental error envelope. One should note that this envelope describes only statistical errors and that it is subject to the important caveat in Section 4 regarding the omission of systematic errors.

2.3.1. Electron Scattering

Elastic electron scattering cross sections have been calculated for the charge density distributions reviewed in Section 2.1 and compared with the 52.9-MeV data from Darmstadt, 124.0- and 167.0-MeV data from Mainz, and 248.2- and 502.0-MeV data from Stanford. The data and corrections are treated exactly as in the ^{208}Pb analysis by Friar and Negele (FN 73a): These corrections have been reviewed in Section 1 and the data are discussed in Section 4. The results of these calculations are presented in Table III and Figures 19–22.

The theories are tabulated in Table III in the same order as they were discussed in Section 2.1 and entered in Table II. The first six columns of χ^2 give the χ^2 per data point for each of the five sets of electron scattering data and the total χ^2 per data point for all the electron scattering data taken together. Since we have noted that the average interior density, or, equivalently, rms radius, is not strongly determined theoretically, we have also rescaled the radial parameter for each distribution to yield a density of the same shape with the experimental rms radius of 5.502 fm. That is, $\varrho(r)$ has been replaced by $(1/\lambda^3)\varrho(r/\lambda)$, which simulates, at least in the cases we have tested where λ is close to unity, the result of repeating the Hartree–Fock calculations for an interaction adjusted for a slightly different saturation density of nuclear matter. The next four columns show the χ^2 per data point for these scaled distributions for the two lowest energy data sets, for the high-energy Stanford data, and for the entire set of data. For the low-momentum-transfer Darmstadt data, one might expect scaling to the proper rms radius to systematically improve the χ^2, from the fact that the rms radius determines the coefficient of q^2 in the expansion of the form factor in powers of the momentum transfer. From Table III one observes that this is not the case. Whereas decreasing the rms radius from values greater than the experimental value of 5.502 down to the experimental

TABLE III

Electron Scattering and Muonic X-Ray Results for Theoretical Charge Distributions

The inclusion of proton and neutron form factors is denoted by P and N, respectively, and all other quantities are defined and discussed in the text. The entry $>1.6 \times 10^4$ indicates that the muonic χ^2 for distributions with unrealistic radii yielded a χ^2 larger than 1.6×10^4 which was not recorded. The values for $4\pi\varrho'_{HD}$ and $\langle r^2_{ch}\rangle^{1/2}$ given in parentheses are calculated from the best fit distribution in Section 4 and the combined experimental and theoretical errors for muonic transition energies are indicated in parentheses.

Theory	Form factors	$4\pi\varrho'_{HD}$, fm^{-4} (0.338 ± 0.008)	$\langle r^2_{ch}\rangle^{1/2}$, fm (5.502 ± 0.006)	52.9 MeV Darmstadt χ^2/N	124.0 MeV Mainz χ^2/N	167.0 MeV Mainz χ^2/N	248.2 MeV Stanford χ^2/N	502.2 MeV Stanford χ^2/N
RBHF	P	0.293	4.868	652	151	362	1320	978
DDHF	PN	0.331	5.495	2.17	1.54	0.97	14.9	31.2
DME	PN	0.332	5.490	1.38	0.61	1.18	4.04	11.8
CS	PN	0.350	5.429	3.61	1.12	6.12	50.0	123
FN-β	P	0.327	5.504	3.32	2.60	2.35	21.5	39.8
CK-1	PN	0.320	5.490	4.12	3.54	4.23	18.7	18.2
CK-2	PN	0.324	5.481	3.54	2.57	3.41	12.6	19.7
SKM-I	PN	0.396	5.420	9.84	14.2	74.4	459	1085
SKM-II	PN	0.333	5.531	2.29	3.80	8.74	70.1	143.3
SKM-III	PN	0.338	5.563	4.13	6.52	20.9	155	405
SKM-V	PN	0.333	5.459	1.73	0.56	2.27	9.03	23.6
MDI-2	P	0.323	5.513	2.98	2.93	1.44	19.7	29.3
MDI-4	P	0.317	5.509	3.57	2.85	2.72	15.4	20.2
NESTOR	P	0.327	5.294	35.9	10.6	35.4	130	145
BL-F1	P	0.334	5.489	1.15	1.01	2.04	20.4	55.4
BL-F2	P	0.331	5.498	1.49	1.40	1.36	16.3	40.6
SRP	P	0.342	5.497	1.23	1.13	3.25	27.8	227
CMTW	P	0.323	5.540	6.04	4.61	2.50	29.2	51.7
OBEP-II	P	0.323	5.360	14.7	4.59	16.4	49.8	50.1
RH-1	PN	0.393	5.495	2.58	18.8	96.5	756	1950
RH-2	PN	0.330	5.534	1.97	1.72	4.03	15.5	20.5

TABLE III (continued)

Theory	Total χ_e^2/N	Scaled Darmstadt χ^2/N	Scaled Mainz 124 MeV χ^2/N	Scaled Stanford 502 MeV χ^2/N	Scaled χ_e^2/N	χ_μ^2/N	Scaled muonic energy deviations, keV		
							$2p_{1/2} - 1s_{1/2}$ (± 0.50)	$2s_{1/2} - 2p_{3/2}$ (± 0.46)	$E_\infty - 2p_{3/2}$ (± 0.50)
RBHF	796	26.7	28.8	245	169	$>1.6\times10^4$	-0.49	9.79	6.55
DDHF	14.8	2.42	1.77	30.9	14.9	141	0.14	1.13	-0.08
DME	5.35	1.70	0.91	11.0	5.14	204	0.18	0.42	-0.61
CS	55.3	2.20	3.08	82.7	39.0	5920	0.17	0.63	-0.48
FN-β	19.8	3.24	2.53	39.7	19.7	37	0.10	1.65	0.32
CK-1	12.4	4.89	4.26	20.2	14.3	580	0.03	2.70	1.12
CK-2	11.0	4.92	3.87	19.7	13.0	1080	0.02	2.67	1.11
SKM-I	495	1.83	9.22	805	359	4950	0.44	-3.16	-3.36
SKM-II	67.8	1.05	2.95	151	71.0	1293	0.30	-1.28	-1.94
SKM-III	180	1.21	5.41	453	202	5202	0.38	-2.33	-2.74
SKM-V	10.8	2.05	1.82	16.4	8.10	2110	0.18	0.51	-0.56
MDI-2	15.7	2.29	2.36	28.8	14.8	72	0.16	0.75	-0.38
MDI-4	11.9	3.15	2.51	18.2	10.5	2.6	0.11	1.53	0.23
NESTOR	91.6	6.60	6.08	78.9	47.0	$>1.6\times10^4$	-0.07	3.90	2.08
BL-F1	24.2	1.38	1.31	53.0	23.4	162	0.22	-0.11	-1.03
BL-F2	18.2	1.69	1.59	39.9	17.7	22	0.20	0.15	-0.83
SRP	84.3	1.38	1.26	217	80.4	21	0.22	-0.14	-1.05
CMTW	26.2	2.64	2.35	57.2	25.2	1101	0.13	1.20	-0.02
OBEP-II	34.0	8.16	8.26	87.7	56.8	$>1.6\times10^4$	-0.11	4.67	2.67
RH-1	864	2.40	18.7	1920	850	171	0.49	-3.92	-3.97
RH-2	11.9	1.20	1.56	19.7	13.5	1830	0.37	-2.20	-2.61

value diminishes χ^2, the true minimum for the Darmstadt data lies at a slightly lower value of the rms radius, so that increasing the radius from roughly 5.49 to 5.502 actually increases the χ^2 for this data. The origin of this apparent anomaly is obvious from Fig. 40 in Section 4. The best fit to all five sets of electron data plus muonic data, from which the rms radius was determined, is systematically slightly lower than the Darmstadt data, so that the Darmstadt data alone prefer a slightly smaller rms radius. The total χ^2 for all electron data, however, is systematically improved by scaling to the best-fit rms radius. It is interesting that the qualitative differences among various theories are not significantly altered by such scaling; that is, no shape that yielded a poor fit to the data compared to the best unscaled fits ever surpassed those fits when scaled. Thus, as far as electron scattering is concerned, in contrast to the case of muonic X-ray data discussed below, comparison of the unscaled density provides an adequate means of distinguishing among the various theoretical density distributions.

Additional insight is provided by comparison of the elastic cross sections as a function of momentum transfer for the various theories (Figs. 19–22). To facilitate comparison of distributions in Fig. 17, results for each of the subgroups of densities are shown on a single graph, using the same legend as in Fig. 17. In order to suppress experimental scatter, the theoretical cross sections are compared with the reference cross sections σ_{ref} obtained from the best phenomenological fit to all five sets of electron data plus muonic data. The percentage deviation between the experimental cross sections and $\sigma_{ref}(q)$ is shown in Fig. 40 in Section 4 and allows one easily to assess the significance of deviations in the present graphs. The actual quantity plotted in these graphs is the difference between the reference and theoretical cross sections, divided by the experimental cross section, which differs negligibly from the reference cross section. For small deviations, this definition reduces approximately to the percentage deviation. For large deviations, it is asymmetric in the sense that deviations of theory above experiment are emphasized relative to deviations below experiment; that is, if theory is twice the experimental value, the definition yields -100%, whereas if theory is half the experimental value, the definition yields $+50\%$. This asymmetry is somewhat compensated by our compression of scales for negative deviations, and in any event our primary interest will be in results yielding small fractional deviations.

The theories have been separated into two groups, according to whether the effect of the neutron form factor has been included or not. For those theories for which the effect of neutrons was omitted (Figs. 21 and 22), we have also displayed by a heavy solid line the deviation expected

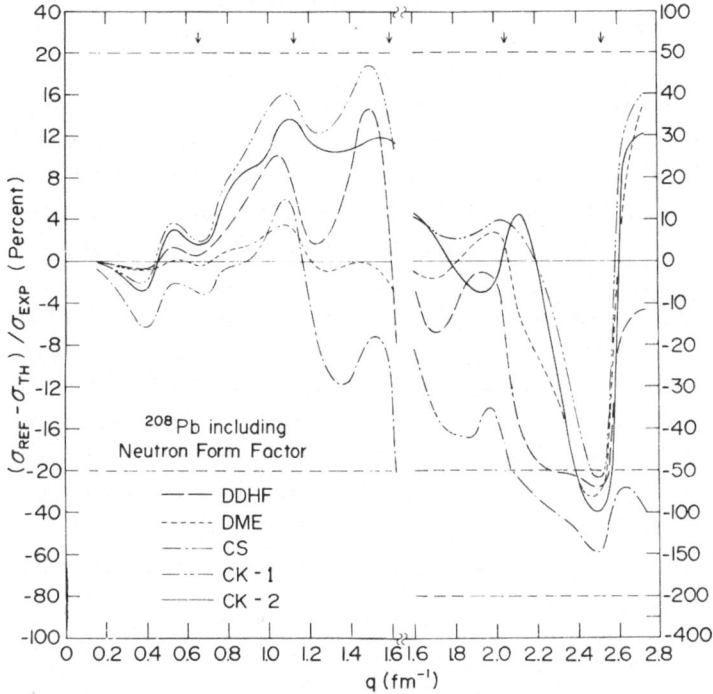

Fig. 19. Deviation of theoretical cross sections from experiment. The positions of the diffraction minima are denoted by the vertical arrows. Note that deviations below $q = 1.6 \, \text{fm}^{-1}$ are to be read on the left scale, those above $1.6 \, \text{fm}^{-1}$ are to be read on the right scale, and that the vertical scales change at the horizontal dashed lines. The abbreviations for the theories are defined in Section 2.1.

purely from the omission of the neutron form factor alone, using the results of Bertozzi *et al.* (Ber+ 73). Thus, although the effect of the neutrons is clearly comparable to many of the theoretical differences under consideration, this effect may be approximately eliminated simply by comparing the curves in Figs. 21 and 22 to the heavy solid curve instead of the horizontal axis.

The complete deviations for all the theories have been presented in Figs. 19–22 so that specialists may compare the various theories in detail. We shall simply comment here on several characteristic features of various density distributions which are readily identified in these figures. The unmistakable signature of an error in rms radius is the large oscillations resulting when the theoretical diffraction minima are out of phase with the experimental minima (denoted by arrows in these figures). The most dramatic cases are the RBHF curve in Fig. 22, corresponding to a charge radius of 4.87 fm, and the Nestor and OBEP II curves in Fig. 21, with rms radii of 5.29 and 5.36 fm, respectively. An unphysically steep surface, as occurs for SKM-I and RH-1, introduces unphysically large higher Fourier components into the density distribution, and these are reflected in cross sections at higher momentum transfer which are larger than experiment. This corresponds to a negative fractional deviation, and in Fig. 20 we observe that from 1.6 to 2.6 fm^{-1}, SKM-I and RH-1 yield deviations which are systematically significantly more negative than any other theories.

One systematic discrepancy with experiment occurs for all the theoretical distributions considered in this work and thus warrants comment. At the last diffraction minimum, in the region of 2.5 fm^{-1}, virtually every theory yields cross sections which are at least 50% larger than the experimental values, which have experimental statistical errors on the order of 10%. The only exception is the MDI-4 curve, which yields large, positive fractional deviations above and below this momentum transfer where other theories produce better fits. If this curve were displaced downward

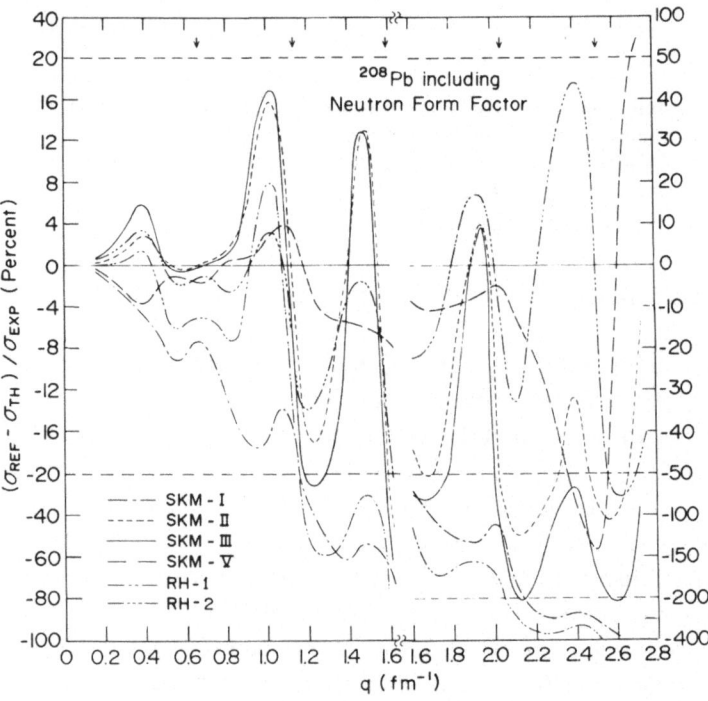

Fig. 20. Deviation of theoretical cross sections from experiment. The notation is the same as in Fig. 19.

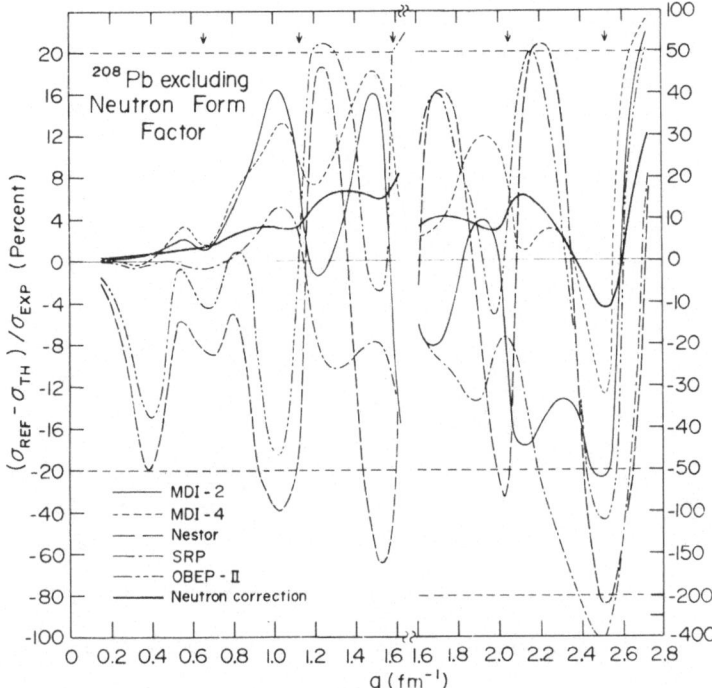

Fig. 21. Deviation of theoretical cross sections from experiment. Since these distributions did not include corrections for the neutron form factor, the deviations should be measured from the heavy solid line denoted "neutron correction" rather than the horizontal axis. All other notation is the same as in Fig. 19.

by 20%, it would also yield a 50% discrepancy at 2.5 fm⁻¹ and give significantly better results throughout the rest of the high-momentum-transfer region. This systematic discrepancy provides the tantalizing possibility that there are distinctly different physical processes becoming important in this region, such as dispersion effects or meson exchange current contributions. Dispersion corrections are a particularly good candidate since they increase with q^2 and, at least in the approximation of Bethe and Molinari (BM 71), the correction amplitude is predominantly imaginary. Thus the contribution is largest in the last diffraction minimum where the real amplitude vanishes.

2.3.2. Muonic X-Rays

Because of the high experimental resolution, muonic X-ray transition energies impose several very stringent constraints on theoretical charge

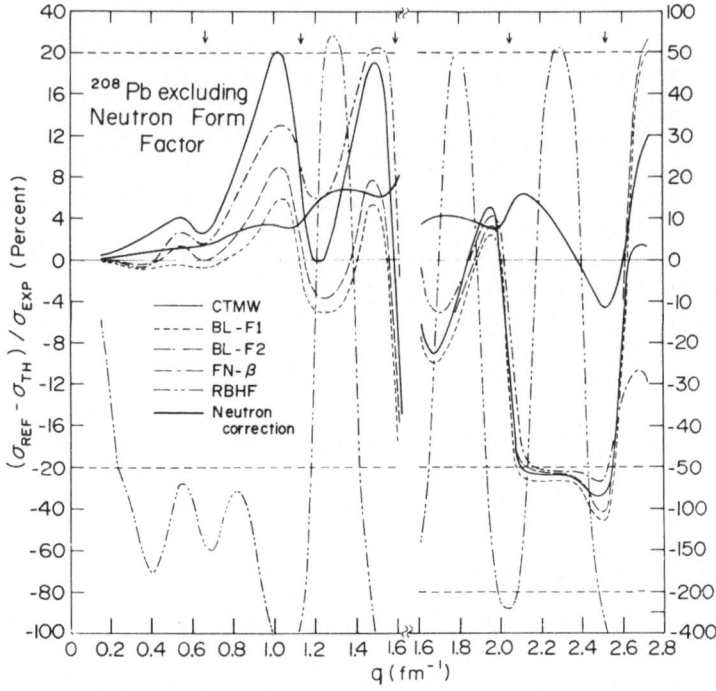

Fig. 22. Deviations of theoretical cross sections from experiment. The notation is the same as in Fig. 21.

distributions. The precise nature of these constraints will be explored in Section 3. For our present purposes, we simply note the sensitivity of the lowest six energy levels by citing the energy shift produced by a 1% change in rms radius given by Friar and Negele (FN 73a): 55 keV for $1s_{1/2}$, 11 keV for $2s_{1/2}$, 7.6 keV for $2p_{1/2}$, 5.3 keV for $2p_{3/2}$, 0.13 keV for $3d_{3/2}$, and 0.06 keV for $3d_{5/2}$. Compared with experimental errors of the order of 0.1–0.5 keV, it is evident that only transitions involving the lowest four levels yield significant constraints. We have calculated the $2p_{3/2}$ and $2p_{1/2}$ energies, corresponding to transitions from an infinitely high orbital $E_\infty \to 2p$ which can be inferred from cascades from high orbitals with point Dirac energies, as well as the redundant transitions $2p_{1/2} \to 1s_{1/2}$, $2p_{3/2} \to 1s_{1/2}$, $2s_{1/2} \to 2p_{3/2}$, and $2s_{1/2} \to 2p_{1/2}$ using the data and corrections of (FN 73a). The resulting values of χ_μ^2 are tabulated in Table III, and because of the extreme sensitivity to the rms radius, yield virtually no information about the shape of the density distribution besides its radius. This is indeed unfortunate, since we have argued above that the radius is very weakly determined theoretically and thus the total muonic χ^2 simply measures the value of the saturation density of nuclear matter used to adjust the phenomenological component of the effective interaction.

Meaningful information concerning the shape of the charge density distributions can still be recovered by scaling the distributions as in Section 2.3.1. Since the muonic transitions do not exactly measure any particular moment of the charge density, it is not sufficient to rescale the distribution to reproduce a precise value of the rms radius. Rather, we have scaled the distributions to fit exactly the $2p_{3/2} \rightarrow 1s_{1/2}$ transition, and then examined the consistency of all the other transitions· with this same scale change. As before, this scaling is intended to simulate the readjustment of the effective interaction to yield a Hartree–Fock distribution which exactly fits the $2p_{3/2} \rightarrow 1s_{1/2}$ transition, and in the case of small changes where the approximation has been checked, it is very accurate. Given that a cascade from high muonic orbitals with essentially point Dirac energies establishes absolute energies of the lowest four orbitals and that we have adjusted the size of the nucleus to exactly reproduce the difference between the $2p_{3/2}$ and $1s_{1/2}$ energies, there are three remaining combinations of energies to compare with experiment.

The results for three convenient combinations are tabulated in the last three columns of Table III. Comparison of these deviations, which are defined as the (positive) experimental transition energy minus the theoretical transition energy, with the quoted experimental and theoretical error estimates discussed in Section 4 indicates that the muonic data pose two very significant constraints on the shapes of the theoretical distributions. The $2p_{1/2} \rightarrow 1s_{1/2}$ transition does not provide significant additional information, and the reason for this will become evident in Section 3.1.2, where we show that the $2p_{3/2} \rightarrow 1s_{1/2}$ transition, which we have fit exactly, and the $2p_{1/2} \rightarrow 1s_{1/2}$ transition specify very similar constraints on the charge distribution. In contrast to the case of electron scattering, where we were guided by Born approximation arguments, the relation of muonic energy deviations in Table III to specific features of the density distributions is not immediately evident, and thus motivates the analysis to be presented in Section 3.

2.3.3. Conclusions from Direct Comparison

The preceding direct comparison of theoretical cross sections and muonic X-ray energies with experimental data effectively discriminates among the various theoretical charge density distributions. From the tabulated values of the total electron scattering χ_e^2, DME yields an excellent fit, DDHF, CK-1, CK-2, SKM-V, MDI-2, MDI-4, and RH-2 result in good fits, and FN-β, BL-F1, BL-F2, and CMTW yield fair fits. Based on χ^2 calculated from the tabulated scaled muonic energy deviations, DME, CS, SKM-V, and BL-F2 produce excellent fits and DDHF, MDI-2, BL-F1, SRP, and CMTW yield good fits. Taken together, the muonic and electron scattering data provide complementary constraints. The very best simultaneous fit is obtained by DME, which also yielded the best separate χ^2 for electron scattering and scaled muonic transitions. Fits which were at least good for both electron and muonic data were produced by DDHF, SKM-V, and MDI-2. The complementary nature of the contraints is emphasized by cases such as CS, which gave a poor fit to electron data and an excellent fit for muonic data, and RH2, which yielded a good fit to electron scattering data and a poor fit for muonic data.

The importance of scaling distributions to extract shape information from muonic transition energies is emphasized by the comparison of MDI-2 and MDI-4. Naively comparing the total χ_e^2 and χ_μ^2 from Table III would lead one to conclude that lowering the compression modulus from MDI-2 to MDI-4 was actually preferred by the data, since χ_e^2/N decreases from 15.7 to 11.9 and χ_μ^2 decreases from 72 to 2.6. In fact, this is not the case, since the change in compression modulus also changed the rms radius by 0.004 fm and this radius change accounts for the drastic difference in χ_μ^2. Since the radius is adjusted in the theory, the only meaningful test of the theory is to compare scaled densities, in which case MDI-4 yields significantly worse agreement for muons than MDI-2, even though it yields slightly better agreement for electrons. Note, as mentioned previously, that scaling does not drastically modify the conclusions drawn from electron scattering. The unscaled χ^2 results for MDI-2 and MDI-4 of 15.7 and 11.9 yield the same conclusions as the scaled results 14.8 and 10.5.

Unfortunately, in spite of the strong discrimination among various theoretical distributions provided by the electron and muonic data, it is not really possible to translate this directly into discrimination among the various theories. A large part of the difficulty arises from the presence of the phenomenological parameters in the theories, which, if chosen fortuitously, can produce accidental agreement for an invalid theory or, if chosen on the basis of an incorrect prejudice concerning the saturation density or compression modulus of nuclear matter, can produce poor numerical results from an otherwise valid theory. Also, of course, a correct theory must reproduce the systematics of energies and shapes throughout the periodic table, and we have restricted our attention to only the density distribution of ^{208}Pb. Perhaps the most dramatic indication of the difficulty in directly evaluating theories on the basis of the results presented in this section is the fortuitous superiority of the DME results to the DDHF results, in spite of the fact emphasized in Section 2.1.2 that DME is an approximation to DDHF. Nevertheless, one should not permit the lack of absolute discrimination among theories to obscure the fact that the direct comparison of theoretical predictions has already provided a great deal of theoretical insight. Two particularly significant results are the extremely good absolute-χ^2 fits which can presently be obtained in the context of the central field approximation, and the fact that the details of the nuclear interaction are immaterial as far as electron scattering and muonic X-rays are concerned, since simple interactions like SKM-V and MDI-2 yield results comparable to the best results obtained with a realistic effective interaction determined from a nucleon–nucleon potential.

The direct comparison of theoretical predictions with experiment raises several questions which require further consideration. Uncertainties arising from dispersion corrections and meson exchange effects have been discussed in Section 1.3 and experimental errors will be treated in Section 4. One additional significant question concerns higher order contributions to the nuclear density distribution. The lowest order contributions to the perturbation expansion of the ground-state expectation value of the one-body density operator are shown in Fig. 23, taken from (Neg 71), using the notation of Thouless (Tho 61), where the heavy dot denotes the density operator and the $-\times$ denotes the one-body potential defining the basis. Only the Hartree–Fock density, given by the diagram of Fig. 23a, has been included in the preceding analyses. The diagram of Fig. 23d has been treated by many investigators (SBK 73, TFM 73, Dav+ 74, RPS 74) and by itself yields a significant contribution to the density expansion. By defining the single-particle potential to include the diagrams of Figs. 18a and 18c, the diagrams of Figs. 23b and 23d are identically cancelled by the diagram of Fig. 23c. Also, the diagrams of Figs. 23g–23j sum to zero. Thus the only second-order correction is the sum of the diagrams of Figs. 23e and 23f.

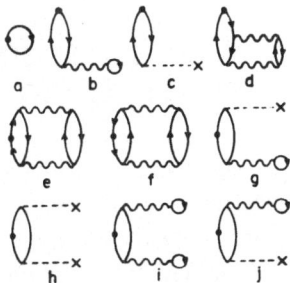

Fig. 23. Contributions to the density expansion. The solid dot indicates the density operator, the wavy line represents the G-matrix, and the cross indicates the one-body potential. Graphs (g)–(j) are intended to represent insertions of the density operator in both particle and hole lines.

The diagram of Fig. 23e counts the probability in the normally unoccupied particle states and the diagram of Fig. 23f subtracts an equal amount of probability from the normally occupied hole states. For high-lying excitations, this correction just describes the change in the one-body density due to short-range correlations. Physically, we expect this correction to be small, since short-range correlations excite pairs to high-energy intermediate states, and by the uncertainty principle, the excitation does not last long enough for the particles to go very far. Furthermore, the particles move in opposite directions, so very little spatial redistribution of probability is expected, and calculations in ^{40}Ca (Neg 70) bear this out. In contrast, low-lying excitations can last long enough for particles to move a large distance and thus significantly alter the one-body density. One of the simplest ways to approximately evaluate processes such as the graphs of Figs. 23e and 23f for states in the neighborhood of the Fermi surface is via BCS pairing theory. When this is done for an open-shell nucleus like ^{90}Zr (Neg 71), a significant correction occurs, but for closed-shell nuclei like ^{40}Ca and ^{208}Pb, the same calculation yields no pairing and thus no correction. A numerical evaluation of the graphs of Figs. 23e and 23f with the Tabakin potential for ^{40}Ca (SBK 73) also yields a very small correction. Thus, present evidence indicates that the omitted corrections in ^{208}Pb are very likely small, but given the extreme sensitivity of present data, the diagrams of Figs. 23e and 23f should definitely be calculated in ^{208}Pb, and higher order terms should at least be estimated.

Another problem raised by the direct comparison of theoretical results with experiment is the obvious difficulty in relating discrepancies in cross sections and muonic energies to errors in the shape of the charge density distribution. It would clearly be of great value to the theorist to know precisely where in coordinate space and by how much the charge density should be modified in order to simultaneously minimize the discrepancies for all the electron scattering and muonic data. It is this problem to which the next section is addressed.

3. SPECIFICATION OF THE INFORMATION ACTUALLY DETERMINED BY AVAILABLE DATA

One of the primary limitations of the early pioneering treatments of electron scattering and muonic X-rays was the lack of any precise specification of what information about the nuclear charge density was actually determined by the data. Data were generally analyzed in terms of some convenient function with variable parameters, such as a Gaussian, a Fermi function, a Fermi function multiplied by a parabolic term, or some smooth function plus "wiggles." The parameters were then varied to minimize χ^2 and one never knew for certain which features of the resulting density were actually determined by the data and which were simply the result of the arbitrary choice of parameterization.

Recently, there has been considerable effort devoted to this problem, resulting in substantial progress, which is reviewed in this section. The first step is the determination of the linear constraint imposed on the charge density by each experimental datum. This can be done in a precise and virtually model-independent way described in Section 3.1. Even with these linear constraints, the determination of the density distribution from the experimental data is not possible without the introduction of additional model assumptions. Various procedures for specifying model assumptions are reviewed in Section 3.2. In the context of a specific set of assumptions, the determination of the density distribution and the errors and error correlations in it arising from the independent statistical errors in the data are obtained straightforwardly.

3.1. Determination of Constraints Imposed by Experimental Data

The starting point for a precise formulation of the constraint imposed by a given measurement is the determination of the kernel $\varkappa_\alpha^{(l)}$ such that the experimental observable $\sigma_\alpha^{(l)}$ is obtained from a linear functional of the density:

$$\sigma_\alpha^{(l)} = \int \varkappa_\alpha^{(l)}(r)\varrho(r)r^2\,dr + c_\alpha^{(l)} \tag{76}$$

where the lepton label l denotes either electrons or muons and α denotes all the quantum numbers of the bound states in a muonic transition or the energy and scattering angle of a scattering cross section. The choice of removing the phase space weighting r^2 from the kernel is purely a matter of convenience, and we will use the normalization

$$\int \varrho(r)r^2\,dr = Z \tag{77}$$

In practice, the constant $c_\alpha^{(l)}$ can be subtracted from the experimental observable $\sigma_\alpha^{(l)}$ to yield an effective observable, or $c_\alpha^{(l)}/Z$ may be added to the kernel by virtue of Eq. (77).

In simple ideal cases the kernel in (76) is trivially obvious. For example, in the case of electron scattering in first Born approximation, if we take $\sigma_q^{(e)}$ as the form factor at momentum transfer q (i.e., the square root of the cross section with appropriate sign and normalization), then $\varkappa_q^{(e)}$ is just $j_0(qr)$ and $c_q^{(e)}$ is zero. In general, however, the observable is related to the density through some nonlinear functional

$$\sigma_\alpha = F_\alpha[\varrho(r)] \tag{78}$$

which must then be appropriately linearized to obtain the form (76). Borysowicz and Hetherington (BH 73) discuss the linearization in terms of expansion about a "base" or reference charge distribution $\varrho_0(r)$. In practice, $\varrho_0(r)$ will be the best fit to all available data, subject to physically motivated model assumptions. A fundamental assumption is necessarily introduced at this stage, which applies to all but the Monte Carlo approach described subsequently. This assumption is that there are no discrete ambiguities in the sense that no two significantly different choices of ϱ_0 yield indistinguishable values of all the observables. In Born approximation, for example, such an ambiguity would arise in the case that only a few cross sections were known in the region of a minimum, and one could fit the cross sections with either a positive or negative sign for the form factor of the point nearest the minimum. In practice, for cases with a large quantity of accurate, complementary data, such ambiguities seldom arise. One interesting case, however, is described by Sick (Sic 75). Linearization of (78) about ϱ_0 yields

$$\sigma_\alpha = F_\alpha[\varrho_0(r)] + \int_0^\infty \frac{\delta F_\alpha[\varrho(r)]}{\delta \varrho(r)}\bigg|_{\varrho_0(r)} (\varrho(r) - \varrho_0(r))\, dr \tag{79}$$

so that $r^2\varkappa_\alpha$ is just the functional derivative of F_α and c_α is $F_\alpha[\varrho_0(r)] - \int \varkappa_\alpha\varrho_0(r)r^2\, dr$.

This is as far as one can go in general terms. To reduce (79) to a practical form, we make use of the separation of the lepton–nucleus problem into the solution of the Dirac equation in a static Coulomb field plus systematic correction terms described in Section 1. The correction terms are small and relatively insensitive to $\varrho(r)$, and are thus included only in $F_\alpha[\varrho_0(r)]$. [In practice they are taken from the literature, and may not even be calculated for a realistic $\varrho_0(r)$, but rather a simple parameterization of ϱ or even a point nucleus.] The functional derivative of the full F_α is replaced by the functional derivative of the solution of the Dirac equation in the static Coulomb field generated by $\varrho(r)$. This functional derivative for the Dirac equation may easily be calculated in perturbation theory, and yields a simple physical interpretation of the kernels $\varkappa_\alpha^{(l)}$.

3.1.1. Perturbation Theory

For simplicity, we first treat the case of the binding energy $\varepsilon_{nlj} = m - E_{nlj}$ of a muonic bound state with principal quantum number n, angular momentum j, and orbital angular momentum (of the large component) l, following the treatment of Friar and Negele (FN 73a). By linearity, the kernel for a transition energy is just the difference between the kernels for each state. The Dirac equation resulting from $\varrho(r)$ may be written

$$(2m - V(r) - \varepsilon)F - \frac{dG}{dr} - \varkappa \frac{G}{r} = 0 \qquad (80a)$$

$$(-V(r) - \varepsilon)G + \frac{dF}{dr} - \varkappa \frac{F}{r} = 0 \qquad (80b)$$

where G and F are the large and small components normalized such that $\int_0^\infty (F^2(r) + G^2(r))\, dr = 1$, $\varkappa = 2(l-j)(j+\frac{1}{2})$, $V(r) = -\alpha \int (1/r_>)\varrho(r')(r')^2\, dr'$, $r_>$ denotes the larger of r and r', and the quantum numbers (nlj) are temporarily suppressed. The corresponding Dirac equation resulting from the density $\varrho_0(r)$ has solutions G^0, F^0, and ε^0. Multiplication of Eq. (80a) by F^0, of Eq. (80b) by G^0, of the analogous equations for ϱ_0 by $-F$ and $-G$, and integration over r yields the familiar identity

$$\varepsilon - \varepsilon^0 = \frac{\int_0^\infty \{V_0(r) - V(r)\}\{F(r)F^0(r) + G(r)G^0(r)\}\, dr}{\int_0^\infty \{F(r)F^0(r) + G(r)G^0(r)\}\, dr} \qquad (81)$$

Considering the limit of an infinitesimal difference between $\varrho(r)$ and $\varrho_0(r)$ yields

$$\varepsilon - \varepsilon^0 = \int_0^\infty dr' \int_0^\infty dr\, \{F^2(r) + G^2(r)\} \frac{\alpha}{r_>} \{\varrho(r') - \varrho_0(r')\}r'^2 \qquad (82)$$

The quantity multiplying $\{\varrho(r') - \varrho_0(r')\}$ is the desired functional derivative, so, changing variables and restoring the (nlj) subscripts, we obtain the appropriate kernel

$$\varkappa_{(nlj)}^{(\mu)}(r) = \alpha \int_0^\infty \frac{dr'}{r_>} \{F_{nlj}^2(r') + G_{nlj}^2(r')\} \qquad (83)$$

This kernel has obvious physical significance since it is just the Coulomb potential seen by the nucleus generated by the charge distribution of the orbiting muon. Thus each muonic energy essentially measures the overlap of the nuclear charge density with the Coulomb potential generated by the muon wave function, and transition energies measure differences between these overlaps.

For electron scattering, the theory is much simpler if one treats the case of high electron energy so that the electron mass is negligible compared to its momentum. In practice, this restriction poses no difficulty, since with a finite-mass partial wave analysis, one can always transform data below 50 MeV into "equivalent" data at the same momentum transfer at higher energy. Following Yennie *et al.* (YRW 54), a partial wave decomposition neglecting the electron mass yields

$$\frac{dG_j}{dx} - \frac{j + \frac{1}{2}}{x} G_j + (1 - v(x))F_j = 0$$

$$\frac{dF_j}{dx} + \frac{j + \frac{1}{2}}{x} F_j - (1 - v(x))G_j = 0 \tag{84}$$

where $x = Er$, $v(x) = (1/E)V(x/E)$, and F_j and G_j are the radial functions for the jth partial wave. Equations (84) are easily obtained from Eqs. (80a) and (80b) by substituting $\varepsilon = m - E$, $x = Er$, neglecting terms containing m, and noting that solutions for $l = j \pm \frac{1}{2}$ are transformed into each other by the substitution $F \to G$, $G \to -F$. In the asymptotic region $x > ER$, where R is assumed to contain the entire nuclear charge,

$$G_j \sim \sin[x - \tfrac{1}{2}(j - \tfrac{1}{2})\pi + \eta_j + Z\alpha \ln(2x)] \tag{85}$$

where η_j is the phase shift for the jth partial wave. The scattering amplitude is

$$f(\theta) = (1/2iE) \sum_j e^{2i\eta_j}(j + \tfrac{1}{2})[P_{j-(1/2)}(\cos\theta) + P_{j+(1/2)}(\cos\theta)] \tag{86}$$

and the cross section for scattering electrons of energy E through an angle θ is

$$d\sigma(\theta)/d\Omega = \sec^2(\theta/2)\,|f(\theta)|^2 \tag{87}$$

Considering two different nuclear charge densities $\varrho(r)$ and $\varrho_0(r)$ and performing the same manipulations which lead to Eq. (81), we obtain

$$\sin(\eta_j - \eta_j^0) = \int_0^R \{V_0(r) - V(r)\}\{F_j(r)F_j^0(r) + G_j(r)G_j^0(r)\}\,dr \tag{88}$$

For infinitesimal differences between $\varrho(r)$ and $\varrho_0(r)$,

$$\eta_j - \eta_j{}^0 = \int_0^R dr' \int_0^R dr \, \{F_j{}^2(r) + G_j{}^2(r)\} \frac{\alpha}{r_>} \{\varrho(r') - \varrho_0(r')\} \, r'^2 \qquad (89)$$

This is completely analogous to Eq. (82), and if phase shifts were actually observable, then the phase shift in each partial wave would measure the overlap of the nuclear density with the Coulomb potential generated by the electron charge distribution in that particular partial wave. Infinitesimal changes in phase shifts are straightforwardly related to infinitesimal changes in cross sections through Eqs. (86) and (87), with the result that the kernel for the electron scattering cross section $d\sigma/d\Omega(\theta)$ at angle θ and energy E is

$$\varkappa_{\theta,E}^{(e)}(r) = \frac{2\alpha}{E} \sec^2 \frac{\theta}{2} \sum_j \{\mathrm{Re}\,[f^*(\theta)\,e^{2i\eta_j}(j + \tfrac{1}{2})(P_{j-(1/2)}(\cos\theta)$$

$$+ P_{j+(1/2)}(\cos\theta))] \int_0^R \frac{dr'}{r_>} [F_j{}^2(r') + G_j{}^2(r')]\} \qquad (90)$$

Given a density distribution $\varrho_0(r)$, the kernels (83) and (90) are straightforward to evaluate numerically. If $\varrho_0(r)$ were the true density distribution, these kernels would exactly specify the constraints imposed by each measurement. In practice, it is possible to implement an iterative procedure in which ϱ_0 is continually improved until it yields an optimal χ^2 fit to all available data (FN 73a). At the last iteration, one has the best possible ϱ_0, subject to the experimental errors and model assumptions, and one can verify that the constraints obtained in earlier iterations with less realistic choices of ϱ_0 differ negligibly from the final values. Thus we conclude that the determination of the linear kernels is quite unambiguous and virtually model independent.

3.1.2. Muonic X-Ray Kernels

The insensitivity of the muonic kernels to the detailed shape of $\varrho_0(r)$ is simple to understand physically (Neg 73). By normalization, ϱ_0 fluctuates about ϱ and if it fits available muonic and electron scattering data, the discrepancy $\varrho(r) - \varrho_0(r)$ only contains significant Fourier components beyond the maximum experimental momentum transfer. Consider a discrepancy comprised of a single Fourier component $\Delta\varrho/\varrho_0 = \lambda[\sin(kr)]/kr$. Then, because the long range of the Coulomb force averages out the positive and negative deviations, the resulting error in the Coulomb potential seen by the muon is very much smaller. Calculating the ΔV arising from $\Delta\varrho$ and comparing with the average Coulomb potential $V_0 \sim Ze^2/R$ of a distribution of radius R, we obtain

$$\frac{\Delta V}{V_0} = \frac{\lambda}{(Rk)^2} \frac{\sin{(kr)}}{kr}$$

so that the attenuation is a factor $\sim 1/(Rk)^2$ in going from the density to the potential. A rough perturbation estimate of the error in the muon density induced by this error in the Coulomb potential, using hydrogenic wave functions, yields $\Delta\varrho_\mu/\varrho_\mu < \lambda/(kR)^3$. For ^{208}Pb, $k > 3$ fm^{-1} and $R = 7$ fm, so that $\Delta\varrho/\varrho_0 < 10^{-4}\lambda$, that is, the fractional error in the muon density is four orders of magnitude smaller than the fractional error in the nuclear density. The error in the kernel is reduced still further, however, because it is determined by the convolution of the error in the muon density with the long-range Coulomb force, which again smooths out fluctuating errors.

By the preceding argument, muon kernels calculated numerically from a density distribution optimally fit to electron scattering data and muonic X-ray data solve the problem of specifying the experimental constraints in an extremely precise and unambiguous manner. Such muon kernels for the transitions in ^{208}Pb treated in Section 2 are shown by the solid lines in Fig. 24, taken from (FN 73a). The curves labeled $2p_{1/2}$ and $2p_{3/2}$ are for

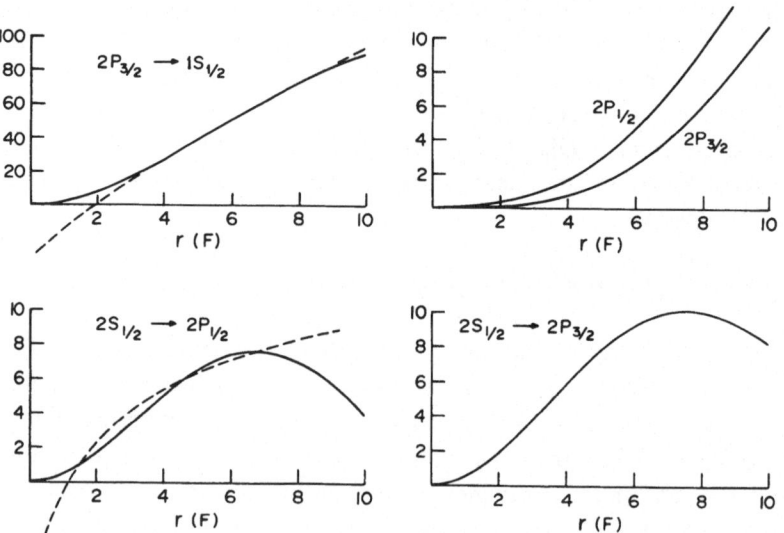

Fig. 24. Muonic kernels $\varkappa_\alpha^{(\mu)}$ (solid curves) on a scale such that energies are in keV if the density is normalized according to Eq. (77). The dashed curves correspond to the approximation of Eq. (94).

the binding energies of those levels or equivalently $E_\infty - E_{2p}$, and the other curves are for the indicated transition energies. The redundant $2p_{1/2} \rightarrow 1s_{1/2}$ result is not graphed, but may easily be reconstructed from the other information in Fig. 24. The constant $c_\varkappa{}^\mu$ in Eq. (76) has been defined such that all kernels are zero at $r = 0$. Given the ease of the numerical calculation of the kernels, we strongly endorse this as the most precise method of defining what is measured. For a variety of reasons, however, a number of simpler approximations to these exact kernels have been developed and are widely used in the literature.

The insensitivity to the shape of the nuclear charge distribution has been exploited by Bethe and Negele (BN 68) to derive semianalytic expressions for the muonic kernels. In this theory, muon wave functions are calculated in a uniform charge distribution of radius R, where R is chosen to reproduce the experimental binding energy. (The theory is only semianalytic since R and the wave function normalization are determined numerically.) Inside R, the Dirac wave functions are easily obtained as power series:

$$\frac{G_{nlj}}{r} = \sum_s a_s \left(\frac{r}{R}\right)^s, \qquad \frac{F_{nlj}}{r} = \sum_s b_s \left(\frac{r}{R}\right)^s \qquad (91)$$

The coefficients are expressed in terms of a_l, determined from normalization, by the recursion relations

$$b_{l-1} = \begin{cases} 0, & \text{for } j = l + \tfrac{1}{2} \\[2mm] \dfrac{[l+1+\varkappa]a_l}{R(2m - \varepsilon_{nlj}) + \tfrac{3}{2}Z\alpha}, & \text{for } j = l - \tfrac{1}{2} \end{cases}$$

$$b_{l+1} = \frac{R\varepsilon_{nlj} - \tfrac{3}{2}Z\alpha}{l + 2 - \varkappa} a_l$$

$$a_{l+2i+2} = \frac{-\tfrac{1}{2}Z\alpha b_{l+2i-1} + [R(2m - \varepsilon_{nlj}) + \tfrac{3}{2}Z\alpha]b_{l+2i+1}}{l + 2i + 3 + \varkappa} \qquad (92)$$

$$b_{l+2i+3} = \frac{\tfrac{1}{2}Z\alpha a_{l+2i} + [R\varepsilon_{nlj} - \tfrac{3}{2}Z\alpha]a_{l+2i+2}}{l + 2i + 4 - \varkappa}$$

where $\varkappa = 2(l - j)(j + \tfrac{1}{2})$, m is the reduced muon mass, ε_{nlj} is the binding energy, i is any positive integer, and all a_s and b_s not specifically defined are identically zero. From these wave functions, the muon kernels are straightforwardly determined:

$$\varkappa_{nlj}^{(\mu)}(r) = -\alpha R^2 \sum_{t=0}^{\infty} \left(\frac{r}{R}\right)^{2t+2} \frac{1}{(2t+2)(2t+3)} \sum_{s=0}^{2t} (a_{2t-s}a_s + b_{2t-s}b_s) \qquad (93)$$

Although this result is strictly applicable only for $r < R$, the error for r slightly greater than R is very small due to the small deviation of the $1/r$ potential from the interior parabolic potential. The chief advantages of this theory are that Eq. (93) provides a simple expression for the linear combination of the even moments of the charge distribution measured by a given energy level and that it is a convenient starting point to explore the sign and magnitude of shifts in transition energies due to various changes in the nuclear shape. Numerical values of the coefficients in (93) for ^{208}Pb are tabulated by Negele (Neg 69). Unfortunately, many moments contribute to each transition energy, so that one cannot conclude that each transition measures a particular moment. Equation (93) is the most accurate of the approximations to the true kernels discussed in this section, and on the scale of Fig. 24, this approximation cannot be distinguished from the exact plotted results.

Ford and Wills (FW 69) essentially approximate the kernels by the parameterization

$$\varkappa(r) = A + Br^k \tag{94}$$

The constant A is the usual inessential additive shift. The value of k, not necessarily an integer, is determined by calculating moments $\langle r^k \rangle$ for a variety of empirical charge distributions constrained to reproduce the exact transition energy and choosing the moment that is least model dependent. Given k, B is found by relating small changes in transition energies to small changes in $\langle r^k \rangle$. From Eq. (93), we know that it is not correct in principle to relate the kernels to a single moment. Barrett (Bar 70) has interpreted the parameterization (94) as a best fit to the slope and curvature of the correct kernel in the region of the nuclear surface. Then, given the limited range of empirical charge distributions treated by Ford and Wills (FW 69), the variations in density were strongly peaked in the surface and only sampled the approximate kernel in the region where it reproduces the true kernel fairly well. Following Barrett (Bar 70), we have displayed the approximation (94) by dashed curves in Fig. 24 for the $2p_{3/2} \to 1s_{1/2}$ and $2s_{1/2} \to 2p_{1/2}$ transitions. Here one observes that whereas the approximation yields a fair qualitative fit for the $2p_{3/2} \to 1s_{1/2}$ transition, it is quite inadequate for the $2s_{1/2} \to 2p_{1/2}$ transition.

Because of the serious shortcoming of Eq. (94), Barrett (Ba 70) suggested parameterization in terms of a generalized moment

$$\varkappa(r) = A + Br^k e^{-\alpha r} \tag{95}$$

A single value of α is chosen for all the transitions in a given nucleus, and

k is optimized for each transition. This still enables one to characterize each transition by a single parameter k and yields much more accurate fits than Eq. (94).

The accuracy of the parameterizations in Eqs. (94) and (95) is indicated by a series of analyses in the Pb region. In calculating ^{209}Bi and ^{207}Pb isomer shifts, Rinker (Rin 71a) found absolute errors of the order of 0.5 keV using Eq. (94). Deliberately searching over a wide range of somewhat unrealistic charge distributions, Rinker (Rin 71b) found much larger errors, of up to 3.5 keV for Eq. (95) and errors an order of magnitude larger for Eq. (94). From the results of a study of Pb isotopes by Ford and Rinker (FR 73), one finds errors of 0.7 and 1.4 keV for the $3p_{3/2} \rightarrow 2s_{1/2}$ and $2p_{3/2} \rightarrow 1s_{1/2}$ transitions using Eq. (94) and errors of 0.10 and 0.15 keV using Eq. (95). From these numbers, one concludes that Eq. (94) is inadequate and that Eq. (95) is adequate only for a restricted set of phenomenological density distributions. Because of the convenience of characterizing muonic transitions by a single parameter, Engfer *et al.* (Eng+ 74) have compiled a tabulation of generalized moment parameters for nuclei throughout the periodic table. They adopt a linear parameterization for α as a function of Z and determine k and B in Eq. (95) by fitting to the exact kernel calculated using a Fermi distribution for the nuclear charge density.

An alternative approximation,

$$\varkappa_\alpha^{(\mu)}(r) = A + Br^2 + Cr^3 \tag{96}$$

has been suggested by Kankeleit (Kan 73). In principle, the exact kernels may contain odd powers of r, even though such terms are absent in the simple model leading to Eq. (93). Since the actual kernels contain many higher moments, Eq. (96) should be regarded, like Eqs. (94) and (95), as a best fit in the region of maximum overlap with the nuclear charge distribution. The added parameter C increases one's flexibility in reproducing the shapes in Fig. 24, but Eq. (96) has the practical disadvantage relative to Eq. (95) that it yields a two-parameter rather than one-parameter characterization of all the transitions in a given nucleus.

3.1.3. Electron Scattering Kernels

The kernels defined in Eq. (90) are straightforward to calculate numerically by using a standard partial wave analysis. The ^{208}Pb results obtained by Friar and Negele (FN 73a) for several angles at 248.2 and

502.0 MeV are shown in the left and right portions of Fig. 25, respectively. The angles were chosen to correspond to comparable momentum transfer at high and low energy, and span the range over which actual data are available. The positions of the selected angles relative to diffraction minima are indicated by the differential cross sections in the lower left and upper right of Fig. 25. The exact kernels, arbitrarily normalized to unity at the

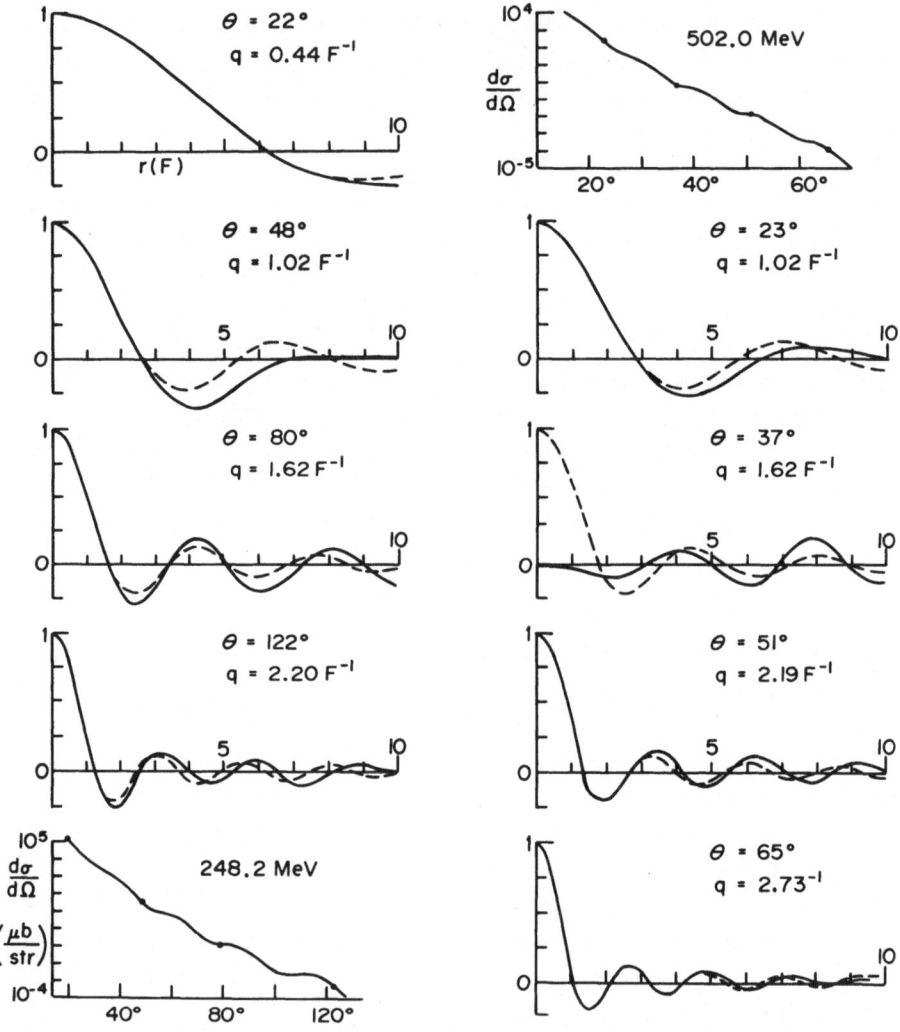

Fig. 25. Comparison of exact kernels $\varkappa_\alpha^{(e)}(r)$ (solid curves) with first Born approximation (dashed curves). The left column corresponds to 248.2 MeV and the right to 502 MeV. The angles selected are indicated by heavy dots on the differential cross-section curves at the lower left and upper right.

origin, are denoted by solid lines. For purposes of comparison, the simple Born approximation kernels

$$\varkappa^{(e)}(r) = j_0(q_{\text{eff}}r) \qquad (97)$$

where

$$q_{\text{eff}} = q\left(1 + \frac{Z\alpha}{E_0 R}\right)$$

are shown by dashed lines. The effective momentum transfer q_{eff} essentially specifies the local wave number of an electron of energy E_0 in the region of the nuclear surface.

One observes from this figure that the essential structure of the exact kernels, which are the Coulomb potentials generated from the exact distorted wave functions, is generally well reproduced by the spherical Bessel functions. The most conspicuous failure occurs at $37°$ at 502 MeV. This point was deliberately chosen at a diffraction minimum where the real amplitude vanishes and the Born approximation is inapplicable. Notice, however, that the small shift in q_{eff} between 502 and 248.2 MeV is sufficient to move the corresponding $q = 1.62$ fm^{-1} low-energy point far enough off the diffraction minimum that the Born approximation is again reasonably accurate. Thus, in the context of the entire angular distribution, the inaccuracy at the diffraction minima is rather insignificant.

The Born approximation kernels are certainly not accurate enough for quantitative calculations of the information content of a single cross section. However, they are very important insofar as they indicate that even in a heavy nucleus, electron scattering essentially specifies the Fourier–Bessel transform of the nuclear charge distribution $\varrho(r)$ [or equivalently the Fourier transform of $r\varrho(r)$].

3.1.4. Definitions of Model-Independent Functionals

As a result of the valid desire on the part of the experimentalists to present results which are free of theoretical bias, there has been considerable discussion recently of "model-independent" analyses of data. For our present purposes, it is useful to distinguish between the attempt to define model-independent functionals of the density, and the attempt to perform a model-independent determination of the density. The latter is manifestly impossible, and the specification of model assumptions for determination of the density will be discussed in detail in Section 3.2.1. The definition of model-independent functionals of the density is closely related to the preceding derivation of kernels for muonic X-ray transitions and electron scattering, and will be briefly discussed in this section.

The basic goal is to write down some set of convenient functionals of the density, such that given a set of electron scattering and muonic X-ray data, these functionals are uniquely determined even though the density is not uniquely determined. If one were to drop the requirement of convenience, the problem is essentially solved by the exact kernels given in Eqs. (83) and (90), one for each measurement. Even a discrete set of the approximate kernels, Eqs. (95) and (97), are not of optimal convenience in compactly representing experimental results, because people prefer to deal with sets of functionals labeled by a single continuous parameter so that the final results can be presented on a graph in terms of a single variable.

One definition of a model-independent functional, advocated by Friedrich and Lenz (FL 72), is equivalent to Eq. (94):

$$M(k) = \left[\int_0^\infty r^k \varrho(r) r^2 \, dr \right]^{1/k}, \qquad k \neq 0 \tag{98}$$

$$M(0) = \exp \left[\int_0^\infty ln(r) \, \varrho(r) r^2 \, dr \right]$$

It is evident that, in principle, even if Eq. (94) were exact for N specific values of k, N infinitely precise measurements could at most only determine N values of $M(k)$. In practice, tacit assumptions regarding sufficiently smooth behavior of $M(k)$ and thus the behavior of $\varrho(r)$ are always made. Friedrich and Lenz consider a very general form of charge distribution expressed as the sum of S δ-functions

$$\varrho(r) = \sum_{i=1}^{S} C_i \, \delta(r - R_i) \tag{99}$$

where S ranges from 7 to 13 for ^{208}Pb. Random values of C_i and R_i are generated and the total χ^2 for all observables predicted by the resulting distribution is calculated. The moment function $M(k)$ is calculated for all distributions yielding a χ^2 per data point less than 1, and the resulting envelope of curves specifies $M(k)$ and the errors in $M(k)$. Since no muonic data were treated by Friedrich and Lenz (FL 72), there is no theoretical reason why $M(k)$ is really model independent for any value of k. (Certain limits in which electron scattering could determine moments for specific k's are not realized in the experimental data.) Thus, there is no unambiguous way to assess the model dependence of the resulting envelope of $M(k)$ values.

A second prescription for an integral functional suggested by Friedrich and Lenz (FL 72) is the average radius $R(Q)$ of the interior fraction Q of the total charge:

$$R(Q) = (1/Q) \int_0^{\bar{r}} r\varrho(r)r^2 \, dr, \qquad Q(\bar{r}) = \int_0^{\bar{r}} \varrho(r)r^2 \, dr \qquad (100)$$

It is evident that this functional is no less model dependent than the density, since $\varrho(r)$ may easily be obtained from Eq. (100) by differentiating twice. When $R(Q)$ is graphed, the errors appear smaller than for $\varrho(r)$ simply because integration damps noise, whereas differentiation amplifies it.

The most obvious candidate for a model-independent function parameterized by a single continuous parameter is simply the form factor:

$$F(q) = \int_0^{\infty} \varrho(r)j_0(qr)r^2 \, dr \qquad (101)$$

The validity for electron scattering has been discussed in connection with Eq. (97), and from Fig. 24 or Fourier transformation of Eq. (93) it is evident that the muonic kernels specify only a limited range of very low Fourier components of the density. Taking the approach of Friedrich and Lenz and constructing the envelope of form factors of all distributions of the form (99) that fit all available muonic and electron scattering data should yield a highly model-independent result between $q = 0$ and the maximum experimental momentum transfer.

In summary, it appears that attempts to define model-independent functionals parameterized by a single, continuous variable are of very limited utility. One already knows from Eqs. (83) and (90) how to accurately specify the information content of each measurement. To define an auxiliary model-dependent functional of the density is useless, since the ultimate intent is to make meaningful statements about the density. It is far better to make a direct model-dependent analysis of the density, focusing on imposing physical model constraints, rather than first determining an auxiliary model-dependent functional which is only indirectly related to the density and then attempting in a second step to obtain the density from this functional. The one sufficiently model-independent function to be of use is the form factor $F(q)$. Its full utility is not realized, however, unless all available electron scattering data and muonic X-ray data for a given nucleus are *simultaneously* analyzed, and traditionally, experimentalists in the fields of electron scattering and muonic X-rays have analyzed their data separately.

3.2. Determination of Densities, Errors, and Error Correlations in Coordinate Space

One of the primary aims of studying the nucleus with electromagnetic probes is to make definitive statements about the charge density in coordinate space. Unfortunately, given a finite number of measurements of elastic electron scattering cross sections and muonic transition energies, the density distribution cannot be determined in a model-independent way. Although this is obvious from the Born approximation, where lack of data beyond the maximum momentum transfer of the experiment q_{max} immediately implies total ignorance about all Fourier components beyond q_{max}, it is perhaps useful to consider a concrete example simply to dramatize the problem. Figure 26 shows two distributions for ^{208}Pb that yield identical observables for the six lowest muonic transition energies and for the energies and angles of the Stanford electron scattering data, in the sense that the total χ^2 calculated from the ratio of the deviations between the results for the two densities divided by the corresponding experimental errors is less than 0.005. The wildly fluctuating density is, of course, physically unreasonable, but the wavelength of the undetermined fluctuation is not

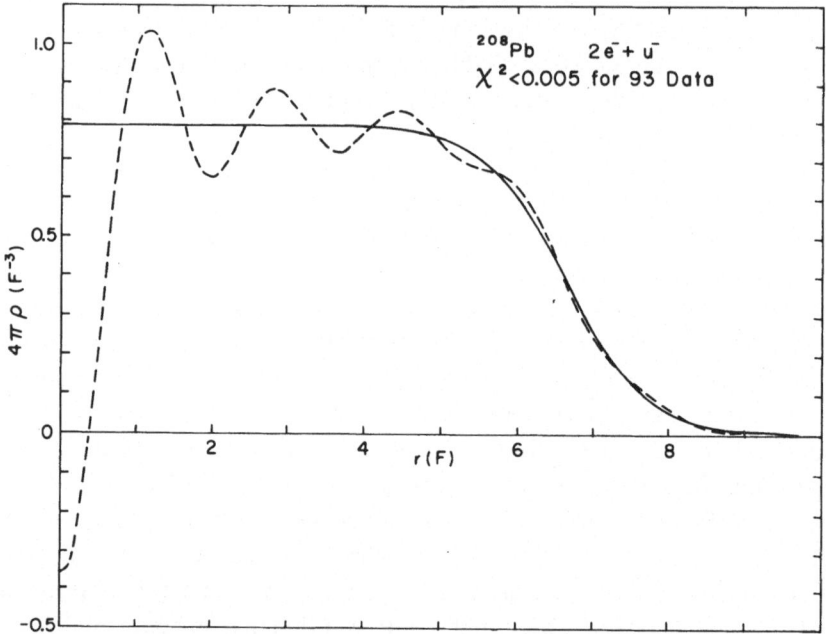

Fig. 26. Two ^{208}Pb distributions which are equivalent for muonic X-ray transition energies and electron scattering cross sections up to $q_{max} = 2.73$ fm^{-1}.

that much smaller than that of the realistic quantum density fluctuations in Fig. 17 which we would like to study experimentally. Also, as long as such large-amplitude fluctuations cannot be excluded, the density $\varrho(r)$ at any particular radius r is very poorly determined. This emphasizes the fact that we must give up all pretense of model independence, and rather address the difficult problem of formulating the most physical model dependence.

3.2.1. Specification of Model Assumptions

There are several alternative philosophies underlying the model assumptions which have been used in analyzing electron scattering and muonic data. One extreme view is to focus on the success of the best theoretical distributions reviewed in Section 2. Thus, one simply asks, given those features of the density that are actually determined by the data, e.g., the low Fourier components, how much these features must be changed to bring the theoretical density into optimal agreement with the data and what are the uncertainties in these changes due to statistical experimental errors. All the features not explicitly measured are then still specified by the theory. The opposite extreme view is to free the analysis from the quantitative predictions of any specific theory, and rather attempt to guess the behavior of the form factor beyond q_{max}. There are, of course, intermediate possibilities between these extremes, such as using theoretical calculations to guide the parameterization of the form factor beyond q_{max}.

In discussing specific implementation of model constraints, it is useful to think of the problem in terms of knowing the Fourier transform of $r^2\varrho(r)$ at a finite number of momenta. Although we will eventually formulate the problem in terms of the exact linear kernels, Eqs. (83) and (90), some general features are more obvious in terms of the Fourier transform, and we have argued above that the Fourier transform embodies the physical information in an essentially model-independent way.

Borysowicz and Hetherington (BH 73) emphasized three relevant general qualitative relations for Fourier transforms. Assume that the Fourier transform of $\varrho(r)$ is known in increments of Δq up to a maximum momentum transfer of q_{max}. Then the maximum distance probed by the experiment is

$$r_{max} = \pi/\Delta q \qquad (102)$$

Conversely, if $\varrho(r)$ is only known out to r_{max}, the smallest momentum scale over which we have significant resolution is Δq. Thus, if we assume that

$\varrho(r)$ vanishes for $r > R$, and R is less than r_{max}, we are essentially assuming a smooth interpolation of the form factor between the discrete values of q for which it is known. This is a specific example of how a tacit assumption about the density manifests itself in smooth interpolation between discrete values of a functional of the density, as mentioned in connection with Eq. (98).

The complimentary relation

$$q_{max} = \pi/\Delta r \qquad (103)$$

specifies the maximum momentum component that can be described on a mesh of spacing Δr or the smallest spatial scale over which a form factor terminating at q_{max} provides resolution. Of special interest is the wavelength corresponding to the last determined Fourier component $\lambda = 2\pi/q_{max}$ compared with the theoretically predicted wavelength of quantum density fluctuations $\lambda = \pi/k_F$. Taking $k_F = 1.35 \text{ fm}^{-1}$ as a characteristic value from Table II and noting that $\lambda \sim 2.3 \text{ fm}$ agrees well with the fluctuations shown in Fig. 17, we find that these quantum density fluctuations begin to be determined only if $q_{max} > 2k_F = 2.7 \text{ fm}^{-1}$. This is very close to the maximum momentum transfer $q_{max} = 2.73 \text{ fm}^{-1}$ of the presently available ^{208}Pb data.

The final general relation specifies the number M of independent functions of r on the interval $(0, R)$ that contribute to $F(q)$ below q_{max}:

$$M = q_{max} R/\pi \qquad (104)$$

This can be shown from Eqs. (102) and (103) and is obvious in the case of a Fourier–Bessel expansion on the interval $(0, R)$ discussed subsequently.

In practice, model constraints are imposed by specifying restrictions on the density $\varrho(r)$ and the form factor $F(q)$ defined in Eq. (101). The first extreme philosophy mentioned above of only changing those features of the charge distribution that are actually determined by the data is implemented by Friar and Negele (FN 73a, FN 75) by writing

$$\varrho(r) = \varrho_{th}(r) + \sum_{i=1}^{M} C_i \, \Delta\varrho_i(r) \qquad (105)$$

where $\varrho_{th}(r)$ is a theoretical charge density distribution and $\Delta\varrho_i(r)$ are M independent functions which vanish beyond $r = R$ and are strongly determined by the available experimental data. The selection of R, M, and $\Delta\varrho_i$ is discussed in detail in Section 3.2.2. The alternative extreme philosophy of imposing minimal restrictions directly on $\varrho(r)$ and instead specifying the behavior of $F(q)$ beyond q_{max} is most fully implemented by Hetherington and Borysowicz (HB 74). In this method, discussed in Section 3.2.3, one assumes an error bound on $F(q)$ from q_{max} to infinity that decreases exponentially or as some power of $1/q$. Although one knows that asymptotically $F(q)$ must decrease at least as fast as q^{-4} for physically acceptable density distributions, there is, of course, no guarantee that this behavior begins at q_{max}. Because the actual data below q_{max} exist only at a finite number of discrete points, according to the relation (102) some restriction must still be placed directly on $\varrho(r)$, such as requiring that ϱ vanishes or decreases exponentially beyond some R. These two extreme philosophies approach the same result in the limit in which q_{max} goes to infinity, and fortunately, even for the values of q_{max} realized in present experiments, the densities and error envelopes do not differ drastically.

Model constraints intermediate between these extremes involve expanding $\varrho(r)$ in a restricted set of functions and, in some cases, adding fictitious data beyond q_{max}. Sick (Sic 74) replaced the δ-shell expansion (99) by expansion in a sum of Gaussians (SOG). To ensure that the slope vanishes at the origin, identical Gaussians are placed at the positions $\pm R_i$:

$$\varrho(r) = \sum_{i=1}^{N} A_i \{\exp[-(r - R_i)^2/\gamma^2] + \exp[-(r + R_i)^2/\gamma^2]\} \quad (106)$$

As with the δ-shell expansion, A_i and R_i are generated as random numbers and only distributions yielding χ^2 per data point less than one are considered. The model dependence is contained in the restriction $R_i < R$, $A_i > 0$ which enforces smooth interpolation between discrete values of q_i, and in the width γ of the Gaussians, which strongly restricts the high Fourier components that can appear in $\varrho(r)$. In contrast to the δ-shell case, where the number of terms N actually limited the maximum Fourier components that could be generated, N does not specify any model restrictions in Eq. (106) as long as it is large enough. The most conservative value for γ is the width imposed by the proton size, but in practice given the values of q_{max} in present experiments, a larger value must be assumed. Since Hartree–Fock densities are generated by sums of squares of single-particle wave functions, it is argued that larger values of γ characteristic of the width of lobes in single-particle wave functions are justified. Unfortunately, it is difficult to quantify any more precisely the theoretical validity of a specific value of γ, either with respect to the ability to accurately reproduce theoretically reasonable density distributions or the ability to reject theoretically unreasonable ones. Nevertheless, the self-contained prescription offered by (106) is very con-

venient, and has been applied to a variety of cases reviewed by deJager
et al. (JDD 74).

Similar model constraints may be imposed by a truncated expansion in
a relevant set of basis functions. The most natural choice, given that one
measures the form factor $F(q)$, is a Fourier–Bessel series:

$$\varrho(r) = \sum_{n=1}^{M} C_n j_0\left(\frac{n\pi r}{R}\right)\theta(R - r) \tag{107}$$

As with the other procedures, the θ function requires that the density
vanish beyond a maximum radius R, and M is essentially specified by Eq.
(104). This truncated expansion assumes that essentially no Fourier components beyond q_{max} exist. If q_{max} is so low that the truncation of higher
Fourier components is unrealistic, one may either introduce them explicitly
from a theoretical model by using Eq. (105) with $\Delta\varrho_i = j_0(n\pi r/R)\theta(R - r)$,
as in (FN 73a), or by constructing enough fictitious data beyond q_{max} to
build in the desired Fourier components, as in (Dre+ 74). Besides the
Fourier–Bessel series in Eq. (107), Borysowicz and Hetherington (BH 73)
have investigated a variety of alternative expansion functions on the interval
$(0, R)$, including cosines, spline, and unspline functions.

3.2.2. Linear Expansion in a Truncated Basis

We now assume that the density is expanded in terms of M linearly
independent functions $f_i(r)$

$$\varrho(r) = \sum_{i=1}^{M} C_i f_i(r) \tag{108}$$

and that there are N experimental constraints

$$\sigma_\alpha = \int \varkappa_\alpha(r)\varrho(r)r^2\,dr \tag{109}$$

with experimental values σ_α^{exp} corresponding to the measured values of
normally distributed variables with standard deviations ε_α. The linear
kernels $\varkappa_\alpha$ are given by Eqs. (83) and (90) and we adopt the convention
that Latin indices refer to expansion coefficients and Greek subscripts
label data points. Expansions of the form of Eq. (105) are trivially included
by solving for $\varrho(r) - \varrho_{th}(r)$ and subtracting σ_α^{th} from σ_α^{exp}.

For simplicity, following Borysowicz and Hetherington (BH 73), we
first treat the case in which the constraint $\int \varrho(r)r^2\,dr = Z$ is not imposed
exactly, but rather is regarded as one of the N constraints with a fictitious

finite but very small standard deviation. Substituting Eq. (108) into Eq. (109) and defining

$$W_{\alpha i} \equiv \int \varkappa_\alpha(r) f_i(r) r^2 \, dr \tag{110}$$

we find the total χ^2 to be minimized is

$$\chi^2 = \sum_{\alpha=1}^{N} \left[\left(\sigma_\alpha^{\exp} - \sum_{i=1}^{M} W_{\alpha i} C_i \right)^2 \Big/ \varepsilon_\alpha^{\;2} \right] \tag{111}$$

The coefficients C_i that minimize χ^2 are solutions to the equation, obtained from varying Eq. (111),

$$\sum_{k=1}^{M} b_{jk} C_k = d_j \tag{112}$$

where

$$b_{jk} = \sum_{\alpha=1}^{N} \frac{W_{\alpha j} W_{\alpha k}}{\varepsilon_\alpha^{\;2}} \qquad \text{and} \qquad d_j = \sum_{\alpha=1}^{N} \frac{\sigma_\alpha^{\exp} W_{\alpha j}}{\varepsilon^{\;2}}$$

This equation is trivially solved by matrix inversion.

The covariance matrix of the C-coefficients is determined by considering an ensemble of measurements of the observables $\sigma_\alpha^{\exp}$, which are each independently normally distributed about their mean values $\langle \sigma_\alpha \rangle$. Using the linear relation (112) between the C's and the σ's and denoting mean values by brackets, the covariance matrix is

$$\delta^2 C_{ij} \equiv \langle (C_i - \langle C_i \rangle)(C_j - \langle C_j \rangle) \rangle$$
$$= \sum_{k,l=1}^{M} b_{ik}^{-1} b_{jl}^{-1} \langle (d_k - \langle d_k \rangle)(d_l - \langle d_l \rangle) \rangle \tag{113}$$

where

$$\langle (d_k - \langle d_k \rangle)(d_l - \langle d_l \rangle) \rangle$$
$$= \sum_{\alpha,\beta=1}^{N} \frac{W_{\alpha k}}{\varepsilon_\alpha^{\;2}} \frac{W_{\beta l}}{\varepsilon_\beta^{\;2}} \langle (\sigma_\alpha^{\exp} - \langle \sigma_\alpha^{\exp} \rangle)(\sigma_\beta^{\exp} - \langle \sigma_\beta \rangle) \rangle = b_{kl}$$

It follows that the correlated error in the density is

$$\langle (\varrho(r) - \langle \varrho(r) \rangle)(\varrho(r') - \langle \varrho(r') \rangle) \rangle$$
$$= \sum_{i,j=1}^{M} f_i(r) f_j(r') \, \delta^2 C_{ij} = \sum_{i,j=1}^{M} f_i(r) f_j(r') b_{ij}^{-1} \tag{114a}$$

and that the error in any integral of the density $\mathcal{M} = \int_0^\infty M(r)\varrho(r)r^2\,dr$ is

$$\langle(\mathcal{M} - \langle\mathcal{M}\rangle)^2\rangle = \sum_{i,j=1}^M \left[\int M(r)f_i(r)r^2\,dr\right]\left[\int M(r)f_j(r)r^2\,dr\right]b_{ij}^{-1}$$

(114b)

The diagonal elements of this function yield a statistical estimate of the error in $\varrho(r)$, and it is convenient to define

$$\Delta\varrho(r) = [\langle(\varrho(r) - \langle\varrho(r)\rangle)^2\rangle]^{1/2}$$

(115)

which yields an envelope enclosing all distributions that would be generated by fitting data that were normally distributed about each of the mean values $\langle\sigma_\alpha\rangle$ with standard deviation ε_α. Not all curves within this envelope correspond to physical densities, however (for example, they may not have charge Z), so the error correlations described by the off-diagonal elements in Eq. (114) are also significant.

There is some advantage to changing to a representation that diagonalizes the matrix b_{ij}. If U_{mn} is an orthogonal transformation that diagonalizes b_{ij}, then the new basis

$$g_i(r) = \sum f_j(r)U_{ji}$$

(116)

has two extremely useful properties. One is that the diagonal error matrix (115) is simply the sum of the squares of the basis functions g_i weighted by the inverse of the eigenvalue. The second is that this basis is the optimal one in which to state unambiguously what is actually determined by a given set of data and to formulate the truncation of the expansion. If the functions g_i are normalized consistently, then ordering these functions in order of decreasing eigenvalues of b automatically orders them in decreasing degree of experimental determination. Furthermore, these functions are statistically independent, so that if the basis size M is sufficiently large, $g_1(r)$ is the very best determined component of the density, $g_2(r)$ is the next best independently determined component, and so on. For some i, the error associated with g_i becomes prohibitively large, indicating that no further information is provided concerning the density. The $i-1$ functions below this point provide the optimal basis for the expansion of the density.

The preceding discussion did not rigorously enforce the normalization of the density to charge Z, although the assignment of a sufficiently small error to the normalization may be adequate for many practical applications.

To enforce the constraint exactly, we require

$$\sum_{k=1}^{M} \alpha_k C_k = \beta \tag{117}$$

where $\alpha_k = \int f_k(r)r^2\,dr$, and where β is equal to Z if we are considering the expansion of $\varrho(r)$ in Eq. (108) and β is zero if we are expanding the difference between ϱ_{th} and ϱ in Eq. (105). Then, introducing a Lagrange multiplier λ and minimizing $\chi^2 + \lambda \sum_{k=1}^{M} \alpha_k C_k$, we obtain the equation

$$\sum_{k=1}^{M} b_{jk} C_k + \lambda \alpha_j = d_j \tag{118}$$

which must be satisfied simultaneously with Eq. (117). These equations may be compactly expressed in the form of an $M + 1$ by $M + 1$ system of equations by defining

$$C_0 \equiv \lambda, \qquad d_0 \equiv \beta, \qquad b_{0i} \equiv b_{i0} \equiv \alpha_i, \qquad b_{00} \equiv 0 \tag{119}$$

With this notation, we obtain

$$\sum_{k=0}^{M} b_{jk} C_k = d_j \tag{120}$$

which is of the same form as Eq. (112) with one additional dimension, and may straightforwardly be solved by matrix inversion. It is shown by Friar and Negele (FN 73a) that the formula for the error matrix in this case is identical to Eq. (114), where b_{ij}^{-1} is now the inverse of the $M + 1$ by $M + 1$ augmented b_{ij} matrix defined above and the sums still run from 1 to M.

Depending upon the method by which model assumptions are introduced, the optimal truncation of the basis and the physical meaning of the error envelope differ significantly. Although we have deferred a review of results to Section 4, two concrete examples from the literature will serve to clarify these differences.

First, consider the case in which the deviation $\varrho(r) - \varrho_{th}(r)$ is expanded in a Fourier–Bessel series, for which there are two parameters to be determined: R, the distance beyond which $\varrho(r)$ is required to be equal to $\varrho_{th}(r)$, and M, the number of terms in the expansion. For given values of M and R, there are three distinct kinds of errors present in the density extracted from the analysis. One source of error is systematic errors in the data, and we know of no satisfactory general means of treating such errors. In special cases, much may be learned by simulating various likely forms of systematic errors, and examples will be discussed in Section 4. A second source of error is uncertainty in the expansion coefficients C_i originating from statistical errors in the data, and this statistical error is quantitatively described by the statistical error envelope $\Delta\varrho(r)$ in Eq. (115). A third source of error is what we shall refer to as the completeness error originating from the fact that the true density cannot be represented exactly by $\varrho_{th}(r)$ plus a finite Fourier–Bessel series. If $\varrho_{th}(r)$ is a realistic approximation to $\varrho(r)$, then we may obtain an estimate of the completeness error by evaluating the error resulting from approximating $\varrho_{th}(r)$ by an M-term series. Thus we define

$$\Delta\varrho_M{}^c(r) = \left| \varrho_{th}(r) - \sum_{I=1}^{M} C_I \frac{1}{r} \sin\frac{I\pi r}{R} \right|$$

(121)

$$C_I = \frac{2}{R} \int_0^R \left(\sin\frac{I\pi r}{R} \right) r\varrho_0(r)\, dr$$

This definition actually expresses the error introduced by attempting to expand the entire density, instead of $\varrho(r) - \varrho_{th}(r)$ in a Fourier–Bessel series. In cases in which $\varrho_{th}(r)$ accurately describes the high Fourier components of $\varrho(r)$, the true completeness error is, in fact, much less than this estimate. Alternatively, one could imagine cases in which $\varrho_{th}(r)$ actually has higher Fourier components out of phase with the true $\varrho(r)$, so that in some places, the true error is of the order of twice the estimate in Eq. (121). In any event, the zeros of $\Delta\varrho^c(r)$ have no significance, and we shall regard a smooth envelope enclosing the maxima of $\Delta\varrho^c(r)$ as a rough estimate of expected completeness error.

The statistical error envelope, Eq. (115), and the completeness error, Eq. (121), are shown in Fig. 27, taken from the ^{12}C analysis of Friar and Negele (FN 75) for various values of M with $R = 6$ fm. Considering first the statistical errors, it is evident that for small enough M, the available data will determine the C's extremely accurately, yielding a very small error envelope. For values of M in the range between 3 and 7 the resulting error envelopes are contained between the solid curves labeled 3 and 7 in the figure, yielding a broad plateau over which the statistical error is very slowly increasing. Adding one additional coefficient, by increasing M to 8, significantly increases the statistical error. This is easily understood in terms of Eq. (114), since $Rq_{max}/\pi = 6.9$, indicating that although the seventh coefficient is well determined, the eighth coefficient is not. Thus the plateau terminates abruptly beyond the maximum value specified by Eq. (114). The completeness error, on the other hand, steadily decreases

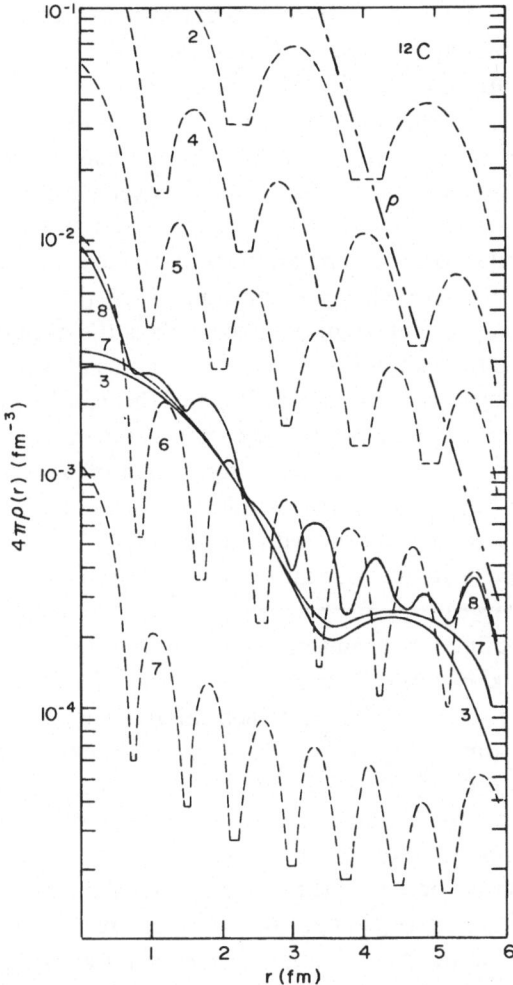

Fig. 27. Statistical error envelopes (solid curves) and completeness error (dashed) curves for ^{12}C using different numbers of Fourier–Bessel coefficients and $R = 6$ fm. The statistical error envelopes for 4, 5, and 6 coefficients are bounded by the curves for 3 and 7 coefficients, and the total density is indicated by the dash-dot curve.

as more terms are included in the Fourier–Bessel expansion, as indicated by the dashed curves in Fig. 27, where the minima have been artificially filled in for the sake of clarity.

The optimal choice of M is determined by the balance between uncertainties arising from statistical and completeness errors. Clearly, too small a value of M introduces unreasonably large completeness errors, whereas too large a value of M introduces unreasonably large statistical errors. The natural choice of $M = 7$ from Eq. (114) is seen to be ideal for the case represented in Fig. 27. The completeness errors for $M = 7$ are roughly an order of magnitude smaller than the statistical errors, yielding a wide margin of safety in the event that the estimate using $\varrho_{th}(r)$ in (121) is not truly representative of the errors in expanding the true density.

The optimal value of R is also apparent from this analysis. Clearly, once the statistical errors are comparable to $\varrho(r)$, the available experimental data are incapable of providing any meaningful information other than the fact that the density is consistent with zero. Thus the minimum value of R for which useful information is not lost is the radius at which the statistical error becomes comparable to the density, and this can be estimated by the point at which the statistical error becomes comparable to $\varrho_{th}(r)$. From Fig. 27, we observe that this occurs in the region of 6 fm and in practice, results are rather insensitive to reasonable variations of R about this value.

It was noted in connection with Eq. (116) that expansion in eigen-functions of the error matrix was in principle preferable to a Fourier–Bessel expansion. In terms of Fig. 27, this means that the statistical error envelope obtained by an optimal M-term series would always be smaller than that of the corresponding M-term Fourier–Bessel series. This has been checked by the authors for the case of ^{208}Pb, which should be a much more stringent test than ^{12}C, by comparing the Fourier–Bessel error envelopes in the region of the optimal value of M, which was 11, with the corresponding error envelopes obtained by retaining the same number of terms in the basis obtained by diagonalizing the error matrix in the space of 20 Fourier coefficients. The qualitative nature of the plateau remains the same in the optimal basis, and the quantitative changes for $M = 11$ are negligible. This is expected by virtue of the fact that the higher Fourier components are so poorly determined and constitutes a very strong argument for the use of the Fourier–Bessel basis. Other bases, such as the expansion in terms of Laguerre polynomials multiplied by $e^{-\lambda r}$ discussed by Friar and Negele (FN 73a), turn out to be intractable unless the error matrix is explicitly diagonalized.

The situation is quite different in the case in which one introduces fictitious data beyond q_{max}. As in the previous case, the systematic error still cannot be described in a satisfactory general way. The statistical error and completeness error, however, merge into a single composite error as more and more data are introduced. This is particularly clear in the results of the ^{3}He analysis of Borysowicz and Heatherington (BH 73), shown in Figs. 28 and 29. Using only the actual experimental data extending to

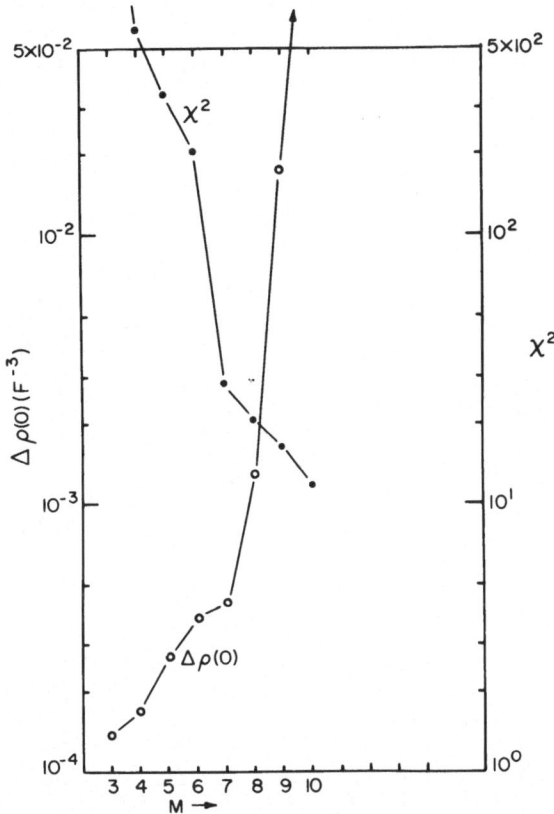

Fig. 28. Plots of χ^2 and $\Delta\varrho(0)$ for ^{3}He as a function of M, using only the experimental data extending to $q_{max} = 4.5$ fm^{-1}.

$q_{max} = 4.5$ fm^{-1} and choosing a cosine basis with $R = 5$ yields the values of $\Delta\varrho(0)$, the statistical error envelope at the origin, and total χ^2 shown in Fig. 28. For this case, $q_{max}R/\pi = 7.1$, and we observe that the end of the $\Delta\varrho$ plateau occurs at $M = 7$, as expected, and that the total χ^2 is of the order of the number of degrees of freedom. Fictitious data are now introduced between $q_{max} = 4.5$ fm^{-1} and $q_{fict} = 8.8$ fm^{-1} defined such that $F(q_i) = 0$ to within a standard deviation $\varepsilon_i = F_0 \exp[-A(q_i - q_{max})]$. The coefficient F_0 is chosen comparable to the form factor at q_{max} and A continues the average exponential decay of the measured form factor. The results, shown in Fig. 29, are virtually identical to the previous results up through $M = 7$. Beyond this point, however, $\Delta\varrho(0)$ approaches a new plateau, which extends up to $M = 12$. Using Eq. (114), we would not necessarily expect the plateau to terminate before $M = q_{fict}R/\pi = 14$, but since the cosine

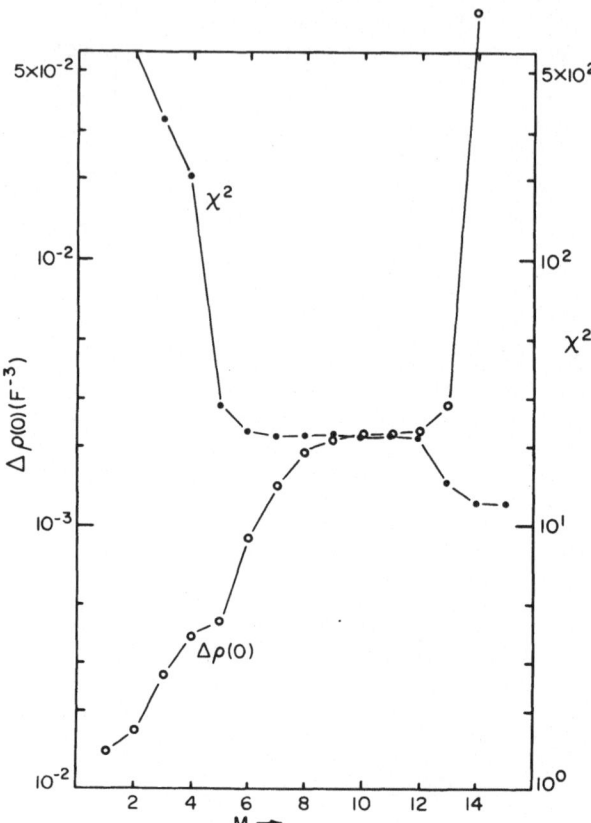

Fig. 29. Plots of χ^2 and $\Delta\varrho(0)$ for ^{3}He as a function of M, with fictitious data added up to $q_{\text{fict}} = 8.8$ fm^{-1}.

basis is not quite optimal, it actually terminates at 13. The strong implication of Fig. 29 is that if q_{fict} is chosen large enough, the completeness error from truncation of the expansion at $q_{\text{fict}}R/\pi$ can be made arbitrarily small while $\Delta\varrho(0)$ approaches some finite asymptotic value. That this is indeed the case is shown rigorously in Section 3.2.7.

Two distinctly different error statements can be deduced from Fig. 29. The error envelope $\Delta\varrho_7(r)$ for $M = 7$ is the same as the statistical error envelope discussed previously, and specifies the errors associated with those components of the density that are actually determined by the available data and arising from the actual statistical errors in that data. The increase in the error envelope from $M = 7$ to $M = 12$ reflects the additional uncertainty specified by the assumed exponentially decreasing errors, and, clearly, different assumptions will yield different values of $\Delta\varrho_{12}$. In practice, if the fictitious data roughly correspond to the form factor calculated from $\varrho_{th}(r)$, then the difference between these two different error envelopes will be comparable to the completeness error estimate given in Eq. (121). Since this correspondence is not exact, it would be extremely desirable for analyses making uses of fictitious data to calculate both error envelopes, so that one could accurately assess the size of the errors actually specified by real data and the size of errors arising from arbitrary assumptions.

3.2.3. Basis-Independent Analyses

The results in Fig. 29 are highly suggestive that sufficiently strong assumptions on $\varrho(r)$ and $F(q)$ enable formulation of the problem in a basis-independent fashion. Such a rigorous treatment has been given by Hetherington and Borysowicz (HB 74). The starting point is the definition of an augmented χ^2:

$$\chi^2 = \sum_\alpha \frac{(\sigma_\alpha - \langle \varkappa_\alpha | \varrho \rangle)^2}{\varepsilon_\alpha^2} + \int_0^\infty V(r')\langle r' | \varrho \rangle^2 \, dr' + \int_0^\infty T'(q)\langle q | \varrho \rangle^2 \, dq \quad (122)$$

where elements in the linear vector space are denoted $| x \rangle$ and the density $| \varrho \rangle$ can be represented either in coordinate space by $\langle r | \varrho \rangle = \varrho(r)$ or in momentum space by $4\pi F(q) = \langle q | \varrho \rangle$, using the normalization in Eq. (101). The error functions $V(r)$ and $T'(q)$, the notation for which will become obvious soon, are continuous generalizations of the constraints used previously. Since the data provide only discrete constraints below q_{max}, by Eq. (102) some restriction must be placed on $\varrho(r)$ for large r. Defining $V(r)$ to be zero inside R and infinite outside R reproduces the old constraint that $\varrho(r)$ vanish beyond R since any finite $\varrho(r)$ outside of R would cause infinite χ^2. Continuous cutoffs, such as exponential increase of $V(r)$, are also possible and are physically more reasonable. The previous introduction of fictitious data corresponds to defining $T'(q) = \sum_j [\delta(q - q_j)/\varepsilon_j^2]$, where ε_j decreases exponentially or as some power of $1/q_j$. This requirement is generalized by defining $T'(q)$ as a continuous function which increases exponentially or as some power. In general, one does not have complete freedom to independently specify $T'(q)$ and $V(r)$, and general compatibility requirements are not presently known. Two sets of compatible conditions, sufficient for cases of practical interest, are given (HB 74): (i) If $V(r)$ equals infinity beyond $r = R$, then $T'(q)$ must go to infinity slower than an exponential, but may increase faster than any power of q. (ii) If $V(r)$ goes to infinity exponentially or slower, then $T(q)$ may go to infinity exponentially or slower.

Variation of (122) with respect to $\langle r \mid \varrho \rangle$ and insertion of integrals over $\mid r' \rangle \langle r' \mid$ where necessary yields

$$\sum_\alpha \frac{\sigma_\alpha \langle r \mid \varkappa_\alpha \rangle}{\varepsilon_\alpha^{\,2}} = \sum_\alpha \frac{\langle r \mid \varkappa_\alpha \rangle \langle \varkappa_\alpha \mid \varrho \rangle}{\varepsilon_\alpha^{\,2}} + V(r)\langle r \mid \varrho \rangle$$

$$+ \int_0^\infty \langle r \mid q \rangle T'(q) \langle q \mid \varrho \rangle \, dq \qquad (123)$$

Since this must hold for all r, it is equivalent to the operator equation

$$\sum_\alpha \mid \varkappa_\alpha \rangle \frac{\sigma_\alpha}{\varepsilon_\alpha^{\,2}} = H \mid \varrho \rangle \qquad (124)$$

where

$$H = \sum_\alpha \frac{\mid \varkappa_\alpha \rangle \langle \varkappa_\alpha \mid}{\varepsilon_\alpha^{\,2}} + T' + V \equiv \sum_n \frac{\mid n \rangle \langle n \mid}{S_n^{\,2}}$$

In the case that the kernels K_i are just Fourier components, then $\sum_\alpha (\mid q_\alpha \rangle \langle q_\alpha \mid / \varepsilon_\alpha^{\,2}) + T'$ is represented in momentum space, reminiscent of the kinetic energy, and V is represented in coordinate space, in analogy with the potential energy, thus motivating the notation H, T', and V. Diagonalization of H yields the eigenfunctions of the error operator denoted $\mid n \rangle$ above. Inversion of H yields the density

$$\mid \varrho \rangle = H^{-1} \sum_\alpha \frac{\mid \varkappa_\alpha \rangle \sigma_\alpha}{\varepsilon_\alpha^{\,2}} = \sum_n \mid n \rangle \Big(\sum_\alpha \langle n \mid \varkappa_\alpha \rangle \frac{\sigma_\alpha}{\varepsilon_\alpha^{\,2}} S_n^{\,2} \Big) \qquad (125)$$

and the correlated error in the density, analogous to Eq. (114), is given by

$$\langle [\varrho(r) - \langle \varrho(r) \rangle][\varrho(r') - \langle \varrho(r') \rangle] \rangle = \sum_n \langle r \mid n \rangle S_n^{\,2} \langle n \mid r' \rangle \qquad (126)$$

The conceptual advantage of this formulation is its independence of basis, even though in (HB 74), Eq. (124) is solved in a suitable large basis. One disadvantage is that statistical errors arising from the data are not distinguished from those arising from the assumptions. In fact, this could be done by repeating the calculation with drastically unrealistic estimates of the errors beyond q_{max}, so that the error envelope arises essentially only from the actual experimental error.

The other basis-independent approach is the sum of Gaussian (SOG) expansion of Sick (Sic 74) mentioned previously. The parameterization (106) effectively requires that $F(q)$ is dominated by $\exp(-\gamma^2 q^2)$, and this, coupled with the requirement that $r < R$, is stronger* than the conditions imposed by Hetherington and Borysowicz (HB 74), so that it is reasonable that the theory does not require truncation in N. The interpretation of the error envelope containing all results yielding χ^2 less than 1 per degree of freedom is the same as that in (HB 74), and does not distinguish uncertainty arising from the data from that arising from the arbitrary choice of γ. This method has two significant disadvantages relative to that of Hetherington and Borysowicz (HB 74). First, whereas error correlations are readily calculated from Eq. (126), they are much more difficult to obtain from a set of distributions generated by random numbers. Second, if one really performs partial wave analyses, rather than simply calculating Fourier transforms, the former method requires only several partial wave calculations, even if the kernels are determined iteratively, whereas the latter requires a very large number to acquire satisfactory statistics.

4. RESULTS FROM ANALYSES OF DATA

The application of the methods of analysis discussed in Section 3 is considered in this section. The results for the charge densities of two nuclei will be discussed in detail. The nucleus ^{208}Pb has been selected, as in Section 2, because a heavy closed-shell nucleus provides a good testing ground for many-body techniques and ^{208}Pb has been extensively studied theoretically. Because of the high Z of this nucleus, muonic X-ray transitions impose strong constraints on the charge density, so this nucleus provides an ideal opportunity to simultaneously analyze electron scattering and muonic data.

The other nucleus we will discuss in detail, ^{12}C, provides a complementary case. Because it is a low-Z nucleus, muonic X-rays do not presently provide useful constraints. The electron scattering data for ^{12}C exist for quite large momentum transfer, in contrast to ^{208}Pb, and the ^{12}C cross sections exhibit a much more detailed structure than the relatively smooth ^{208}Pb cross sections. This difference is again a result of the high Z of Pb, which produces a large Coulomb distortion of the electron's wave function, and the higher order terms in the Born series fill in the deep diffraction minima which exist in first Born approximation.

---

* The incompatibility of a sharp cutoff in R with exponential limits on the density is irrelevant, since Eq. (124) is never solved.

4.1. Pseudodata

In order to demonstrate that a particular method of analysis is practical from an operational point of view, it is desirable to check the analysis procedure using cross-section and muonic atom "data" generated from a known charge distribution. Such pseudodata were used by Friar and Negele (FN 73a, FN 75) for ^{208}Pb and ^{12}C, by Dreher *et al.* (Dre+ 74) in their treatment of ^{208}Pb, and by Sick (Sic 74) in his analysis of ^{12}C and ^{32}S.

4.1.1. ^{12}C Pseudodata

The ^{12}C pseudodata used by Friar and Negele (FN 75) were generated using a density whose basic component was the two-parameter modified harmonic oscillator density (MHO) of Sick and McCarthy (SM 70). This MHO density was further modified by replacing the unphysical Gaussian tail of the distribution with a realistic tail. Outside the nuclear potential in the single-particle approximation, the wave function of a proton orbital is a Whittaker function (Sic 74, FN 75). This is approximately an exponential for the *p*-states of ^{12}C and the parameter in the exponential is simply related to the separation energy of the protons. The Whittaker tail was matched smoothly to the MHO density at 3.95 fm and the resulting density is denoted MHOT. Using the MHOT density, cross sections were calculated using a phase-shift code at angles and energies corresponding to real data taken at Stanford (SM 70) and Amsterdam (JPD 72), which will be discussed later. These cross sections were randomized by means of a random number generator, each of the pseudodata being chosen from a set of Gaussian-distributed numbers with a mean equal to the calculated cross section and with a variance equal to the standard deviation of the real data point. Altogether 25 independent sets of these pseudodata were generated with 62 data points in each set. The Fourier–Bessel expansion of $\varrho - \varrho_{\text{th}}$ described in Section 3 was then used on each of the data sets with a starting density ϱ_{th} equal to ϱ_{MHOT}. The truncation parameters $R = 6$ fm and $M = 7$ derived in Section 3 were used. The deviations of resulting densities from the original density ϱ_0 are plotted in Fig. 30, together with the statistical error estimate $\Delta\varrho(r)$ from Eq. (115), which is denoted by two solid lines. The error envelope is roughly constant beyond 3.5 fm, although the density is exponentially decreasing. On a statistical basis one would expect 31% of all densities (8 out of 25) to deviate from $\varrho(r)$ by more than one standard deviation [i.e., $\Delta\varrho(r)$] at any value of *r*. Furthermore, approximately 5% (1 out of 25) should lie outside two standard deviations. A check at intervals of 1 fm from the center of the nucleus

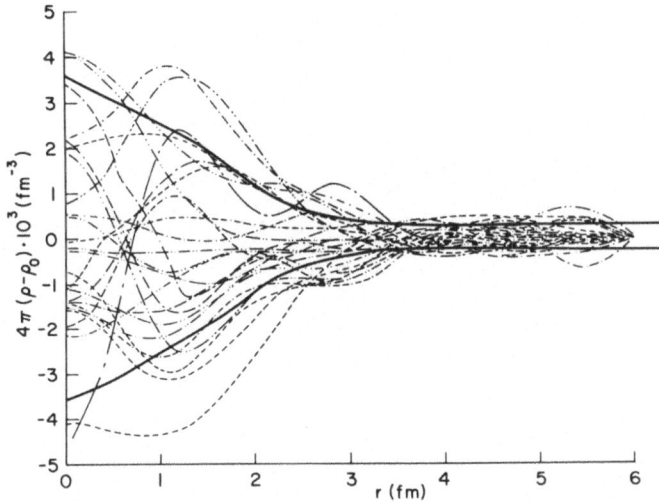

Fig. 30. Deviation of densities calculated for 25 sets of ^{12}C pseudodata from the density that generated the pseudodata. The solid lines denote the statistical error envelope $\Delta\varrho(r)$.

indicates that both of these expectations are fulfilled. One can also check the distribution of rms radii of these distributions and compare them with the rms radius of the original density ϱ_0. One finds that the average rms radius of these densities deviates only slightly from the original one and a calculation of the actual standard deviation of the rms radii differs by only 10% from the value obtained directly from the statistical analysis. Other tests also prove satisfactory, indicating that the error estimates do precisely what they are supposed to do: They provide an estimate of errors due to statistical fluctuations in the data.

Further information is provided by Fig. 31. The curves labeled ϱ_0 and $\Delta\varrho_0$ are typical results obtained from pseudodata. The solid curve labeled ϱ_0 is the density MHOT used to generate the pseudodata, and is indistinguishable on this graph from the fit to the pseudodata for r less than 4 fm. The centers of the error bars indicate the density calculated from the pseudodata. The statistical error envelope is denoted both by the solid curve marked $\Delta\varrho_0$ and by the error bars. Note that the density and error become roughly equal at 6 fm and that the error envelope is flat between 3.5 and 6 fm. We see that the density is quite accurately determined in regions where the density is 10^{-2}–10^{-3} of the central density. The same reproducibility of the low-density region is also obtained using other choices for ϱ_{th}. Thus, as expected from the estimates of the very small completeness error shown in Fig. 27, the Fourier–Bessel expansion works well even in

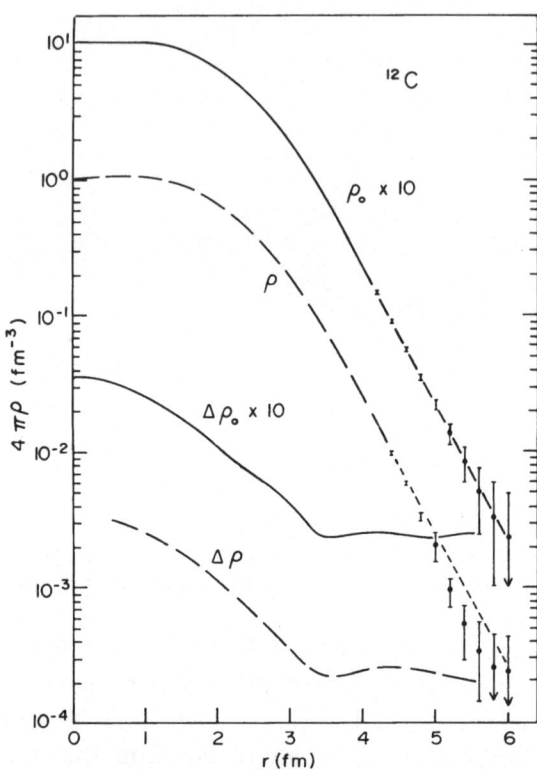

Fig. 31. Carbon densities and error envelopes obtained from a typical set of pseudodata (ϱ_0, displaced one decade) and from the experimental data with $\Delta\phi$ corrections (ϱ) using seven coefficients and a radius of 6 fm. The calculated density ϱ_0 is indicated by the centers of the error bars and inside 4 fm is indistinguishable from the exact density denoted by the solid line. The experimental density ϱ is indicated by the long-dashed curve inside 4 fm and by the centers of the error bars beyond 4 fm. The short-dashed curve is a Whittaker tail of the proper energy joined at 4 fm. The error envelopes are denoted both by the $\Delta\varrho$ curves and the magnitude of the error bars.

the low-density region where the density decreases approximately exponentially. Similar results are obtained for ^{208}Pb.

Sick generated pseudodata for the nucleus ^{32}S to test his sum of Gaussians (SOG) method. The cross sections were calculated using a phase shift code and a simple model density, and errors were assigned to these data corresponding to the experimental errors of real ^{32}S data. Unlike the pseudodata discussed above, his cross sections were not randomized and in this respect were similar to the ^{208}Pb pseudodata used by Friar and Negele (FN 73a), which will be discussed subsequently. Roughly half of the densities obtained by fitting the strengths of 12 Gaussians at randomly chosen positions had a χ^2 of less than 0.1 for 50 degrees of freedom. The distribution of the positions of the Gaussians for the remaining cases was unfavorable for a good fit. This test shows that Sick's density parameterization had sufficient flexibility to reproduce cross sections calculated from the model density. The test also confirmed the accuracy of various technical aspects of his procedure.

4.1.2. ^{208}Pb Pseudodata

An analogous analysis of electron and muon pseudodata was made for ^{208}Pb by Friar and Negele (FN 73a). The pseudodata were generated using a particular choice of Fermi distribution for ϱ_0. Electron scattering cross sections corresponding to the energies and angles of the Stanford experiment (Hei+ 69) were calculated, as were the energies of the six muonic atom transitions, $2s_{1/2} \rightarrow 2p_{1/2}$, $2s_{1/2} \rightarrow 2p_{3/2}$, $2p_{3/2} \rightarrow 1s_{1/2}$, $2p_{1/2} \rightarrow 1s_{1/2}$, $E_\infty \rightarrow 2p_{3/2}$, and $E_\infty \rightarrow 2p_{1/2}$, for a total of 63 pseudodata. Errors were assigned to these pseudodata corresponding to the fractional experimental errors assigned by Heisenberg *et al.* (Hei+ 69) to the scattering data and from Table III for the muonic atom data. These data were *not* randomized and a good fit would correspond to $\chi^2 = 0$. On the basis of the maximum momentum transfer in the high-energy Stanford data set, one would expect M, the highest Fourier component, to be 10. Many different starting densities ϱ_{th} were tried, ranging from Fermi distributions to results from density-dependent Hartree–Fock calculations. From this investigation it was found that $R = 11.0$ fm satisfied the criterion of Section 3, and that $M = 11$ was a marginally better choice of cutoff than $M = 10$, since the increase in statistical error by adding the 11th coefficient was more than compensated by the significant decrease in completeness error. The 11th coefficient, whose error was roughly equal to its size, was fairly large and was an important contribution to the density. The need for including this coefficient results from the lack of data at sufficiently high q, since the data for Pb terminate at 2.73 fm^{-1}, in contrast to the higher q ^{12}C data. With these parameters, in all cases the resulting fitted densities $\varrho(r)$ were contained within the error envelopes shown in Fig. 32. Three representative choices of Fermi distribution parameters used in specifying ϱ_{th}, together with the original "exact" density used in generating the pseudodata, are shown in this figure. The oscillation in the error envelope is caused by the marginally determined 11th coefficient. The densities determined from the pseudodata have the same fluctuation as well, for the same reason, and the phase may have either sign. The case with the

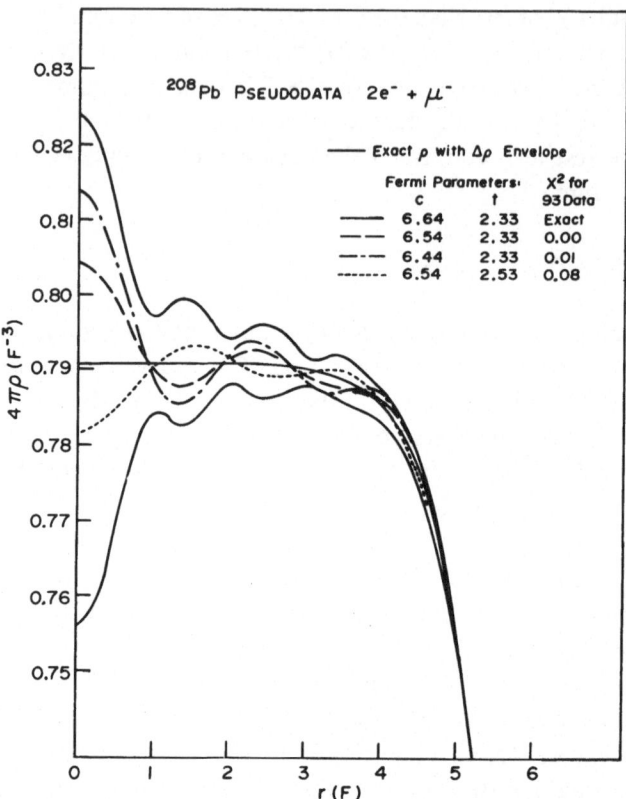

Fig. 32. Comparison of various fits to Pb pseudodata with the Fermi function that generated the pseudodata. In this and subsequent density graphs the normalization factor of 4π is written explicitly so that ϱ is the conventional density instead of that given in Eq. (31).

negative fluctuation at the origin is produced by an 11th coefficient considerably smaller than the other two cases and this shows up as a fluctuation of the opposite sign. These calculations suggest that the statistical error envelope actually bounds the completeness error resulting from our lack of knowledge of the higher Fourier components, and evaluation of Eq. (121) yields the same conclusion. The total χ^2, furthermore, is essentially zero, indicating a near perfect fit to the data.

The error envelopes for ^{208}Pb exhibit the same type of behavior as those of ^{12}C; the Pb envelopes are roughly constant for $r > 5$ fm and depend only weakly on the number of coefficients until one too many coefficients is added (i.e., M bigger than $q_{max}R/\pi$), when ringing begins to develop. As discussed in Section 3, these two features allow a consistent and reasonable determination of R and M.

The complementarity of data sets and the effect of systematic errors are most easily investigated with pseudodata. Table IV shows the χ^2 values for each of the two Stanford pseudodata sets, corresponding to electron energies of 248.2 and 502.0 MeV, and the muonic pseudodata set separately for fits using a variety of choices of ϱ_{th} and for a variety of cross-section normalization factors. It is quite difficult to experimentally determine the absolute cross section in an electron scattering experiment at the percent level, so that a given data set may have cross sections which are uniformly too high or low by a few percent. If this were the case in an experiment, published data would differ from the "correct" data by a normalization factor, if we ignore statistical effects. Six of the cases listed in Table IV have used the correctly normalized pseudodata and were used to investigate the way the various data sets complement one another. The remaining cases have had the pseudodata cross sections multiplied by the tabulated normalization factors before a fit was attempted.

The first three cases in Table IV are the three densities shown in Fig. 32 and lead to rms radii which are statistically indistinguishable from the original "exact" distribution. The effect of eliminating the muon data set

TABLE IV

Tabulation of χ^2 and rms Radii Obtained in Various Fits to the ^{208}Pb Pseudodata Using Fermi Distributions for ϱ_{th}

The distribution from which the pseudodata were generated is given in the first line of the table.

	c	t	Norm (248)	Norm (502)	$\chi_e^2(248.2)$ (37 data)	$\chi_e^2(502.0)$ (50 data)	χ_μ^2 (6 data)	rms radius
	6.64	2.33	1.00	1.00		Exact distribution		5.5071
1	6.54	2.33	1.00	1.00	0.00	0.00	0.00	5.5071 ± 0.0013
2	6.44	2.33	1.00	1.00	0.00	0.00	0.00	5.5071 ± 0.0013
3	6.54	2.53	1.00	1.00	0.02	0.04	0.02	5.5074 ± 0.0013
4	6.54	2.33	1.00	1.05	18.82	30.84	0.76	5.5061 ± 0.0013
5	6.54	2.33	1.05	1.00	46.82	23.50	0.35	5.5066 ± 0.0013
6	6.54	2.33	1.05	1.05	25.25	6.33	1.99	5.5057 ± 0.0013
7	6.54	2.33	1.00	1.00	0.00	0.00	—	5.5069 ± 0.0120
8	6.54	2.33	1.05	1.05	1.39	1.07	—	5.4425 ± 0.0118
9	6.54	2.33	1.00	1.00	0.00	—	—	5.5070 ± 0.0174
10	6.54	2.33	1.00	1.00	—	0.00	—	5.5068 ± 0.0323

and one of the two electron scattering data sets is demonstrated by cases 7, 9, and 10. Eliminating the muonic data primarily affects the lowest three or four of the Fourier coefficients. As we will show later, the rms radius (and other simple radial moments) depends primarily on the low Fourier coefficients. This is reflected in the tenfold increase in the error in the rms radius indicated for case 7. The radius itself has scarcely changed, however. Eliminating the high-energy data set leads to a somewhat greater uncertainty in the radius, while the effect on the error of the rms radius caused by elimination of the low-energy data is even larger. These results demonstrate the complementarity of the various types of data, and an examination of the Fourier coefficients and their statistical errors confirms this. The muonic data make a profound effect on the lowest few coefficients because of their great accuracy. They do not have a significant effect on the higher coefficients, which, as indicated in Section 3, depend primarily on electron scattering data with the appropriate large values of momentum transfer.

Adjusting the normalization produces the effects shown in Fig. 33 and which correspond to cases 4–6 in Table IV. Although this procedure leads to a substantial χ^2, the χ^2 per degree of freedom is still considerably less than one, and thus would be unimportant in the presence of genuine statistical fluctuations. The effect of increasing the normalization of the cross sections is to produce an overall positive slope to the interior density, which is usually referred to as a "wine-bottle shape." One effect of the normalization change is to decrease the rms radius, but the sensitivity of the muonic atom data to this parameter prevents much of a change. If, however, no muonic atom data are included (case 8 of Table IV), the effect is a sharp decrease of the radius and a very small χ^2. In effect, the muonic atom data provide an absolute scale against which the electron scattering data are measured. With muonic data included, an incorrect normalization does not change the radius much, but it will alter other features of the density in order to lower the χ^2 as much as possible, although the χ^2 still will be fairly high. When there is no such scale, the electron data can easily accommodate a substantial normalization error with an attendant shift in the rms radius, and this fit will be nearly perfect. Because the data sets overlap in momentum transfer, a different normalization for the two sets shows up as an enhanced χ^2. As indicated in the table, adjusting both sets by the same amount gives a lower χ^2.

The error correlation in the Pb density, $\langle \Delta\varrho(r) \Delta\varrho(r') \rangle$, reflects the uncertainty in the last coefficient. This uncertainty, in fact, dominates the correlation. If the density is too high at one maximum of the error envelope,

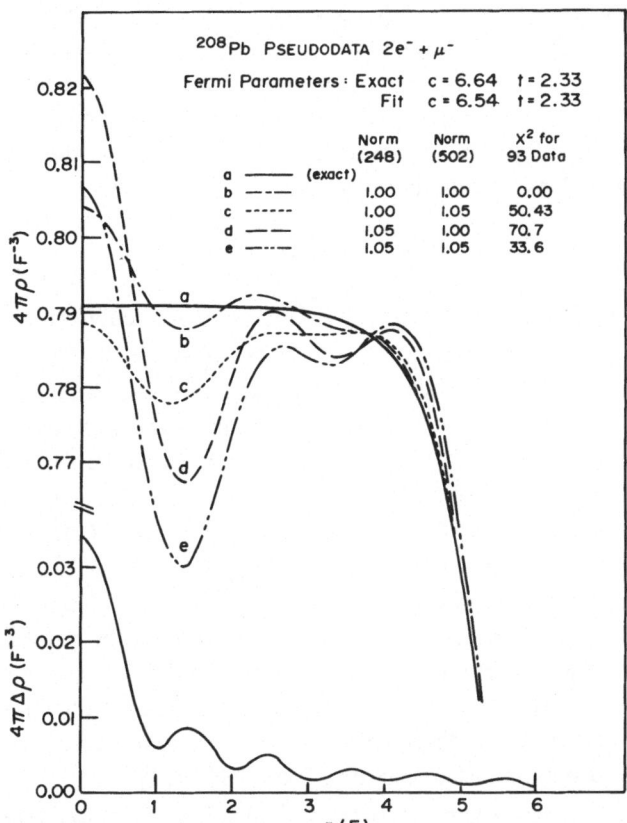

Fig. 33. Effect of normalization changes on the Pb density distribution obtained from the pseudodata. In this and subsequent density graphs, the vertical scale is broken so that the statistical error envelope $\Delta\varrho$ may be displayed at the bottom.

it will be too low at the next, and so on. Such an oscillatory structure is very evident in Fig. 32. Two of the three fits are out of phase with the third fit, but all three display the property discussed above. Since the densities deduced from the pseudodata have an oscillation, while the original "exact" distribution did not, one must be very careful to interpret any density fluctuations with reference to the size of the error envelope.

Dreher *et al.* (Dre+ 74) have investigated the statistical error envelope and the completeness error for ^{208}Pb using electron scattering pseudodata alone. In their investigation they used the simpler procedure of calculating form factors from their model density, rather than cross sections, and then analyzed the form factor pseudodata. The momentum transfers of their data corresponded to the Stanford experiment and unrandomized 1%

errors were assigned to the data. The analysis of such form factor data originally led Meyer–Berkhout *et al.* (MFG 59) to introduce the Fourier–Bessel expansion of the density, in addition to other expansions. The latter group made use of a property of the Fourier–Bessel density expansion, which is easily obtainable from Eq. (121). Evaluating the form factor expression at values of momentum transfer $q_N = N\pi/R$, one finds

$$C_N = 2q_N F(q_N)/R \tag{127}$$

This relationship was used directly (MFG 59) to calculate the density of ^{12}C. The question of errors in the density was not considered.

Because the error is such a crucial aspect of the complete analysis, Dreher *et al.* did not use Eq. (127) but minimized the χ^2 function in order to determine the Fourier coefficients and errors. Since the form factor is a linear function of the coefficients, a simple set of linear equations (normal equations) results for the C_N. For fixed cutoff radius R, they calculated densities corresponding to cutoffs $M = q_{max}R/\pi$, $M + 1$, and $M + 2$. This analysis confirmed the prior results of Friar and Negele (FN 73a). The statistical error band corresponding to M is small, the band for $M + 1$ is somewhat larger, particularly at the origin, and the error envelope for $M + 2$ is extremely large with an oscillatory structure reflecting the last Fourier coefficient. These three cases would correspond to $M = 10, 11, 12$ in the earlier discussion [see also Fig. 3 of (FN 73a)].

The completeness error, of course, is determined by those higher Fourier coefficients, such as C_{M+2} in the previous example, that are unspecified by the data. In order to examine this aspect of the data analysis, Dreher *et al.* specified the maximum size of the form factor F_{max} in the region $q > q_{max}$, and assumed a uniform random distribution of form factors within the limits $\pm F_{max}$. By means of Eq. (127), randomly generated form factors at a momentum transfer q_N generate randomly distributed Fourier coefficients C_N. A set of densities may be generated in this way with the higher Fourier coefficients distributed according to the size of F_{max}, and the distribution of these densities determines an "error" envelope which specifies not only statistical fluctuations, but model dependence as well. This procedure is analogous to the prescription of Borysowicz and Hetherington (BH 73).

It is quite clear that the results of such an analysis depend critically on the values of $F_{max}(q)$. This aspect of the density determination was thoroughly discussed in Section 3, and it was demonstrated in Fig. 26 that a single large Fourier component can make a contribution to the density much larger than the statistical error envelope. Dreher *et al.* assume a smooth dependence of F_{max} on the momentum transfer, which they obtain from asymptotic arguments. If the charge density obtained from the nucleon distribution is continuous, but has a discontinuous second derivative, such as one obtains for a particle in a square-well potential, the asymptotic form of the density for very large $\mathbf{q}^2$ is

$$F(\mathbf{q}^2) \sim C/\mathbf{q}^4 \tag{128}$$

where C is a constant. The maximum value of the form factor was specified by Dreher *et al.* as the product of F in Eq. (128) and a Gaussian nucleon charge form factor, with the constant chosen to produce continuity at $q = q_{max}$. If one further assumes continuous derivatives of ϱ, and that ϱ and all its derivatives vanish at infinity faster than $1/r$, the asymptotic expansion of the form factor can be shown to be (Lig 70)

$$F(\mathbf{q}^2) \sim \frac{-2\varrho'(0)}{\mathbf{q}^4} + \frac{4\varrho'''(0)}{\mathbf{q}^6} + \cdots \tag{129}$$

The reasonable assumption that $\varrho'(0)$ vanishes, which is true in single-particle models with smooth potentials, generates a $1/\mathbf{q}^6$ asymptotic form. This would further reduce the effect of the higher Fourier components on the density. Not too much weight should be given to the distinction between the two asymptotic forms, however. A single Fourier component can make a contribution larger than that of (128) or (129), since the region where asymptotic behavior sets in is not known.

The numerical results of Dreher *et al.* (Dre+ 74) using their pseudodata suggest that the model-dependent part of the complete error envelope is not too important compared to the statistical part, except in the center of the nucleus. A quantitative comparison depends on whether one takes M or $M + 1$ coefficients for the case of statistical error only, as we discussed above.

4.2. ¹²C Results

4.2.1. ¹²C Data

As discussed in Section 1.2, muonic transition data are not sufficiently precise at the present time to provide useful constraints on the charge density of ¹²C, so that information must be obtained from analyzing elastic electron scattering cross-section data alone. Because ¹²C is widely used as a laboratory standard in electron scattering work, there exist a fairly large number of data sets. Among the recent data are the two Stanford data sets of Sick and McCarthy (SM 70). Their low-energy set for 374.5-MeV electrons is comprised of 38 data with momentum transfers ranging from 1.04 to 2.77 fm^{-1}, while the high-energy set at 747.2 MeV contains 15 data which extend from 2.08 to 3.62 fm^{-1} in momentum transfer. These data overlap in q and have a rather high maximum momentum transfer. It is necessary to correct all these data for multiple Coulomb scattering and finite spectrometer acceptance angles. Because carbon is very light, the second Born amplitude is not very large and hence the scattering amplitude is primarily real. Thus the resulting cross sections have a very deep diffraction minimum. The cross sections vary rapidly in the minimum and the acceptance angle corrections can be as large as 10%. Both vertical and horizontal acceptance angles are available for each datum. These data supersede the earlier, less accurate data of Crannell (Cra 66) also taken at Stanford.

Nine extremely accurate low-energy data are available from Amsterdam taken by Jansen *et al.* (JPD 72). Because these data exist for seven different electron energies, they are inconvenient for analysis. It is relatively easy to convert the data to "equivalent" data all at the same electron energy and with corresponding momentum transfers, which range from 0.18 to 0.70 fm^{-1}. The conversion was made to 80 MeV electron energy by Friar and Negele (FN 75) and Sick (Sic 74) to facilitate their analyses. The error in the cross section introduced by this process should be considerably less than 1%. The acceptance angle corrections are negligible for these data. The three sets of data are plotted in Fig. 34, which shows the percentage deviation of the data from the best fit by Friar and Negele (FN 75). The arrow indicates the position of the diffraction minimum. Of some importance is the gap in the data between 0.7 and 1.0 fm^{-1}. More recent data, which were not available for analysis by Friar and Negele (FN 75) or Sick (Sic 74), have been reported from Stanford (Kli+ 73), Darmstadt (Fey+ 73), and Mainz (Mer 73).

4.2.2. Analysis of ^{12}C Electron Scattering Data

The analysis of the ^{12}C pseudodata and the discussion of Fig. 27 in Section 3 provided clear criteria for selecting the two parameters R and M which must be specified before data analysis begins. To quantitatively justify the parameters $R = 6.0$ fm and $M = 7$, a variety of fits were made and the χ^2 per degree of freedom for the 62 data was determined for each

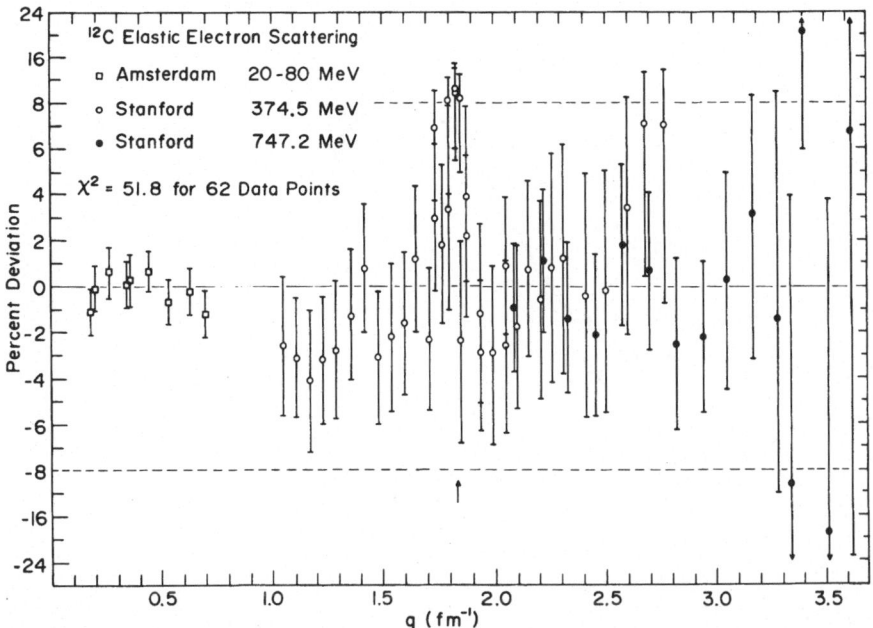

Fig. 34. Composite plot of the percentage deviation between experimental ^{12}C cross sections and the cross sections obtained from the optimal fit to the data with $\Delta\phi$ corrections and using seven coefficients and a radius of 6 fm. The arrow denotes the position of the diffraction minimum and the vertical scale changes at 8%.

fit. This quantity was found to be constant for M/R greater than 1.15 fm^{-1} and to rise rapidly for smaller values of M/R. This behavior is quite consistant with the value of q_{max}/π, 1.16 fm^{-1}. Values of M/R much smaller than q_{max}/π produce poorer fits, while larger values of M/R do not improve the quality of fit, with χ^2 dropping by one as each new parameter is introduced. The signature of too many coefficients, according to the conventions of Section 3, is a rapid increase in the statistical error envelope, which is most apparent at the origin as illustrated in Fig. 27. This error is roughly constant for M/R less than 1.20 and rises steeply for larger values of this quantity. We have thus confined M/R between 1.15 and 1.20 and, using the value of R determined previously, the choice $M = 7$ gives $M/R = 1.17$, which is quite satisfactory.

The pseudodata analysis suggests that the omitted higher Fourier components will have a very slight effect on the density and this expectation is confirmed in Fig. 35, which depicts the results of five separate fits to carbon data in which multiple Coulomb scattering and horizontal spectrometer acceptance angle corrections have been made where needed.

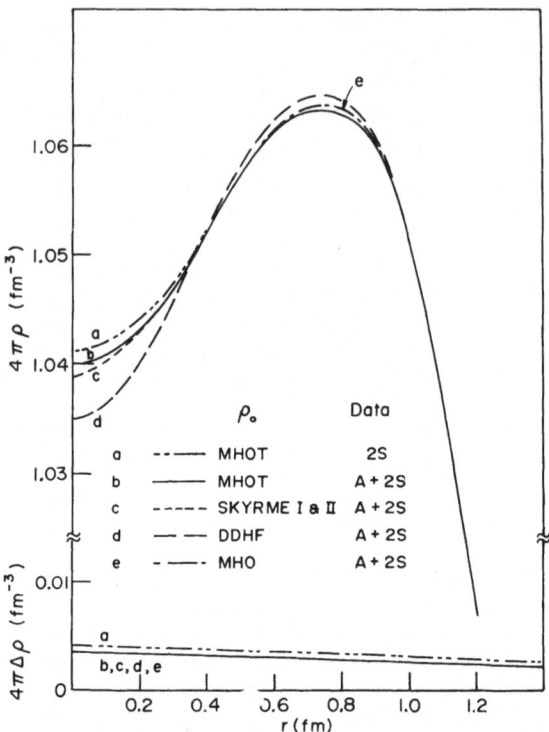

Fig. 35. Density distributions for ^{12}C obtained for different choices of ϱ_{th} using the experimental data, seven coefficients, and a radius of 6 fm. The two sets of Stanford data are indicated by 2S, the Amsterdam data by A, and the label ϱ_0 has been used instead of ϱ_{th}. The statistical error envelopes are shown at the bottom of the graph.

Curve *a* results from fitting only the Stanford data, starting from the density MHOT described earlier, while the remaining cases are fits to all the data, Stanford plus Amsterdam, for four different starting charge distributions. In addition to the phenomenological densities MHO and MHOT, DDHF represents a density-dependent spherical Hartree–Fock calculation and Skyrme I and II depict the numerically equivalent results of two Hartree–Fock calculations using Skyrme forces. The various densities are statistically consistent with one another, as indicated by the error envelope at the bottom of the figure. The low-energy Amsterdam data have a very small effect on the size of the error envelope for the interior density. The resulting densities all have a small, 2%, central depression. Since the agreement of the various cases is attributable to only small contributions to the density from the unfitted Fourier components, an extreme starting density was constructed using only the first seven Fourier components of the MHOT density, renormalized to make the total charge Z. The resulting fitted density differs from curve b by less than one standard deviation and the fit has virtually the same χ^2, as might be expected from Fig. 27. The rms radius is 0.008 fm less, reflecting the dependence of this quantity on charge which is far from the center of the nucleus and which in this case lies outside the 6-fm cutoff radius. Table V lists the χ^2 and rms radii of the densities for the various fits.

The complete density is depicted in Fig. 31 by the curve labeled ϱ, which has a corresponding error envelope $\varDelta\varrho$. The density in the tail of the distribution for $r > 4$ fm is indicated by the individual points with error bars, while the dashed curve is the prediction based on the behavior of a p-wave nucleon's wave function outside a potential well. In principle, there could be two reasons for the observed discrepancy. The first is that the density has not become asymptotic by 5 fm, due to the presence of the nuclear potential and nonvanishing s-state tails. However, estimates (FN 75) show that these effects cannot resolve the discrepancy. The second is the presence of a systematic error, as suggested by Sick (Sic 74). It is apparent from Fig. 34 that the fit to the data is much too low in the diffraction minimum. The reason for this is some unknown systematic error whose origin could be experimental or theoretical, particularly if dispersion corrections are important in the minimum. Sick felt that since it was impossible to fit the data with *any* charge distribution, the six data points in the minimum should be left out of the fitting procedure to avoid any bias in the resulting density. Such a procedure is quite acceptable, but risky, since the

TABLE V

Results Obtained for ^{12}C for Different Choices of ϱ_{th} Using the Real Data, Seven Coefficients, and a Radius of 6 fm

The two sets of Stanford data are indicated by 2S, the Amsterdam data by A, and the various choices of ϱ_{th} are explained in the text.

ϱ_{th}	Data	Number of data	χ^2	$\langle r^2 \rangle^{1/2}$, fm
MHOT	2S	53	57.68	2.449 ± 0.008
MHOT	A + 2S	62	61.13	2.449 ± 0.006
SKM I	A + 2S	62	61.52	2.453 ± 0.006
DDHF	A + 2S	62	61.49	2.447 ± 0.006
SKM II	A + 2S	62	61.72	2.453 ± 0.006
MHO	A + 2S	62	61.36	2.442 ± 0.006

excluded data may play a significant role. In the present case they define
the position of the diffraction minimum and this is a dominant feature of
the form factor. For the ^{12}C data, however, Sick's argument is a strong one.
Eliminating the six points in the minimum lowers the χ^2 of the MHOT fit
drastically, to 26.95, while raising the rms radius slightly, as tabulated in
the last row of Table VI. More importantly, the experimentally determined
charge density points are now in good agreement with the dashed line in
Fig. 31. Other features of the density are scarcely altered.

Our discussion of the density has centered on the error envelope,
which provides bounds on acceptable density variations. This envelope is
actually the diagonal part of the error correlation function. The full error
correlation function contains useful information about constraints on the
density, an example of which is the requirement that all charge distributions
satisfy the same normalization condition.

The square root of the magnitude of the error correlation function
for the analysis of the complete set of ^{12}C data is plotted in Fig. 36. Since
the function can have either sign, the positive contours are indicated by
solid lines and negative values by dashed lines. The gross structure of this
function may be described by the four regions of constant sign, which can
be roughly separated by vertical and horizontal lines at 3.4 fm. Thus if the

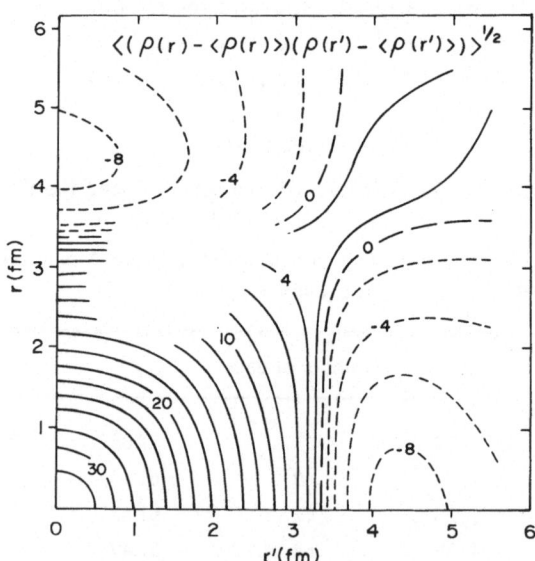

**Fig. 36. Contour plot of the square root of the magnitude of the error correlation function
for ^{12}C times the sign of the correlation function.** Integers labeling contours should be
multiplied by 10^{-4} to yield $4\pi \Delta\varrho$ in fm^{-3}.

data allow a statistical fluctuation in the density of some sign anywhere inside 3.4 fm, it will have the same sign everywhere inside 3.4 fm and opposite sign outside 3.4 fm. This structure of half-wavelength 3.4 fm corresponds to a momentum of 0.93 fm^{-1}. We remarked earlier that there was a gap in the data between 0.7 and 1.0 fm^{-1}, so it is quite natural that this should produce uncertainty in the Fourier components of this magnitude. The same effect is reflected in the diagonal elements of the error correlation function, the error envelope, where the second Fourier coefficient with a wavelength of 1.05 fm^{-1} is found to be the most uncertain. Obviously it will not always be possible to understand the correlation in such a simple fashion in terms of the data.

Traditionally, fits to charge densities have been made assuming smooth model densities with only a few parameters. If the data are not too extensive, fits of varying degrees of quality are possible. This is not possible if many data are available over a broad range of momentum transfer. Sick and McCarthy (SM 70), whose high-energy ^{12}C data are currently the best available, attempted to fit their data with simple models and found it necessary to modify the simple parameterizations by assuming additional components of the density which changed the tail of the charge distribution and which added fluctuations. In spite of using nine parameters, their χ^2 per degree of freedom was substantially greater than one.

Recently Sick (Sic 73, Sic 74) analyzed ^{12}C electron scattering data using the sum of Gaussians expansion discussed in Sections 3.2.1 and 4.1.1. Analyzing only the low-q Amsterdam data, Sick found a considerable range in the values of the rms radius of the charge distributions he obtained, and concluded that the radius was not precisely determined by low-q data as had been suggested (JPD 72). In the limit of small momentum transfer, only the mean square radius contributes to the form factor, although this contribution vanishes as the momentum transfer vanishes. Because of the experimental errors of the data, the most accurate information on the nuclear size is not obtained from the very low-q data but from a somewhat larger range of values, where higher moments of the charge distribution can also contribute. These higher moments have a considerable model sensitivity, and the nuclear size is actually best determined from a combination of both high-q and low-q data.

The ^{12}C charge density (with error envelope) determined by Sick is shown in Fig. 37. Only the Stanford data were used in the fit, and the points indicate the model density determined by Sick and McCarthy (SM 70). The possibility of a small central depression compatible with the densities shown in Fig. 35 is indicated by the solid lines, which delimit

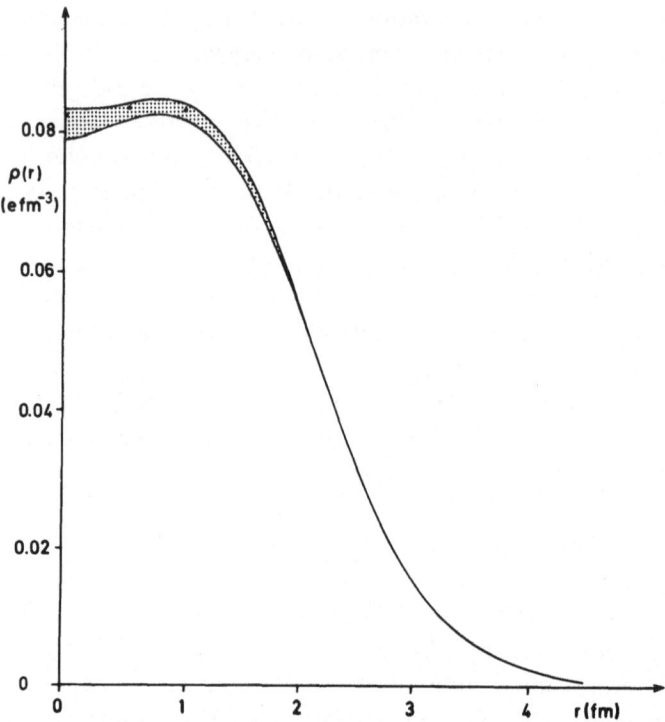

Fig. 37. Carbon density derived from the Stanford data using the sum of Gaussians method. The points indicate the best-fit model density of Sick and McCarthy (SM 70).

the error envelope, whose interior is shaded. The size of this envelope, which is approximately $\pm 3\%$ of the central density, precludes a more definite statement about the possible existence of such a depression, since a flat density is also possible in the region inside 1 fm. The size of this error envelope should be contrasted with the envelopes of Fig. 35, which are approximately $\pm 1/3\%$ of the central density. Although the SOG method and the Fourier–Bessel expansion method rely heavily on Hartree–Fock calculations for their realization, the resulting error envelopes are quite different for the two methods.

The tail of Sick's ^{12}C density is shown in Fig. 38, with the model density of Sick and McCarthy (SM 70) again indicated by points. The dashed curve is the Whittaker tail, the dotted lines delimit the error envelope obtained using the Stanford data alone, and the shaded area indicates the error envelope obtained using both high-q and low-q data. The only appreciable difference in the density produced by inclusion of the Amsterdam data is in the tail. The rms radius was determined to be 2.468 ± 0.016 fm.

It should be reemphasized that the two approaches to density fitting we have discussed in this section have a fundamental difference and that this difference is reflected in the magnitude and interpretation of the error envelopes. Clearly the introduction of higher Fourier components can increase the error envelope significantly, and the size of Sick's error envelope for ^{12}C is essentially determined by the allowable variation, within the framework of his model assumptions, of those Fourier components of the density that are not sensitive to the data. The Fourier–Bessel expansion method has a much smaller error envelope because it assumes ignorance of these components and specifies them using a model. In future SOG investigations it would be highly desirable to display error envelopes for a range of values of the parameter γ in Eq. (106), so that the sensitivity to

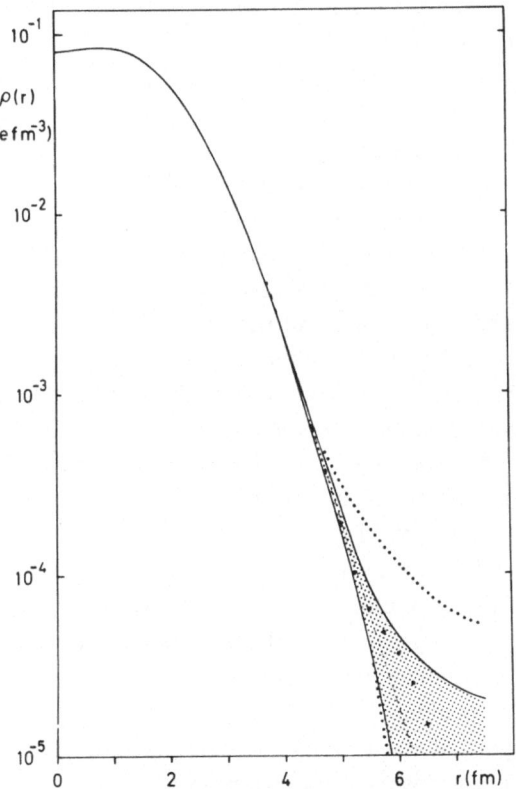

Fig. 38. Carbon density determined from the Stanford data (dotted lines) or Stanford plus Amsterdam (shaded area) using the sum of Gaussians method. The model density of Sick and McCarthy (SM 70) is indicated by points. The dashed curve represents the theoretical Whittaker function tail of the density at large radii.

model assumptions, which this parameter specifies, is more readily apparent.

In addition to the ^{12}C analysis, Sick has applied his method to ^{32}S (Sic 74), ^{40}Ca, ^{48}Ca, and ^{39}K (Sic 75). Of particular interest is the latter paper, where densities are found which have form factors that exhibit minima without sign change. This type of form factor behavior is not very common.

4.2.3. Systematic Errors in the ^{12}C Analysis

In the previous section we found strong evidence for systematic errors of unknown origin, which had an appreciable effect on the tail of the density distribution. In that case eliminating the error meant eliminating data and one cannot be entirely certain that other data than those deleted could be appreciably affected by the systematic error. One strong possibility for a sizable effect which was left out of the analysis and could possibly raise the theoretical cross sections in the diffraction minimum is dispersion corrections. These corrections, which were discussed in Section 1.3, are poorly known at present. Diffraction minima are filled in by the imaginary part of the scattering amplitude, which for light nuclei is small compared to the real part. The real part is dominated by the first Born amplitude and this is proportional to the form factor. Dispersion corrections have both a real and an imaginary part, and the real part can be expected to play the largest role in ^{12}C because it interferes with the dominant real Coulomb amplitude. In the diffraction minimum the real part is of no importance and the imaginary Coulomb and dispersion amplitudes are dominant. For the harmonic oscillator calculation of deForest (Def 70), which is exact within the framework of his model, and the approximate calculations of Friar and Rosen (FR 72), using the same model and in agreement with deForest, the imaginary part of the dispersive amplitude is small near the minimum and the corrections are 1 or 2%, (Fig. 8), not the 10% needed to bring the fits into agreement with the data. This could simply be a defect of the harmonic oscillator shell model, and the resolution of this anomaly is an important problem.

The real part of the dispersion correction affects the data at all momentum transfers and is most important in ^{12}C at high momentum transfers. The effect of adding in the complete dispersive amplitude of Friar and Rosen (FR 74) was investigated and a charge density was determined that allowed for dispersive effects on the cross section. The change in density which was found (FN 75) is plotted in Fig. 39, together with the statistical

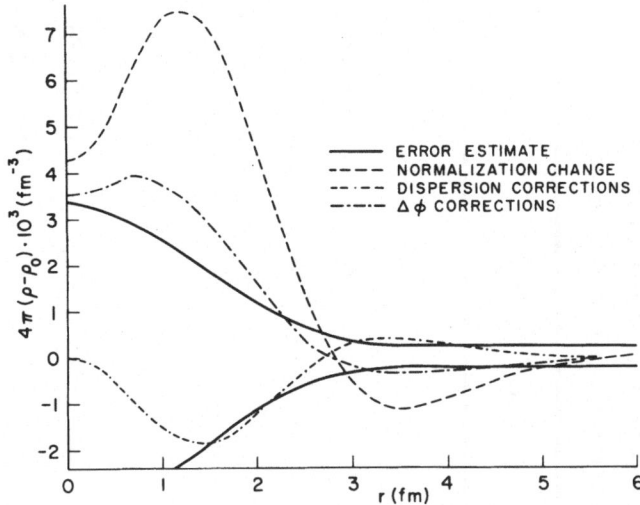

Fig. 39. Carbon density changes from systematic effects described in the text, compared with the statistical error envelope $\Delta\varrho(r)$ **denoted by solid lines.** The label ϱ_0 is used for ϱ_{th}.

error envelope. The effect is barely more than one standard deviation and this is the result of the relatively small size of the dispersion corrections in relation to the experimental errors in the data. For the present, these corrections do not seem to be important unless the imaginary part is much larger than calculations indicate.

The fits described in Section 4.2.2 included the effect of Coulomb multiple scattering in the target and the horizontal part of the spectrometer acceptance angle correction. If these effects are neglected, the χ^2 is 108.96, instead of 61.13, and the effect on the density is appreciable, being of the order of three or four standard deviations. This correction is obviously far too important to ignore. The vertical acceptance angle corrections are generally smaller, except at small scattering angles. The result of including these corrections is denoted in Fig. 39 and Table VI as $\Delta\phi$ corrections. The effect on the density is rather small, although the χ^2 of the fit drops considerably.

Allowing the cross-section normalizations to vary produces the density change labeled normalization change in Fig. 39. The effect is substantial, and the parameters resulting from the fit are listed in Table VI. Most of the change is produced in the fit to the first few points in the low-energy Stanford data set, which is too low, as indicated in Fig. 34. These points are the ones most affected by the $\Delta\phi$ corrections and if one includes these

TABLE VI

Tabulation of χ^2 Obtained Including Various Corrections Using the Distribution MHOT with Seven Coefficients and a Radius of 6 fm

The normalization factors that minimized χ^2 are shown in parentheses for the two cases of free normalization.

	Amsterdam	Stanford 374.5 MeV	Stanford 747.2 MeV	Total χ^2	Degrees of freedom	$\langle r^2 \rangle^{1/2}$
Number of data points	9	38	15	—	—	—
Original data	3.45	51.82	5.86	61.13	56	2.449 ± 0.006
Free norm	4.80 (1.004)	44.41 (1.032)	5.64 (1.053)	54.85	53	2.437 ± 0.006
$\Delta\phi$ Corrections	4.02	42.30	5.51	51.82	56	2.444 ± 0.006
$\Delta\phi$ Corrections with free norm	6.03 (1.005)	36.93 (1.030)	5.61 (1.046)	48.58	53	2.430 ± 0.006
$\Delta\phi$ Corrections + dispersion corrections	3.63	36.79	5.45	45.86	56	2.447 ± 0.006
$\Delta\phi$ Corrections (minus six data)	3.63	17.79	5.52	26.95	50	2.455 ± 0.006

corrections and then varies the normalization, the change in χ^2 is substantially less than before, although the resulting density shift is similar to the previous case. Parameters for an additional fit, which includes $\Delta\phi$ corrections and dispersion corrections, are tabulated in the next to last line of Table VI.

It is apparent from Table VI that systematic effects are more important than statistical errors. Although the spectrometer acceptance angle corrections are not in doubt and should be used in every analysis, the appropriate parameters $\Delta\phi$ and $\Delta\theta$ are not always published or available. Not using these corrections biases the results. Of similar size are corrections arising from uncertainty in the overall cross-section normalization, which are very difficult to determine experimentally. Any conclusions which we wish to draw on the value of the rms radius or the density error envelope are largely influenced by these systematic effects. It is more important to eliminate these systematic experimental uncertainties than to lower the statistical errors.

4.3. ^{208}Pb Results

4.3.1. Pb Data

There exist a number of recent sets of electron scattering data for ^{208}Pb. Low-energy scattering experiments (52.9-MeV electrons) at Darmstadt (Van 69) provide 12 data in which q ranges from 0.15 to 0.53 fm^{-1}. Medium-energy data from Mainz (FL 72) are comprised of 24 points at 124.0 MeV with q ranging from 0.45 to 1.10 fm^{-1} and 26 points at 167.0 MeV with q varying from 0.85 to 1.63 fm^{-1}. Two high-energy data sets from Stanford (Hei+ 69, FN 73b) have 37 data at 248.2 MeV with q varying from 0.44 to 2.20 fm^{-1} and 50 data at 502.0 MeV with q varying from 0.67 to 2.73 fm^{-1}. These two sets provide high-momentum-transfer information, although not nearly as high as for ^{12}C. It is necessary to correct the Stanford data for finite acceptance angles, but only the horizontal angle $\Delta\theta$ is available. Recently data have become available from Tohoku (NT 71) which overlap the Stanford data. There also exist older data from Stanford (BV 67) which are largely superseded.

These data have considerable overlap in momentum transfer, which is important because any errors in overall normalization will be most evident in that case. If the data sets do not overlap and there are normalization errors, charge distributions may be obtained which deviate considerably from the real charge distribution, as we discussed in Section 4.1. It is quite unlikely that all the data sets from different experimental facilities will have identical systematic normalization errors. Five of these six data sets [the Tohoku data were not available for the analysis in (FN 73a)] are displayed in Fig. 40, which shows the deviation of the data from the final fit to all the data to be discussed in Section 4.3.2.

The muonic X-ray data for the low-lying states come from two sources: the VPI–William and Mary collaboration (Jen+ 71) and the Chicago–Ottawa collaboration (Kes 71, And+ 69). These two data sets show a slight systematic disagreement, which is negligible, however, compared to the theoretical uncertainties in the corrections. At the time the analysis of Friar and Negele (FN 73a) was being performed, only the data of Jenkins *et al.* (Jen+ 71) were widely available and these were used in that analysis. In contrast to older muonic data, these data have one new ingredient. Transitions that involve the $2s_{1/2}$ level of the atom ($2s_{1/2} \rightarrow 2p_{3/2}$, $2s_{1/2} \rightarrow 2p_{1/2}$) were observed, and these transitions contain information about the nuclear charge density not available in other data. The absolute position of the $2p$ levels also poses a significant constraint, and should not be omitted from the analysis. As discussed in Section 2, the positions of these two levels were determined from cascades from higher levels and used as additional data. Because of uncertainties in the theoretical corrections to the $2p$ energy levels, it is necessary to increase the experimental errors in the position of these levels so that we include estimates of the theoretical uncertainty in order to avoid a serious bias in the χ^2-minimization procedure. At the present time, theoretical uncertainties are somewhat larger than the experimental errors and merit renewed theoretical attention.

4.3.2. Simultaneous Analysis of ^{208}Pb Muonic and Electron Scattering Data

In Section 1 we discussed the physics of obtaining elastic electron scattering cross sections from nuclear wave functions and a systematic and critical review of results for such wave functions for ^{208}Pb, obtained by means of Hartree–Fock techniques, was presented in Section 2. Procedures by which a density could be deduced, given the experimental data, were developed in Section 3. In this section we present the results for the charge density of ^{208}Pb obtained from these procedures, first treating the simultaneous analysis of electron and muonic data by Friar and Negele (FN 73a).

As discussed in Section 1.3.6, the vacuum polarization calculations of Fricke (Fri 69) were in error, and these erroneous numbers were used in certain noncritical parts of the analysis. Where they are used, they are

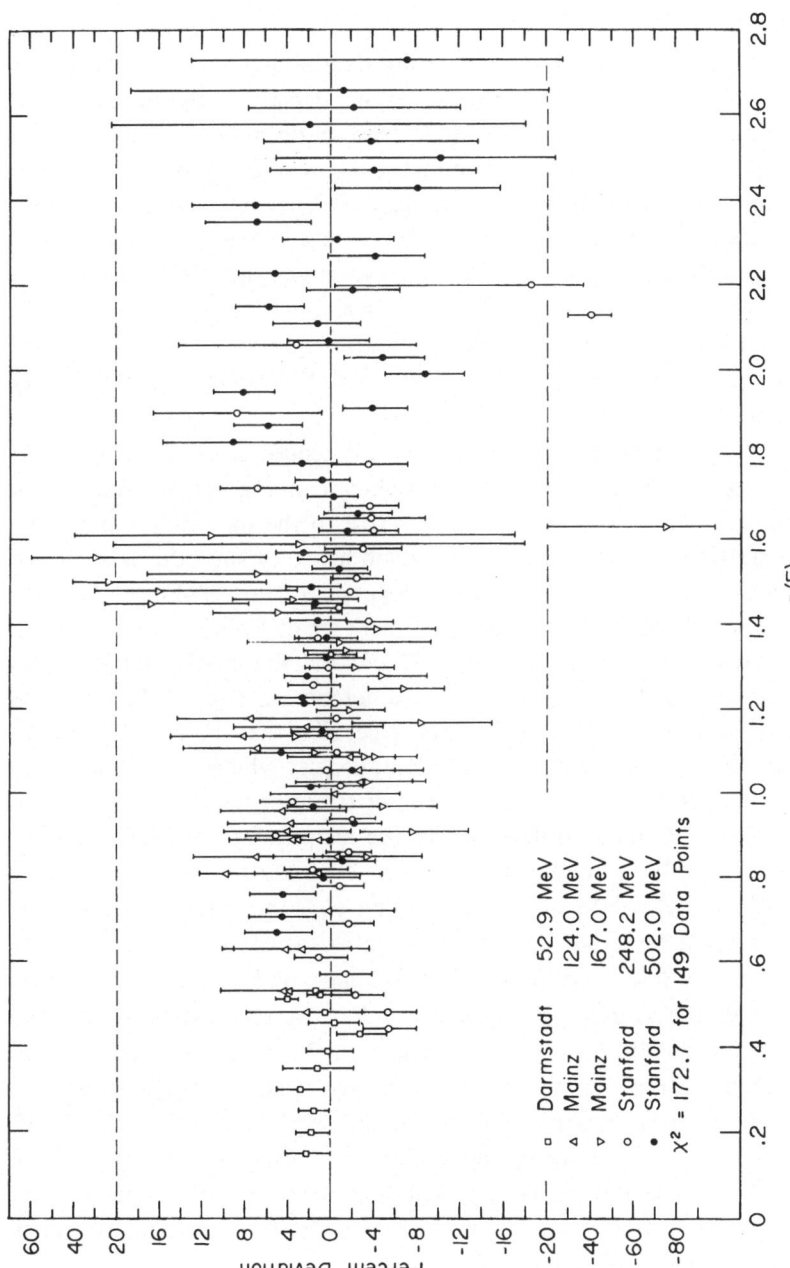

Fig. 40. Composite plot of the percentage deviation between the experimental cross sections for Pb and the cross sections calculated using the optimal fit to both electron scattering and muonic data.

denoted "OldVP." This error actually provides an interesting illustration of the effect of systematic errors on the analysis and we will discuss this in greater detail later.

As discussed earlier, quantum density fluctuations are specified by the Fermi momentum k_F. Components of the density with an oscillation frequency of this order should affect electron scattering corresponding to a momentum transfer $2k_F$, which is numerically very close to the maximum momentum transfer in the data set, $q_{max} = 2.73$ fm^{-1}. Different shapes with fluctuations of this frequency are possible and the data cannot completely discriminate among them. In fact, we saw earlier that starting with pseudodata generated using a smooth density, we could generate fitted densities with apparent fluctuations that were, however, smaller than the statistical error bars. These fluctuations had the frequency of the last fitted Fourier component.

In contradistinction to this case, the differences in the fitted densities determined from the experimental data shown in Fig. 41 are due to the 12th Fourier component, which is insensitive to the data and was not fit. The two starting Fermi distributions, which are very smooth, nevertheless have a sizable C_{12}, while the density-dependent Hartree–Fock (DDHF) density of Negele (Neg 70) has an extremely small Fourier component of this frequency, less than 1/100 of the C_{12} of the Fermi distribution. This component of the density leads to the difference in Fig. 41 between the Fermi and DDHF results. It might appear peculiar that the DDHF density, which contains fluctuations, leads to a final density which is smoother than densities obtained from the smooth Fermi distributions, but this emphasizes the importance of minor differences in the shape of the surface and the difficulty in clearly distinguishing interior fluctuations from the surface shape. Although the data do not determine C_{12}, the solutions in Fig. 41 have overlapping error bars, so that the statistical error exceeds the completeness error, and the oscillations therefore have no statistical significance.

Although the Hartree–Fock densities have a physical basis and the Fermi distributions do not, one should not attribute undue significance to the very high Fourier components. The surface region is vitally important in determining these components, as can be seen by examining the difference of a Fermi distribution and a uniform density distribution chosen to have the same interior density. This difference is quite small except between $r = 5$ and $r = 8$ fm. The uniform density, however, has extremely large high Fourier coefficients $C_m{}^0$,

$$C_m{}^0 \equiv 6j_1(q_m R_0)/R_0{}^2 \sim -6[\cos(q_m R_0)]/q_m R_0{}^3 \tag{130}$$

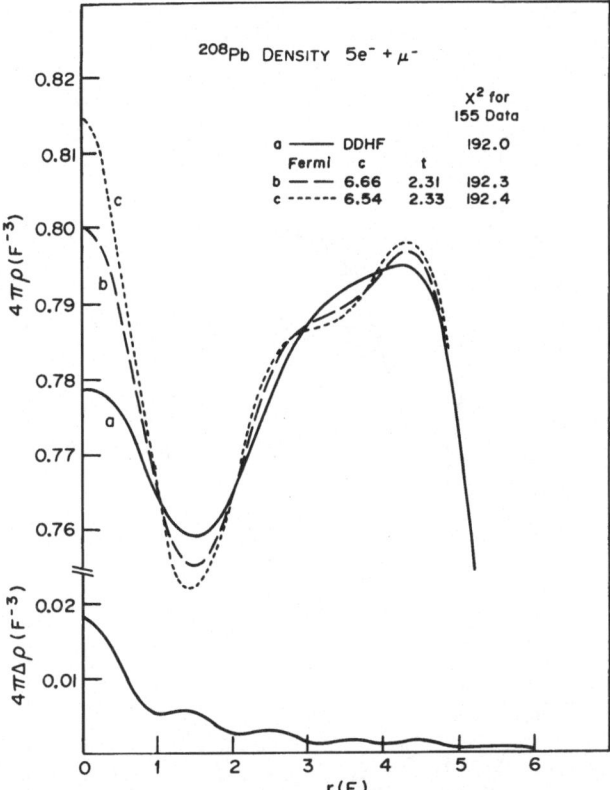

Fig. 41. Comparison of various fits to the entire set of electron scattering and muonic data for ^{208}Pb. These data were analyzed using the old vacuum polarization corrections (OldVP).

where q_m is $m\pi/R$ and R_0 is the radius of the uniform distribution, while the Fermi distribution has coefficients which also oscillate but whose envelope falls off exponentially, in contrast to the linear falloff in Eq. (130). The many-body techniques used to calculate nuclear densities, particularly those based on an expansion about the nuclear matter result, are least accurate in the surface. There is therefore no conclusive evidence as to the validity of the C_{12} determined from the Hartree–Fock density. Since the results differ by less than the statistical error envelope and results are more easily described by the Fermi parameterization, subsequent discussion will treat the expansion about the Fermi function.

If one selects a Fermi distribution ($c = 6.54$, $t = 2.33$) as the density ϱ_{th}, fitted densities corresponding to various combinations of data sets can be obtained as shown in Fig. 42. The corrected vacuum polarization has

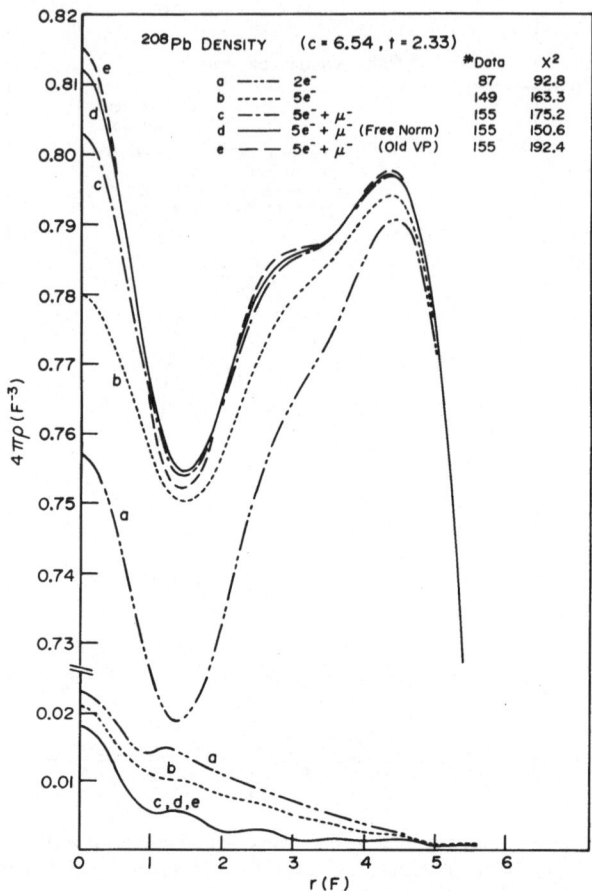

Fig. 42. Comparison of fits to the Pb density using the same Fermi distribution for ϱ_{th} to different sets of data. The Stanford data are indicated by $2e$ and the complete sets of electron data and muon data are indicated by $5e + \mu$. Curves c and d used the revised vacuum polarization corrections and curve e used the old corrections.

been used in conjunction with the muonic data, and case e will be discussed later. Case a is the density fitted from the Stanford data alone, while case b results from fitting all the electron data. Case c results from adding the muonic data to the data set. Note the gradual progression toward smaller errors indicated at the bottom of the figure, and the pronounced wine-bottle shape of all the fits. The progression of the central density upward as the amount of data to be fit is increased is symptomatic of a systematic error. The extreme cases, a and c, do not even lie within one another's error bars. If one adds a properly normalized data set to other data sets with incorrect normalization, there will be a shift in the density away from the result obtained with only the bad set toward the proper result. All the data sets compete against one another in the χ^2-minimization process and the more "correct" data that are added, the weaker will be the effect on the density of the "bad" set. Because the muonic data have such small errors, they play a dominant role in determining certain features of the fitted charge density. The facts that the density obtained with all five data sets agrees quite well with the density obtained when the muonic data are added as well, and that these both disagree with the result from the two Stanford data sets alone, suggest (but certainly does not prove) that there is a systematic error in the latter data. This is explored below.

Curve c was selected (FN 73a) as the optimal fit to all the data. Since the last (11th) coefficient is marginally determined, a ten-coefficient fit was also made. This density is nearly identical to curve c and has a somewhat smaller error envelope, which is consistent with the arguments of the previous section. The fits are statistically equivalent.

Allowing the overall normalization of all of the electron scattering data sets to vary individually so as to minimize χ^2 produces the results labeled d in Fig. 42. The total χ^2 is lowered considerably, which again is indicative of possibly sizable systematic errors in the electron scattering data. The density change is small, however, being rather less than the error envelope. The tendency is to increase the central density. Although the resulting normalization changes were found to be small, this result is inconclusive. Multiple solutions with only slightly different χ^2 are possible for different normalizations, and the method used to do the fitting prefers solutions near normalization factors equal to 1 (i.e., properly normalized data). Indeed, other systematic errors, both experimental and theoretical, may lead us to an absolute best fit (i.e., minimum χ^2) which is not the correct solution as far as the normalization factors are concerned.

This possibility is explored in Fig. 43, where for historical reasons the old vacuum polarization corrections were used. Using a starting Fermi distribution, densities were determined with all data (curve a) and with only the Stanford electron data and the muonic atom data (curve b). These two fits are virtually identical. Using only the Stanford data and lowering the normalization of the lower energy set by 2.5% and the higher energy data set by 5.0% produces the density plotted as curve c. The change in the density is substantial, while the increase in χ^2 is slight. Changing both sets by 5% produces a further reduction in the wine-bottle shape but an unacceptable increase in χ^2. The slight change in χ^2 resulting from the normalization changes in case c of this figure is not very significant in view of all the

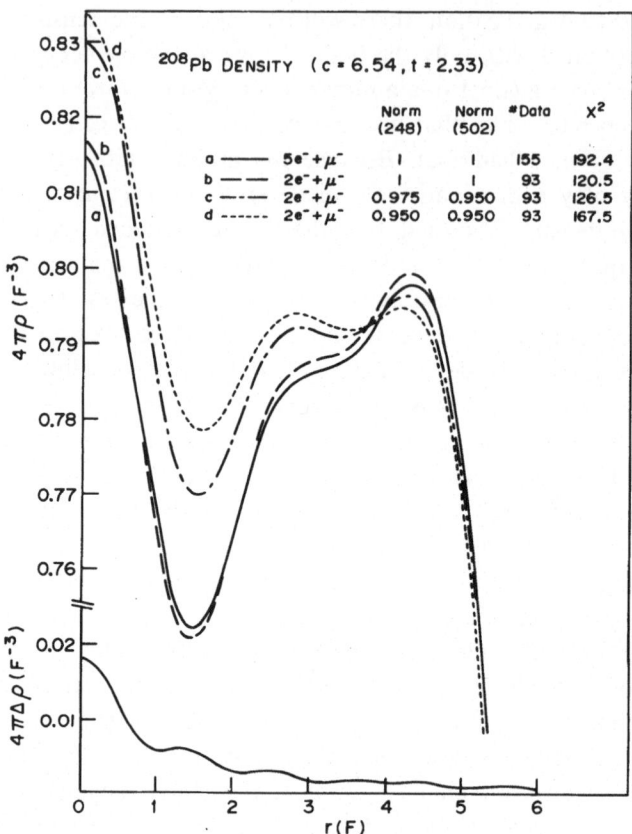

Fig. 43. Effect of normalization changes on fits to the Pb density using the same Fermi distribution for ϱ_{th} . These data include the old vacuum polarization corrections.

other data which must be simultaneously fit (and which may also need some normalization adjustment), and when compared to the large decrease in χ^2 which resulted when all the norms were adjusted in case d of Fig. 42. In any event we do not know what the correct normalization factors should be and the numbers used above were convenient guesses.

The effect of a normalization change on the density is most graphic if the entire density is plotted, not just the central portion. Figure 44 shows two densities plotted with error bars instead of the previously used error envelope. Case a is the OldVP fit to the muonic and Stanford data, which is indistinguishable from the best fit to all the data with corrected VP, and case b is the result of deleting the muonic data, and this was previously presented as case a in Fig. 42. The effect of fitting all the data rather than just the two high-energy electron scattering data sets is observed as a substantial gap between the interior portions of the two densities. The best fit also has a much reduced wine-bottle shape. Of particular note are the central, theoretically expected hump, which is statistically significant, and a secondary and tertiary hump, which are barely statistically significant (and barely perceptible). The two densities in this figure are not statistically compatible. For comparison, the density introduced earlier in case c of Fig. 43, where the Stanford data had been fitted after they had been re-normalized, is plotted in its entirety in Fig. 45. Averaging over the central hump and dip of this density leaves virtually no average interior slope, compatible with the theoretical predictions in Section 2.

Systematic errors do not necessarily produce a sizable density shift. As pointed out earlier, during the course of the work reported by Friar and Negele (FN 73a) it was found that the vacuum polarization corrections were incorrect. Fits were made with both the correct and incorrect (OldVP) corrections. Case c of Fig. 42 shows the density corresponding to the OldVP corrections, and this differs very little from the best fit result (with the VP errors removed) depicted in curve e. The quality of the two fits was different, however, as indicated by the χ^2. Increasing the experimental errors in the muonic data set was found to produce very little change in the

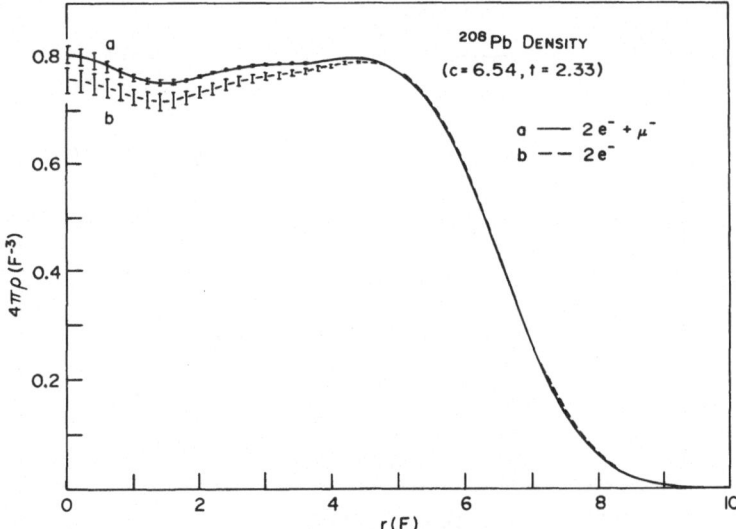

Fig. 44. Plot of entire Pb density distribution obtained from fitting the Stanford data plus muonic data with the old vacuum polarization corrections (a) and the Stanford data alone (b). The statistical error envelope $\Delta\varrho$ is indicated by the error bars.

density, although the error envelope grew. This increases confidence that any remaining systematic errors in the muonic analysis do not seriously distort the resulting densities. The same phenomenon was noted with respect to the density obtained by freeing the normalizations of the electron data sets. There was little change in density but a substantial change in χ^2. Other features of a charge density can be changed more than one standard deviation, however. Table VII lists the rms radii of various fitted densities and one sees that while freeing the norms produces a relatively slight change in the rms radius, the correction of the VP errors produces a shift in the rms radius of between two and three standard deviations. This quantity depends much more on the density in the surface and tail regions than in the interior region, and the two different regions can behave quite differently when a systematic error occurs in the data. If one were interested primarily in the rms radius, the effect of the VP systematic error would be extremely serious.

While a systematic error in one datum may produce a density shift in a given direction, it must compete against the constraints on the density provided by all the other data. If these constraints are sufficiently strong, the result will be a substantial increase in χ^2 but little change in the density. If these constraints do not exist, the density will change. One must also

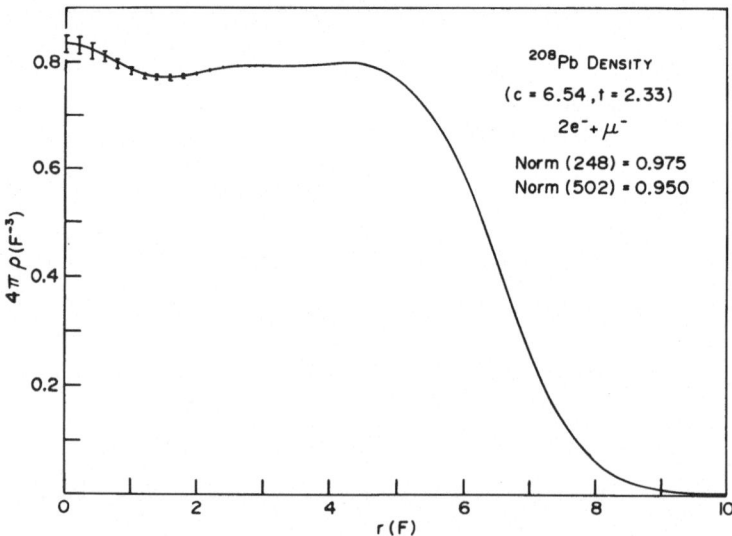

Fig. 45. Plot of entire Pb density distribution obtained from fitting the data of case (a) in Fig. 44 with the electron scattering normalization decreased by 2.5% at 248.2 MeV and 5% at 502 MeV.

TABLE VII

Results Obtained for ^{208}Pb by Fitting Various Combinations of Data Starting with Different Choices for the Starting Density ϱ_{th}

The first two columns specify ϱ_{th} with c and t indicating the parameters of a Fermi distribution and DDHF referring to the theoretical distribution of (Neg 70). The entries u and c in the column under μ^- indicate that muon data were included with the uncorrected vacuum polarization values (Old VP) or the corrected values, respectively. The column under e^- refers to the Stanford data (2) or the entire set of electron scattering data (5). The entry (Free) in the norm column indicates the result of allowing the normalization factors to vary in order to obtain a best fit.

	c	t	μ^-	e^-	Norm	χ_μ^2	χ_e^2	rms radius
1	(DDHF)		u	5	1	18.59	173.35	5.5053 ± 0.0011
2	6.66	2.31	u	5	1	18.93	173.32	5.5052 ± 0.0011
3	6.54	2.33	u	5	1	19.03	173.37	5.5052 ± 0.0012
4	6.54	2.33	u	2	1	13.70	106.84	5.5067 ± 0.0013
5	6.54	2.33	u	2	(0.975, 0.950)	10.21	116.30	5.5082 ± 0.0012
6	6.54	2.33	—	2	1	—	92.84	5.5327 ± 0.0122
7	6.54	2.33	—	5	1	—	163.33	5.5038 ± 0.0076
8	(DDHF)		c	5	1	8.63	167.20	5.5022 ± 0.0011
9	6.54	2.33	c	5	1	8.19	167.03	5.5024 ± 0.0011
10	6.54	2.33	c	5	(Free)	6.43	144.52	5.5032 ± 0.0011

reckon with constraints which are implicit in our method of analysis. For example, if electron scattering data were taken for extremely small values of the momentum transfer and were incorrectly normalized, there would be virtually no chance of obtaining a good fit since the cross sections would be almost independent of the nuclear charge distribution. In fact, the condition that the form factor must equal one at vanishing momentum transfer q is a consequence of the charge normalization condition, and it is this constraint that would have to be broken to obtain a good fit to the bad data. Since we are clearly unwilling to do this, the result will be a high-χ^2 fit. Similar examples can be constructed for muonic atoms.

Generally speaking, the more data sets there are, the smaller the chance that they all require renormalization in the same direction (e.g., all sets having too large cross sections). In this case the "true" density will likely be close to the fitted density, but the χ^2 will be high. This is particularly true if the data sets overlap considerably in momentum transfer, as was true for Pb.

4.3.3. Separate Analysis of ^{208}Pb Muonic or Electron Scattering Data

Although many fits to the experimental electron scattering or muonic atom data have been attempted over the years, most have used model densities of sufficient inflexibility that little is determined but the fitted values of a few parameters which relate to the skin thickness, rms radius, etc. This is particularly true of the muonic data. Most analyses for the case of Pb involve modifications of the Fermi distribution that contain two, three, or four parameters, and the fits are generally fairly good (Jen+ 71, Kes 71, And+ 69). It was shown in Section 3 that the muonic data constrain relatively simple features of the density, such as radial moments. Although the mean square radius is not one of the exact constraints, it is well-constrained by the transition data. If a Fourier–Bessel expansion of the density difference $\varrho - \varrho_{\text{th}}$ is made, the mean square radius of this difference can be related to the Fourier coefficients

$$\delta\langle r^2 \rangle = \frac{6R^4}{Z\pi^3} \sum_{I=1}^{M} \frac{(-1)^I}{I^3} C_I \tag{131}$$

The combination of decreasing Fourier coefficients for large I and the factor of $1/I^3$ shows that small changes in the mean square radius will only affect the lowest few Fourier coefficients and one expects, therefore, that the muonic transitions are only sensitive to the lowest three or four Fourier coefficients. This indicates why several parameter fits to the muonic data can be quite successful.

Fits to all the electron scattering data are more difficult because more parameters are needed, the number being 10 or 11 if one uses the Fourier representation for the density. The Stanford group, when reporting their electron scattering data, fit the density using a smooth Fermi distribution and a rapidly fluctuating part. The reason was that the smooth distribution by itself did not fit the data at high momentum transfer. As we demonstrated in Eq. (127), an oscillation in the density at the appropriate frequency will strongly affect the form factor and the cross section at momentum transfers corresponding to that frequency. The fits obtained with this two-component density were fairly good. The question which immediately arises, and which illustrates the weakness of the conventional approach to charge density fitting that we are describing, is: What does one do next to improve the fit, if indeed it can be improved? In addition we must ask if we have built features into the density which are not actually present and to which the data are completely insensitive.

The results of Section 4.3.2 do not indicate any rapid oscillation in the density, although several bumps, two of which are barely statistically significant, do occur. The apparent disagreement between the original Stanford fits and those of Friar and Negele (FN 73a) can be resolved by examining Fig. 46. The oscillating density component of Heisenberg *et al.* (Hei+ 69) is plotted together with error envelopes obtained by fitting the Stanford data alone and the Stanford data plus the muonic data. The oscillatory density component is not statistically significant inside 5 fm. Beyond this distance the surface region begins, and in this region an oscillation is poorly defined; in fact, such an oscillation can be absorbed into a redefinition of the surface density. A striking feature (a rapid fluctuation) was built into the density to produce a fit to the data, and its only significant effect was to redefine the surface region of the density, which was inadequately described by a Fermi distribution. Other expansion methods, such as the Fourier method or the method of Sick described in Sections 3 and 4.2, are much more flexible and make very weak assumptions about the form of the density; indeed, this is the source of their strength.

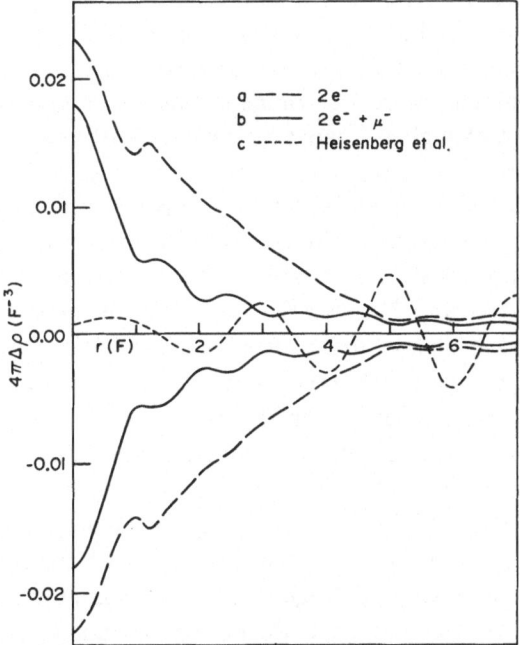

Fig. 46. Comparison of the Pb density fluctuations of Heisenberg *et al.* (Hei+ 69) with the error envelopes obtained from the Stanford data (2e) and from the simultaneous analysis of the Stanford data and the muonic data ($2e + \mu$).

There have been two attempts to obtain the charge density of Pb using models that are sufficiently flexible to reproduce all the data. In addition to the work described in Section 4.3.2, Friedrich and Lenz fitted the Mainz data alone using their δ-shell method discussed in Section 3. They were among the first to point out that very different charge densities could fit the available data. These densities differ, of course, by having different high-frequency components, and consideration of only the relatively low-q Mainz data leaves tremendous freedom in the range of unphysical densities that are consistent with the data. Recently, Dreher et al. (Dre+ 74) examined the high-q Stanford data using a combination of the δ-shell and Fourier–Bessel methods. They first obtained a set of δ-shell distributions compatible with the data and from this set generated a set of form factors. As discussed in Section 3.1.4, the form factor is virtually model independent and contains all the information needed to determine the density and an error envelope using the Fourier expansion method and suitable model assumptions. The assumption that the magnitude of the form factor for $q > q_{max}$ does not exceed the asymptotic estimate discussed in Section 4.1.3 was then invoked and fictitious data were generated for the experimentally unobserved region by assuming a uniform distribution of form factors within the bounds of the asymptotic estimate. The combined data sets were analyzed using the Fourier–Bessel expansion method, and the density and its error estimate were determined. The error contains some estimate of the effect of higher (than experimentally observed) Fourier coefficients.

Since the matching of the asymptotic estimate was made at $q = 2.3$ fm^{-1}, it is not clear that any differences between the analysis of Dreher et al. and those previously discussed in Section 4.3.2 are meaningful, since the analysis of Friar and Negele (FN 73a) used all the data extending to $q = 2.73$ fm^{-1}. Nevertheless, the error envelopes obtained by Dreher et al. (Dre+ 74) appear to be roughly the same size as those of Friar and Negele (FN 73a) except near the origin. In this region the model-dependent part of their complete error dominates the statistical part. For distances greater than 1 fm from the center the statistical error dominates. They also find that varying the cutoff radius R has a weak effect on the error envelope, unless a very large value of R is used. These results confirm similar conclusions reached by Friar and Negele (FN 73a) and previously discussed, in part, in Section 4.3.2. Because their error is enhanced in the central region by the model dependence estimate, Dreher et al. can draw no conclusion about the existence of a central hump.

5. DISCUSSION AND CONCLUSIONS

It is evident that substantial progress has been made recently both in the theoretical calculation of nuclear charge distributions from microscopic many-body theory and in the extraction of detailed information concerning the spatial charge distribution from the simultaneous analysis of all available electron scattering and muonic X-ray data. Rather than reiterate portions of the already lengthy discussion of the recent advances, it is probably more productive to conclude by emphasizing the most critical theoretical and experimental limitations that require attention in future investigations.

The discussion most naturally separates into the treatment of those difficulties relating to the determination of the static nuclear charge distribution from experimental data, and problems associated with theoretical calculations of the total charge distribution. Concerning the former, the effect of systematic experimental errors in the data and the effect of theoretical inadequacies in the various corrections to scattering from the static Coulomb potential are the greatest uncertainties in the analysis. In the analysis of electron scattering, the dominant theoretical uncertainty is the size of dispersion corrections. Considerably more theoretical work is needed on the transverse contributions and Coulomb distortion effects, and additional thought must be given to the elusive two-body correlation function, which is the dominant ingredient in dispersion calculations. The polarization corrections to muonic atom energy levels are better known than the analogous dispersion corrections in the electron scattering case, but they presently set the limit of theoretical accuracy and have uncertainties significantly larger than the experimental errors. The remaining recoil corrections, which have not yet been calculated, are expected to be small, but they must at least be estimated in order to establish this.

Quantum electrodynamic corrections are also in need of attention. Uncertainties in the Lamb shift correction for the $1s$ state of muonic Pb are approximately 1 keV, nearly as large as the polarization uncertainties, and could be improved by numerical calculations using one of the techniques discussed in Section 1.3.6. Almost as important are uncertainties in the vacuum polarization corrections, particularly the contribution to low-lying muonic bound states of the $\alpha^2(Z\alpha)^2$ diagrams and the charge density dependence of the $\alpha(Z\alpha)^N$ diagrams, where N is greater than one. Most of the work done on these graphs has centered on the high-lying states, for which calculations are simpler. Hopefully, these techniques will be extended to include low-lying states in the near future.

Turning to experimental limitations, it is incumbent upon experimentalists to give greater attention to questions involving the absolute cross-section normalization, since this is one of the dominant uncertainties in the analysis of charge densities. It is also desirable that such experimental considerations as spectrometer acceptance angles and precisely what radiative corrections have been applied be explicitly dealt with in the publication of data. When this information is missing and unavailable, the value of data is considerably diminished. The lack of high-q data is a serious limitation, and is, of course, to some extent dependent on the maximum energies and currents available at existing experimental facilities. Nevertheless, the greatest effort should be made to obtain data of the highest possible momentum transfer. As emphasized by Borysowicz and Hetherington (BH 73), even null experiments placing upper bounds on the high Fourier components are of genuine value. If sufficient high-q data were available, the difference between existing methods of analysis would be largely of academic rather than practical interest.

It is regrettable that little experimental information is available on dispersion corrections to aid theoretical analysis. It has been suggested by Borie and Drechsel (BD 71) that 0^+–0^- transitions offer considerable information on these corrections, because such excitations are forbidden to occur by single-photon exchange. Measurements would indeed be helpful, even though they give little information on the dominant Coulomb part of dispersion corrections. Investigations of the diffraction minimum anomalies in ^{12}C and ^{16}O would be even more valuable, since elimination of all experimental uncertainties would force theorists to seriously study any remaining discrepancy. Also, high-precision measurements of the energy dependence of nonstatic corrections and, ultimately, high-resolution companion measurements of e^+ and e^- scattering could provide significant tests of theories of dispersion corrections.

Outstanding problems associated with theoretical calculations of the total charge distribution exist on two levels. The first level is the determination of the point nucleon density distributions from nonrelativistic many-body theory. The most obvious topic for subsequent investigations is the higher order contributions to the one-body density discussed in Section 2.3.3, which are inextricably connected with the calculation of three-body diagrams in finite nuclei.

The second level concerns effects arising from the underlying relativistic meson theory of nuclear forces and nuclear structure. Because of the serious limitation of our present fundamental understanding of strong interactions, the calculations of such effects are ambiguous and difficult. Few calculations of mesonic contributions to charge densities exist and these calculations have centered on the two- and three-body systems. Mesonic and isobaric effects on the density in heavy nuclei are a completely open subject, but they are potentially important enough to require serious consideration. Relativistic effects on the nucleon–nucleon potential, investigated recently by Coester *et al.* (CPS 75), may also require serious consideration.

The neutron charge form factor is not well known experimentally and this is a serious problem, since the neutrons make a significant contribution to the charge form factor. This is also a theoretical problem. Although relativistic corrections to scattering that depend on the total nucleus mass are doubtless unimportant for any but the lightest nuclei, these corrections and mesonic and isobaric corrections to electron–deuteron scattering are important if we wish to extract the neutron charge form factor from this process. In addition, there is still some sensitivity to the model of the nucleon–nucleon potential. In view of these problems associated with electron–deuteron scattering, consideration of other possible alternative experimental methods would also be desirable.

Finally, it is well to bear in mind that the entire discussion in this work has been restricted to elastic electron scattering and muonic X-ray transitions for spherical closed-shell nuclei, and thus we should consider briefly what extensions to inelastic scattering and other nuclei are most likely to have a maximum impact on our understanding of nuclear theory. Virtually all of the techniques of determining density distributions in coordinate space from experimental data generalize straightforwardly, so the primary question is the status of the relevant nuclear theory.

Since our present understanding of the structure of finite nuclei is so firmly rooted in the average field, or Hartree–Fock approximation, it is highly desirable to treat those cases that are still described in this framework. Heavy, well-deformed, rotational nuclei should be adequately described by a Hartree–Fock intrinsic wave function, and the technology of projected Hartree–Fock calculations with the realistic interactions described in Section 2 is tractable. Since the experimental resolution needed to resolve the inelastic scattering to individual rotational states is now available, a coordinated experimental and theoretical effort to treat elastic and inelastic electron scattering and muonic X-ray transitions in heavy rotational nuclei is clearly desirable.

The other category of excited states that is presently amenable to microscopic nuclear theory is the class of collective, low-lying excited states, which may be described by the random phase approximation built on the Hartree–Fock ground state. One recent notable success is the calculation of the transition density to the 3^- state in ^{208}Pb by Bertsch and Tsai (BT 75) using the Skyrme interaction. Clearly, further such tests are highly desirable.

Aside from these two categories, the application of nuclear many-body theory to the experimental data that will become available in the near future will be very difficult and require the development of theoretical techniques which presently do not exist. This is an exciting prospect, and we expect the next generation of high-resolution data to provide a strong impetus for further theoretical development. If such progress comes to pass, the electromagnetic interaction will indeed have served us well.

ACKNOWLEDGMENTS

In addition to the numerous investigators who sent us preprints of work in advance of publication, the authors are particularly indebted to M. Beiner, X. Campi, S. Coon, R. Y. Cusson, K. T. R. Davies, S. Krewald, R. Machleidt, L. D. Miller, S. A. Moszkowski, J. Németh, B. Rouben, and D. Vautherin, who provided tabulations of the ^{208}Pb density distributions analyzed in Section 2. Correspondence and discussions concerning a preliminary draft of this manuscript with S. Coon, A. Faessler, and R. Machleidt are also gratefully acknowledged. Although we have made every effort to eliminate factual errors and distortions of the contributions of various authors whose work is reviewed, we have deliberately attempted to interpret and critically evaluate the work in this field. Thus the philosophy and interpretations expressed in this review are not necessarily those of the original investigators, and the reader is strongly encouraged to consult the original literature.

REFERENCES

AD 64   R. J. Adler and S. D. Drell, *Phys. Rev. Lett.* **13**:349 (1964).
Adl 66   R. J. Adler, *Phys. Rev.* **141**:1499 (1966).
AM 74   H. Arenhövel and H. G. Miller, *Z. Phys.* **266**:13 (1974).
And+ 69   H. L. Anderson, C. K. Hargrove, E. P. Hincks, J. D. McAndrew, R. J. McKee, R. D. Barton, and D. Kessler, *Phys. Rev.* **187**:1565 (1969).
Ara 74   J. Arafune, *Phys. Rev. Lett.* **32**:560 (1974).
AW 72   H. Arenhövel and H.-J. Weber, *Ergeb. Exakten Naturwiss.* **65**:58 (1972).
Bar 68   R. C. Barrett, *Phys. Lett.* **28B**:93 (1968).
Bar 70   R. C. Barrett, *Phys. Lett.* **33B**:388 (1970).
Bar 74   R. C. Barrett, *Rep. Prog. Phys.* **37**:1 (1974).

Bar+ 68   R. C. Barrett, S. J. Brodsky, G. W. Erickson, and M. H. Goldhaber, *Phys. Rev.* **166**:1589 (1968).

Bar+ 73   R. C. Barrett, D. A. Owen, J. Calmet, and H. Grotch, *Phys. Lett.* **47B**:297 (1973).

BB 48   G. Breit and G. E. Brown, *Phys. Rev.* **74**:1278 (1948).

BCM 74a   L. S. Brown, R. N. Cahn, and L. D. McLerran, *Phys. Rev. Lett.* **32**:562 (1974).

BCM 74b   L. S. Brown, R. N. Cahn, and L. D. McLerran, *Phys. Rev. D*, to be published.

BD 65   J. D. Bjorken and S. D. Drell, *Relativistic Quantum Mechanics*, McGraw-Hill, New York (1965).

BD 71   E. Borie and D. Drechsel, *Phys. Rev. Lett.* **26**:195 (1971).

BDL 70   M. A. Braun, Y. Y. Dmitriev, and L. N. Labzovskii, *Sov. Phys.—JETP* **30**:1188 (1970).

Bei+ 75   M. Beiner, H. Flocard, Nguyen Van Giai, and P. Quentin, *Nucl. Phys. A* **238**:29 (1975).

Bel 73   T. L. Bell, *Phys. Rev. A* **7**:1480 (1973).

Ber+ 72   W. Bertozzi, J. Friar, J. Heisenberg, and J. W. Negele, *Phys. Lett.* **41B**:408 (1972).

Bet 47   H. A. Bethe, *Phys. Rev.* **72**:339 (1947).

Bet 71   H. A. Bethe, *Ann. Rev. Nucl. Sci.* **21**:93 (1971).

BG 71   R. Blankenbecler and J. F. Gunion, *Phys. Rev. D* **4**:718 (1971).

BH 73   J. Borysowicz and J. H. Hetherington, *Phys. Rev. C* **7**:2293 (1973).

BK 71   W. D. Brown and E. Kujawski, *Ann. Phys. (N.Y.)* **64**:573 (1971).

BKK 73   V. N. Boytsov, L. A. Kondratyuk, and V. B. Kopeliovich, *Sov. J. Nucl. Phys.* **16**:287 (1973).

BKL 71   S. I. Bilen'kaya, Y. M. Kazarinov, and L. I. Lapidus, *Sov. Phys.—JETP* **33**:247 (1971).

BL 74   M. Beiner and R. J. Lombard, *Ann. Phys. (N.Y.)* **86**:262 (1974).

Blo 72   J. Blomqvist, *Nucl. Phys. B* **48**:95 (1972).

BLS 59   G. E. Brown, J. S. Langer, and G. W. Schaefer, *Proc. Roy. Soc.* **A251**:92 (1959).

BM 59   G. E. Brown and D. F. Mayers, *Proc. Roy. Soc.* **A251**:105 (1959).

BM 71   H. A. Bethe and A. Molinari, *Ann. Phys. (N.Y.)* **63**:393 (1971).

BN 68   H. A. Bethe and J. W. Negele, *Nucl. Phys. A* **117**:575 (1968).

BP 69   S. J. Brodsky and J. R. Primack, *Ann. Phys. (N.Y.)* **52**:315 (1969).

Bre 29   G. Breit, *Phys. Rev.* **34**:553 (1929).

BT 75   G. F. Bertsch and S. F. Tsai, *Physics Reports* **18C**:125 (1975).

BV 67   J. B. Bellicard and K. J. Van Oostrum, *Phys. Rev. Lett.* **19**:242 (1967).

CC 70   F. E. Close and L. A. Copley, *Nucl. Phys. B* **19**:477 (1970).

CG 67   B. M. Casper and F. Gross, *Phys. Rev.* **155**:1607 (1967).

Che 70   M.-Y. Chen, *Phys. Rev. C* **1**:1167 (1970).

Che 75   M.-Y. Chen, *Phys. Rev. Lett.* **34**:341 (1975).

CK 74   S. A. Coon and H. S. Köhler, *Nucl. Phys. A* **231**:95 (1974).

CMR 74   M. Chemtob, E. J. Moniz, and M. Rho, *Phys. Rev. C* **10**:344 (1974).

CO 70   F. E. Close and H. Osborn, *Phys. Rev. D* **2**:2127 (1970).

Col 69   R. K. Cole, Jr., *Phys. Rev.* **177**:164 (1969).

Coo 75   S. A. Coon, in *Hartree–Fock and Self-Consistent Field Theories in Nuclei*, Trieste, Italy (1975).

CPS 75   F. Coester, S. C. Pieper, and F. J. D. Serduke, *Phys. Rev. C* **11**:1 (1975).

CR 71   M. Chemtob and M. Rho, *Nucl. Phys. A* **163**:1 (1971).

Cra 66   H. Crannell, *Phys. Rev.* **148**:1107 (1966).

CS 72   X. Campi and D. W. L. Sprung, *Nucl. Phys. A* **194**:401 (1972).

CSM 74   X. Campi, D. W. L. Sprung, and J. Martorell, *Nucl. Phys. A* **223**:541 (1974).

Cus+ 74   R. Y. Cusson, H. P. Trivedi, H. W. Melner, and M. S. Weiss, to be published.

Dav+ 74   K. T. R. Davies, R. J. McCarthy, J. W. Negele, and P. U. Sauer, *Phys. Rev. C* **10**:2607 (1974).

Def 70   T. deForest, Jr., *Phys. Lett.* **32B**:12 (1970).

Des 55   S. Deser, *Phys. Rev.* **99**:325 (1955).

DF 75   T. deForest, Jr. and J. L. Friar, submitted to *Phys. Lett.*

DJ 71   A. M. Desiderio and W. R. Johnson, *Phys. Rev. A* **3**:1267 (1971).

DKW 73   K. T. R. Davies, S. J. Krieger, and C. Y. Wong, *Nucl. Phys. A* **216**:250 (1973).

DM 71   K. T. R. Davies and R. J. McCarthy, *Phys. Rev. C* **4**:81 (1971).

DMS 72   K. T. R. Davies, R. J. McCarthy, and P. U. Sauer, *Phys. Rev. C* **6**:1461 (1972).

Dre+ 74   B. Dreher, J. Friedrich, K. Merle, H. Rothhaas, and G. Lührs, *Nucl. Phys. A* **235**:219 (1974).

DW 66   T. deForest, Jr. and J. D. Walecka, *Adv. Phys.* **15**:1 (1966).

EH 72   T. E. O. Ericson and J. Hüfner, *Nucl. Phys. B* **47**:205 (1972).

EH 73   T. E. O. Ericson and J. Hüfner, *Nucl. Phys. B* **57**:604 (1973).

EHM 74   K. Erkelenz, K. Holinde, and R. Machleidt, *Phys. Lett.* **49B**:209 (1974).

EM 72   J. W. Ehlers and S. A. Moszkowski, *Phys. Rev. C* **6**:217 (1972).

Eng+ 74   R. Engfer, H. Schneuwly, J. L. Vuilleumier, H. K. Walter, and A. Zehnder, *Nuclear Data Tables* **14**:509 (1974).

Eri 74   G. W. Erickson, *Phys. Rev. Lett.* **27**:780 (1971).

EV 74   R. Engfer and J. L. Vuilleumier, Invited talk given at Fourth International Conference on Atomic Physics (1974).

EY 65a   G. W. Erickson and D. R. Yennie, *Ann. Phys.* (*N.Y.*) **35**:271 (1965).

EY 65b   G. W. Erickson and D. R. Yennie, *Ann. Phys.* (*N.Y.*) **35**:447 (1965).

Fae+ 75   A. Faessler, J. E. Galonska, K. Goeke, and S. A. Moszkowski, *Nucl. Phys. A* **239**:477 (1975).

Fey+ 73   G. Fey, H. Frank, W. Schütz, and H. Theissen, *Z. Phys.* **265**:401 (1973).

FFY 59   L. L. Foldy, K. W. Ford, and D. R. Yennie, *Phys. Rev.* **113**:1147 (1959).

FL 72   J. Friedrich and F. Lenz, *Nucl. Phys. A* **183**:523 (1972).

FM 54   T. Fulton and P. C. Martin, *Phys. Rev.* **95**:811 (1954).

FN 72   J. L. Friar and J. W. Negele, *Comm. Nuclear Particle Physics* **5**:181 (1972).

FN 73a   J. L. Friar and J. W. Negele, *Nucl. Phys. A* **212**:93 (1973).

FN 73b   J. L. Friar and J. W. Negele, *Phys. Lett.* **46B**:5 (1973).

FN 73c   G. Fái and J. Németh, *Nucl. Phys. A* **208**:463 (1973).

FN 75   J. L. Friar and J. W. Negele, *Nucl. Phys. A* **240**:301 (1975).

Fol 58   L. L. Foldy, *Rev. Mod. Phys.* **30**:471 (1958).

FR 72   J. L. Friar and M. Rosen, *Phys. Lett.* **39B**:615 (1972).

FR 73   K. W. Ford and G. A. Rinker, Jr., *Phys. Rev. C* **7**:1206 (1973).

FR 74   J. L. Friar and M. Rosen, *Ann. Phys.* (*N.Y.*) **87**:289 (1974).

Fri 69   B. Fricke, *Z. Phys.* **218**:495 (1969).

Fri 71   J. L. Friar, *Nucl. Phys. A* **173**:257 (1971).

Fri 72	J. L. Friar, *Part. Nucl.* **4**:153 (1972).
Fri 73a	J. L. Friar, *Ann. Phys. (N.Y.)* **81**:332 (1973).
Fri 73b	H. M. Fried, *Functional Methods and Models in Quantum Field Theory*, M.I.T. Press, Cambridge, Massachusetts (1973).
Fri 75a	J. L. Friar, submitted to *Phys. Lett.*
Fri 75b	J. L. Friar, submitted to *Nucl. Phys.*
FS 70	G. Feinberg and J. Sucher, *Phys. Rev. A* **2**:2395 (1970).
FW 50	L. L. Foldy and S. A. Wouthuysen, *Phys. Rev.* **78**:29 (1950).
FW 69	K. W. Ford and J. G. Wills, *Phys. Rev.* **185**:1429 (1969).
GH 74	A. M. Green and P. Haapakoski, *Nucl. Phys. A* **221**:429 (1974).
Gro 65	F. Gross, *Phys. Rev.* **410**:B140 (1965).
Gro 66	F. Gross, *Phys. Rev.* **142**:1025 (1966); **(E)152**:1517 (1966).
Gro 67	F. Gross, in: *Medium Energy Nuclear Physics with Electron Linear Accelerators* (W. Bertozzi and S. Kowalski, eds.), M.I.T. Summer Study (1967).
GS 57	S. Gartenhaus and C. Schwartz, *Phys. Rev.* **108**:482 (1957).
GY 69	H. Grotch and D. R. Yennie, *Rev. Mod. Phys.* **41**:350 (1969).
Gyu 74	H. Gyulassy, *Phys. Rev. Lett.* **32**:1393 (1974).
Gyu 75	M. Gyulassy, *Nucl. Phys. A* **244**:497 (1975).
Hag 63	R. Hagedorn, *Relativistic Kinematics*, W. A. Benjamin, New York (1963).
HB 74	J. H. Hetherington and J. Borysowicz, *Nucl. Phys. A* **219**:221 (1974).
Hei+ 69	J. Heisenberg, R. Hofstadter, J. S. McCarthy, I. Sick, B. C. Clark, R. Herman, and D. G. Ravenhall, *Phys. Rev. Lett.* **23**:1402 (1969).
Jan+ 66	T. Janssens, R. Hofstadter, E. B. Hughes, and M. R. Yearian, *Phys. Rev.* **142**:922 (1966).
JDD 74	C. W. deJager, H. deVries and C. deVries, *Nuclear Data Tables* **14**:479 (1974).
Jen+ 71	D. A. Jenkins, R. J. Powers, P. Martin, G. H. Miller, and R. E. Welsh, *Nucl. Phys. A* **175**:73 (1971).
JLR 75	A. D. Jackson, A. Lande, and D. O. Riska, *Phys. Lett.* **55B**:23 (1975).
JPD 72	J. A. Jansen, R. T. Peerdeman, and C. DeVries, *Nucl. Phys. A* **188**:337 (1972).
Kal+ 74	A. J. Kallio, P. Toropainen, A. M. Green, and T. Kouki, *Nucl. Phys. A* **231**:77 (1974).
Kan 73	E. Kankeleit, *J. Phys. Soc. Japan (Suppl.)* **34**:553 (1973).
Kes 71	D. Kessler, Muon Physics Conference, Fort Collins, Colorado (1971).
KF 70	R. A. Krajcik and L. L. Foldy, *Phys. Rev. Lett.* **24**:545 (1970).
KF 74	R. A. Krajcik and L. L. Foldy, *Phys. Rev. D* **10**:1777 (1974).
Kli+ 73	F. J. Kline, H. Crannell, J. T. O'Brien, J. McCarthy, and R. R. Whitney, *Nucl. Phys. A* **209**:381 (1973).
Koh 65	H. S. Köhler, *Phys. Rev.* **138**:B831 (1965).
KR 66	V. E. Krohn and G. R. Ringo, *Phys. Rev.* **148**:1303 (1966).
KR 74	J. Knoll and R. Rosenfelder, *Nucl. Phys. A* **229**:333 (1974).
KS 55	G. Källen and A. Sabry, *K. Dan. Vidensk. Selsk. Mat.-Fys. Medd.* **29**, No. 17 (1955).
KT 74a	Y. E. Kim and A. Tubis, *Ann. Rev. Nucl. Sci.* **24**:69 (1974).
KT 74b	W. M. Kloet and J. A. Tjon, *Phys. Lett.* **49B**:419 (1974).
Lab 71	L. N. Labzovskii, *Sov. Phys.—JETP* **32**:1171 (1971).
Leh 71	E. Lehman, *Phys. Rev. D* **4**:3324 (1971).
Len 69	F. Lenz, *Z. Phys.* **222**:491 (1969).

Lig 70   M. J. Lighthill, *Introduction to Fourier Analysis and Generalized Functions*, Cambridge University Press (1970).

Lin 72   W.-F. Lin, *Phys. Lett.* **39B**:447 (1972).

Lin 73   W.-F. Lin, *Nucl. Phys. A* **199**:14 (1973).

Lip 58   H. J. Lipkin, *Phys. Rev.* **110**:1395 (1958).

LL 62   L. D. Landau and E. M. Lifshitz, *The Classical Theory of Fields* (transl. by M. Hamermesh), Addison-Wesley, Reading, Massachusetts (1962).

LPD 72   B. E. Lautrup, A. Peterman, and E. deRafael, *Physics Reports* **3C**:193 (1972).

Max 69   L. C. Maximon, *Rev. Mod. Phys.* **41**:193 (1969).

McC+ 70   J. S. McCarthy, I. Sick, R. R. Whitney, and M. R. Yearian, *Phys. Rev. Lett.* **25**:884 (1970).

McK 69   R. J. McKee, *Phys. Rev.* **180**:1139 (1969).

MEH 74   R. Machleidt, K. Erkelenz, and K. Holinde, *Nucl. Phys. A* **232**:398 (1974).

Mer 73   K. Merle, in: *Proc. International Conf. on Photonuclear Reactions and Applications*, Asilomar, California (B. L. Berman, ed.), AEC Office of Information Services (1973).

MFG 59   U. Meyer-Berkhout, K. W. Ford, and A. E. S. Green, *Ann. Phys. (N.Y.)* **8**:119 (1959).

MG 72   L. D. Miller and A. E. S. Green, *Phys. Rev. C* **5**:241 (1972).

MHN 74   R. Machleidt, K. Holinde, and J. Németh, to be published.

Mil 74a   L. D. Miller, *Phys. Rev. C* **9**:537 (1974).

Mil 74b   L. D. Miller, to be published.

Moh 74a   P. J. Mohr, *Ann. Phys. (N.Y.)* **88**:26 (1974).

Moh 74b   P. J. Mohr, *Ann. Phys. (N.Y.)* **88**:52 (1974).

Mos 70   S. A. Moszkowski, *Phys. Rev. C* **2**:402 (1970).

MT 69   L. W. Mo and Y. S. Tsai, *Rev. Mod. Phys.* **41**:205 (1969).

MV 62   K. W. McVoy and L. VanHove, *Phys. Rev.* **125**:1034 (1962).

Neg 69   J. W. Negele, *Nucl. Phys. A* **138**:401 (1969).

Neg 70   J. W. Negele, *Phys. Rev. C* **1**:1260 (1970).

Neg 71   J. W. Negele, *Phys. Rev. Lett.* **27**:1291 (1971).

Neg 73   J. W. Negele, in: *Proc. International Conf. on Photonuclear Reactions and Applications*, Asilomar, California (B. L. Berman, ed.), AEC Office of Information Services (1973).

Neg 74a   J. W. Negele, *Phys. Rev. C* **9**:1054 (1974).

Neg 74b   J. W. Negele, in: *Proc. International Workshop II on Gross Properties of Nuclei and Nuclear Excitations*, Kleinwalsertal, Austria, Technische Hochschule Darmstadt (1974).

Nes+ 68   C. W. Nestor, Jr., K. T. R. Davies, S. J. Krieger, and M. Baranger, *Nucl. Phys. A* **113**:14 (1968).

NR 72   J. Németh and G. Ripka, *Nucl. Phys. A* **194**:329 (1972).

NT 71   M. Nagao and Y. Torizuka, *Phys. Lett.* **37B**:383 (1971).

NV 72   J. W. Negele and D. Vautherin, *Phys. Rev. C* **5**:1472 (1972).

Osb 68   H. Osborn, *Phys. Rev.* **176**:1514, 1523 (1968).

Owe 73   D. A. Owen, *Phys. Rev. D* **8**:424 (1973).

PRS 73   J. M. Pearson, B. Rouben, and G. Saunier, in: *Proceedings of Symposium on Correlations in Nuclei*, Balatonfüred, Hungary (1973).

Rao 72   S. Y. Rao, Ph.D. thesis, University of Maryland (1972).

RG 41   B. Rossi and K. Greisen, *Rev. Mod. Phys.* **13**:240 (1941).

Rin 71a   G. A. Rinker, Jr., *Phys. Rev. C* **4**:2150 (1971).

Rin 71b   G. A. Rinker, Jr., *Muon Physics Conference*, Ft. Collins, Colorado, September 6–10 (1971).

Ros 50   M. N. Rosenbluth, *Phys. Rev.* **79**:615 (1950).

Ros 73   R. Rosenfelder, *Nucl. Phys. A* **216**:477 (1973).

RPS 74   R. Rouben, R. Padjen, and G. Saunier, *Phys. Rev. C* **10**:2561 (1974).

RW 73   G. A. Rinker, Jr. and L. Wilets, *Phys. Rev. Lett.* **31**:1559 (1973).

Sac 62   R. G. Sachs, *Phys. Rev.* **126**:2256 (1962).

Sal 52   E. E. Salpeter, *Phys. Rev.* **87**:328 (1952).

SB 51   E. E. Salpeter and H. A. Bethe, *Phys. Rev.* **84**:1232 (1951).

SB 71   D. W. L. Sprung and P. K. Banerjee, *Nucl. Phys. A* **168**:273 (1971).

SBK 73   M. R. Strayer, W. H. Bassichis, and A. K. Kerman, *Phys. Rev. C* **8**:1269 (1973).

Sic 73   I. Sick, *Phys. Lett.* **44B**:62 (1973).

Sic 74   I. Sick, *Nucl. Phys. A* **218**:509 (1974).

Sic 75   I. Sick, *Phys. Lett.* **53B**:15 (1975).

Sie 70   P. J. Siemens, *Nucl. Phys. A* **141**:225 (1970).

Ska 70   H. F. Skardhamar, *Nucl. Phys. A* **151**:154 (1970).

Sky 56   T. H. R. Skyrme, *Phil. Mag.* **1**:1043 (1956).

SM 70   I. Sick and J. S. McCarthy, *Nucl. Phys. A* **150**:631 (1970).

Spr 72   D. W. L. Sprung, in: *Advances in Nuclear Physics*, Vol. 5, Plenum Press (1972).

SR 75   D. W. L. Sprung and K. S. Rao, *Phys. Lett.* **53B**:397 (1975).

SRP 74   G. Saunier, B. Rouben, and J. M. Pearson, *Phys. Lett.* **48B**:293 (1974).

Str 71   M. R. Strayer, Ph.D. thesis, Massachusetts Institute of Technology (1971).

SW 72   M. K. Sundaresan and P. J. S. Watson, *Phys. Rev. Lett.* **29**:15, 1122 (1972).

SW 73   C. M. Shakin and M. S. Weiss, *Phys. Rev. C* **8**:411 (1973).

Tan 71   K. Tanabe, *Phys. Rev. A* **3**:1282 (1971).

TB 58   L. J. Tassie and F. C. Barker, *Phys. Rev.* **111**:940 (1958).

TD 68   R. M. Tarbutton and K. T. R. Davies, *Nucl. Phys. A* **120**:1 (1968).

TFM 73   R. K. Tripathi, A. Faessler, and A. D. MacKellar, *Phys. Rev. C* **8**:129 (1973).

TFM 74   R. K. Tripathy, A. Faessler, and H. Müther, *Phys. Rev. C* **10**:2080 (1974).

Tho 61   D. J. Thouless, *The Quantum Mechanics of Many Body Systems*, Academic Press, New York (1961).

Ube 71   H. Überall, *Electron Scattering from Complex Nuclei*, Academic Press, New York (1971).

Ueh 35   E. A. Uehling, *Phys. Rev.* **48**:55 (1935).

Van 69   G. J. C. Van Niftrik, *Nucl. Phys. A* **131**:574 (1969).

VB 72   D. Vautherin and D. M. Brink, *Phys. Rev. C* **5**:626 (1972).

WAM 71   H. T. Williams, H. Arenhövel, and H. G. Miller, *Phys. Lett.* **36B**:278 (1971).

WK 56   E. H. Wichmann and N. M. Kroll, *Phys. Rev.* **101**:843 (1956).

WR 75   L. Wilets and G. A. Rinker, Jr., *Phys. Rev. Lett.* **34**:339 (1975).

WS 74   P. J. S. Watson and M. L. Sundaresan, *Can. J. Phys.* **52**:2037 (1974).

WW 69   C. S. Wu and L. Wilets, *Ann. Rev. Nucl. Sci.* **19**:527 (1969).

YLR 57   D. R. Yennie, M. M. Lévy, and D. G. Ravenhall, *Rev. Mod. Phys.* **29**:144 (1957).

YRW 54   D. R. Yennie, D. G. Ravenhall, and R. N. Wilson, *Phys. Rev.* **95**:500 (1954).

Notes Added In Proof

Contributions by C. K. Hargrove and G. Backenstoss at the High Energy Physics and Nuclear Structure Conference, Sante Fe, 1975, show that there is some doubt if a discrepancy actually exists between the high-lying energy level data in muonic atoms and QED. Possible systematic errors in the data exist because of calibration difficulties. More recent measurements are in disagreement with the older data, and the newer data, if accurate, remove the discrepancy discussed in Section 1.3.6.

Recent calculations by E. Borie, M. Sundaresan and P. J. S. Watson, and D. H. Fujimoto have confirmed a small value for the $\alpha^2(Z\alpha)^2$ vacuum polarization correction in high-lying levels. These corrections favor the new data discussed above.

F. Coester and A. Ostebee [*Phys. Rev. C* **11**:1836 (1975)] have examined the assumptions needed to produce the Gross relativistic corrections for elastic electron–deuteron scattering discussed in Section 1.3.5.

The authors would like to thank Prof. D. W. L. Sprung for sending them a list of errors in the manuscript.

INDEX

A

Alpha-cluster model for hypernuclei, 80
Analysis of data, 331-367
Anomalous magnetic moment, 225, 260
Antisymmetric nuclear states, 89
Antisymmetrization (*see also* Pauli principle), 112
Approximations, comparison of, 161-163
Auger effect, nuclear, 100
Augmented χ^2, 329
Average interior density, 282, 291
Average shape of interior density, 283-285

B

$^{12}_{\Lambda}$B, 87-89
  decay rates, 89
Baranger *et al.*, method of, 148
Baranger half-shell function, 142
Baryonic states, 223
Basis, truncated, 320-329
Basis-independent analysis, 329-331
Bethe–Faddeev equations, 26
Bethe–Goldstone equation, 128
Bethe logarithm, 259
Bethe–Salpeter equation, 246
Bhatia and Walker approximation, 156-157
Binding energies, 3, 10, 23, 28, 31, 31-81, 96, 140, 197, 269, 305
  cluster calculations, 37
  Hartree–Fock calculations, 36-37
  for $^{5}_{\Lambda}$He, 42
    overbinding, 43, 45
  muonic atom, 245
  nuclear matter, 187
  nuclear structure effects, 46
  rigid core approximation, 32-35
  shell model calculations, 37-38, 58, 68
  shift of, 248
  s-shell hypernuclei, 38-57
  variational calculations, 35-36

Born approximation, 246, 313, 316
Bound state, 123
  poles, 156
  t_0-matrix pole, 169
  wavefunction, 149
Boundary condition model, 178-180
Breit equation, 246
Bremsstrahlung
  calculation, 191
  experiments, 191
  neutron–proton, 189
  nucleon–nucleon, 189-192
  proton–proton, 189

C

$^{12}_{\Lambda}$C
  excitation spectrum, 99
  formation cross section, 104
^{12}C, charge density, 331
  data, 341-342
Center-of-mass
  coordinate, 251
  corrections, 261-262
  frame, 252
Central depression, 285
Charge
  bare, 254
  physical, 254
Charge-current operator, 224, 225
Charge density
  effective, 231
  exponential, 227
  Gaussian, 227
  for ground state, 265-281
  moments of, 247
  normalization of, 233
  of neutron, 229
  of ^{208}Pb (*see* ^{208}Pb)
  for spherical nuclei, 230, 247